20 044 411 43
University of
WITHDRAWN
AF616319

Experimental protocols for reactive oxygen and nitrogen species

Experimental protocols for reactive oxygen and nitrogen species

Edited by

NAOYUKI TANIGUCHI
Department of Biochemistry
Osaka University Medical School

and

JOHN M. C. GUTTERIDGE
Oxygen Chemistry Laboratory, Directorate of Anaesthesia and Critical Care, Royal Brompton and Harefield NHS Trust

This publication was supported by a generous donation from the Daido Life Foundation

OXFORD
UNIVERSITY PRESS

Great Clarendon Street, Oxford OX2 6DP

Oxford University Press is a department of the University of Oxford
It furthers the University's objective of excellence in research, scholarship,
and education by publishing worldwide in

Oxford New York

Athens Auckland Bangkok Bogotá Buenos Aires Calcutta
Cape Town Chennai Dar es Salaam Delhi Florence Hong Kong Istanbul
Karachi Kuala Lumpur Madrid Melbourne Mexico City Mumbai
Nairobi Paris São Paulo Singapore Taipei Tokyo Toronto Warsaw

with associated companies in Berlin Ibadan

Oxford is a registered trade mark of Oxford University Press
in the UK and in certain other countries

Published in the United States
by Oxford University Press Inc., New York

First published in Japanese 1994
© Shiyusha 1994
Translation © Oxford University Press 2000

The moral rights of the author have been asserted
Database right Oxford University Press (maker)

First published in English 2000

All rights reserved. No part of this publication may be reproduced,
stored in a retrieval system, or transmitted, in any form or by any means,
without the prior permission in writing of Oxford University Press,
or as expressly permitted by law, or under terms agreed with the appropriate
reprographic rights organization. Enquiries concerning reproduction
outside the scope of the above should be sent to the Rights Department,
Oxford University Press, at the address above.

You must not circulate this book in any other binding or cover
and you must impose this same condition on any acquirer

A catalogue record for this book is available from the British Library

Library of Congress Cataloging in Publication Data

Experimental protocols for reactive oxygen and nitrogen species / edited by Naoyuki
Taniguchi and John M. C. Gutteridge.
p. cm.
Includes bibliographical references and index.
ISBN 0 19 850668 6 (Hbk)
1. Active oxygen—Laboratory manuals. 2. Nitric oxide—Laboratory manuals. 3. Free
radicals (Chemistry)—Physiological effect—Laboratory manuals. I. Taniguchi, Naoyuki,
1942– II. Gutteridge, John M. C.

QP535.O1 .E976 2000 572′.53—dc21 00–021378

Typeset by Footnote Graphics,
Warminster, Wilts
Printed in Great Britain
on acid-free paper by
T. J. International Ltd, Padstow

UNIVERSITY OF
NORTHUMBRIA AT NEWCASTLE
LIBRARY
ITEM No. CLASS No.
2004441143 547.024 EXP

Preface for experimental protocols for ROS and RNS

Interest in the importance of free radicals in the biomedical sciences continues to grow, with molecules such as nitric oxide making major impacts on clinical, pharmacological, and medical knowledge.

At the molecular level, the Cu,Zn-superoxide dismutase gene has been identified as a candidate gene for familial amyotrophic lateral sclerosis (ALS) and already over fifty different mutations have been found. Studies such as these have stimulated intense interest in the roles played by free radicals in a range of key cell functions which might be under redox control with a balance between antioxidants and reactive oxygen, nitrogen, and iron species triggering whether cells enter proliferative, apoptotic, or necrotic states.

Many researchers from different disciplines who are not familiar with the specialized techniques used in free radical research, require detailed advice. There are now several excellent texts dealing specifically with such methods, and readers are referred to these for techniques adequately covered by them.

The aims of this publication are to bring to a wider readership a popular methods book previously only available to Japanese readers, and to prepare a book which contains established working methodologies to cover the disciplines of biochemistry, molecular biology, and cell biology in the area of free radical research.

The book has been written in the form of a 'cook book' and it is hoped that researchers will be able to prepare the solutions described and follow recipes to complete their experiments without a need to consult the original research papers.

We would like to thank all contributors and my colleagues in the Department of Biochemistry, Osaka University Medical School, for their admirable efforts and selfless contribution. Special thanks are due to Mrs Tomoko Matsubara, Mr. Shinji Ihara, and Mr. Takeshi Endo for their patience in typing and preparing the drafts for the book. We also thank the Shujunsha Company, who published the original Japanese version, for agreeing with publication of the English version, and the Daido Foundation for the partial support.

Osaka N.T.
London J.M.C.G.
Nov. 1999

Contents

Contributing authors

TETSUO ADACHI (ch.26) Laboratory of Clinical Pharmaceutics, Gifu Pharmaceutical University

TAKAAKI AKAIKE (ch.5,10,36,87) Department of Microbiology , Kumamoto University School of Medicine

KENJI AKI (ch.21) Institute for Health Science, Tokushima Bunri University

SHINYA ASAMI (ch.65) Surgery, Tsubame Rosai Hospital

SHIRO BANNAI (ch.60) Department of Biochemistry, Institute of Basic Medical Sciences, University of Tsukuba

WENYI CHE (ch.83) Department of Pathology, University Rochester Medical Center

YUTAKA EGUCHI (ch.61) Department of Medical Genetics, Biomedical Research Center, Osaka University Medical School

T. W. EVANS (ch.80) Directorate of Anaesthesia and Critical Care, Royal Brompton and Harefield NHS Trust

JUNICHI FUJII (ch.6,27,28,57) Department of Biochemistry, Osaka University Medical School

KOSHI FUJIKAWA (ch.33) Fourth Department of Internal Medicine, Sapporo Medical University School of Medicine

AKI FUJISAWA (ch.37) Research Center for Advanced Science and Technology, University of Tokyo

J. M. C. GUTTERIDGE (ch.1,11,32,46–55, 71–73,75,80) Oxygen Chemistry Laboratory, Directorate of Anaesthesia and Critical Care, Royal Brompton and Harefield NHS Trust

MIKA HAYAKAWA (ch.70) Institute for Biomedical Research

KEIICHI HIRAI (ch.34) Department of Anatomy, Kanazawa Medical University

FUJIKO HIROSAWA (ch.31) Department of Biochemistry, Akita University School of Medicine

YOUNG HO KOH (ch.83) Department of Biochemistry, Osaka University Medical School

SHIRO HOSHIDA (ch.77) Division of Cardiology, Osaka Rosai Hospital

MASUO HOSOKAWA (ch.90) Laboratory of Pathology, Cancer Institute, Hokkaido University School of Medicine

KOHJI ICHIMORI (ch.2) Department of Physiology 2, Tokai University, School of Medicine

YOSHITAKA IKEDA (ch.23) Department of Biochemistry, Osaka University Medical School

KATSUHISA INOUE (ch.5) Department of Microbiology, Kumamoto University School of Medicine

NAOAKI ISHII (ch.85) Department of Molecular Life Science, Tokai University School of Medicine

MUTSUO ISHIKAWA (ch.86) Department Obstetrics and Gynaecology, Asahikawa Medical College

YASUHIRO KAMBAYASHI (ch.3,38) Japan Immunoresearch Laboratories

HIDEKI KANETO (ch.67) Department of Internal Medicine and Therapeutics, Osaka University Graduate School of Medicine

DONGCHON KANG (ch.8) Department of Clinical Chemistry and Laboratory Medicine, Kyusyu University

HIROSHI KASAI (ch.65) Surgery, Tsubame Rosai Hospital

JUNJI KATO (ch.33) Fourth Department of Internal Medicine, Sapporo Medical University School of Medicine

NAOHISA KAWAMURA (ch.19) Osaka Rosai Hospital

SHOSUKE KAWANISHI (ch.66) Department of Hygiene, Mie University School of Medicine

YOSHIRO KAWARADA (ch.31) Department of Biochemistry, Akita University School of Medicine

YOSHIRO KAYANOKI (ch.44) Department of Internal Medicine and Molecular Science , Graduate School of Medicine, Osaka University

TORU KITA (ch.40,82) Department of Geriatric Medicine, Kyoto University Faculty of Medicine

YUKI KITAOKA (ch.13) Department of Obstetrics and Gynaecology, Kyoto National Hospital

TAKAHIKO KONDO (ch.12,24) Department of Biochemistry and Molecular Biology In Disease, Atomic Bomb Disease Institute, Nagasaki University School of Medicine

HIROAKI KOSAKA (ch.4) Second Department of Physiology, Kagawa Medical University

HIDEO KUSUOKA (ch.76) Institute for Clinical Research Osaka National Hospital

TSUNEHIKO KUZUYA (ch.76) Institute for Clinical Research Osaka National Hospital

HIROSHI MAEDA (ch.10) Department of Microbiology, Kumamoto University School of Medicine

HISASHI MAKI (ch.69) Graduate School of Biological Science, Nara Institute of Science and Technology

MASAYASU MATSUMOTO (ch.79) First Department of Medicine, Osaka University Medical School

TATSUYA MATSURA (ch.16) Department of Biochemistry, Faculty of Medicine, Tottori University

YOSHIKI MIYACHI (ch.84) Department of Dermatology, Kyoto University Graduate School of Medicine

YOICHI MIYAMOTO (ch.10,36,87) Department of Microbiology, Kumamoto University School of Medicine

TERUO MIYAZAWA (ch.39) Tohoku University Graduate School of Life Science and Agriculture

TETSUYA MORIUCHI (ch.62) Division of Cell Biology, Cancer Institute, Hokkaido University, School of Medicine

SHARON MUMBY (ch.48–51) Directorate of Anaesthesia and Critical Care, Royal Brompton and Harefield NHS Trust

YUICHIRO NAGATA (ch.37) Research Center for Advanced Science and Technology, University of Tokyo

YUJI NAITO (ch.78) 1st Department of Medicine, Kyoto Prefectual University of Medicine

MASAHIRO NAKAMURA (ch.35) Department of Anatomy, Sapporo Medical University School of Medicine

MASAO NAKAMURA (ch.29) Department of Chemistry, Asahikawa Medical College

MINORU NAKANO (ch.3) Japan Immunoresearch Laboratories

YOSHIRO NIITSU (ch.33) Fourth Department of Internal Medicine, Sapporo Medical University School of Medicine

ETSUO NIKI (ch.43,45) Research Center for Advanced Science and Technology, University of Tokyo

MORIMITSU NISHIKIMI (ch.64) Department of Biochemistry, Wakayama Medical College

HIROKI NOGAWA (ch.35) Department of Anatomy, Sapporo Medical University School of Medicine

NORIKO NOGUCHI (ch.45) Research Center for Advanced Science and Technology, University of Tokyo

KIYOSHI NOSE (ch.56) Department of Microbiology, Showa University

TOHRU OGIHARA (ch.15) Department of Pediatrics, Osaka Medical College

SHINJI OIKAWA (ch.66) Department of Hygiene, Mie University School of Medicine

DAISUKE OKADA (ch.88) Precursory Research for Embryonic Science and Technology (PRESTO)

FUTOSHI OKADA (ch.90) Laboratory of Pathology, Cancer Institute, Hokkaido University School of Medicine

TAKAYUKI OZAWA (ch.70) Institute for Biomedical Research

G. J. QUINLAN (ch.1,11,32,46–55,71-73,75,80) Oxygen Chemistry Laboratory, Directorate of Anaesthesia and Critical Care, Royal Brompton and Harefield NHS Trust

MITSUAKI SANO (ch.18) School of Pharmaceutical Sciences, University of Shizuoka

KENZO SATO (ch.63) Department of Molecular Biology, School of Life Science,

TOMOHIRO SAWA (ch.36) Department of Microbiology, Kumamoto University School of Medicine

FUMI SAWADA (ch.12,24) Department of Biochemistry and Molecular Biology In Disease, Atomic Bomb Disease Institute, Nagasaki University School of Medicine

HIROYUKI SHIMASAKI (ch.41,42) Department of Biochemistry, Teikyo University School of Medicine

TOSHIHIRO SUGIYAMA (ch.31) Department of Biochemistry, Akita University School of Medicine

KEIICHIRO SUZUKI (ch.7,25,58,81,92) Department of Biochemistry, Osaka University Medical School

AKIRA TACHIBANA (ch.89) Graduate School of Science, Kyoto University

KAZUHIKO TAKAHASHI (ch.20) Department of Hygienic Chemistry, Graduate School of Pharmaceutical Science, Hokkaido University

KOICHIRO TAKESHIGE (ch.8) Department of Clinical Chemistry and Laboratory Medicine, Kyusyu University

HIROSHI TAMAI (ch.14,17) Department of Pediatrics, Osaka Medical College

NAOYUKI TANIGUCHI (ch.67,83) Department of Biochemistry, Osaka University Medical School

YOSHIHISA TANIGUCHI (ch.59) Department of Anaesthesia and Intensive Care Medicine, Omiya Medical Center, Jichi Medical School

HARUYUKI TATSUMI (ch.35) Department of Anatomy, Sapporo Medical University School of Medicine

ISAO TOMITA (ch.18) School of Pharmaceutical Sciences, University of Shizuoka

SHINYA TOYOKUNI (ch.91) Department of Pathology and Biology of Diseases, Graduate School of Medicine, Kyoto University

SHIGEKI TSUCHIDA (ch.22) Second Department of Biochemistry, Hirosaki University School of Medicine

YOSHIMOTO TSUJIMOTO (ch.61) Department of Medical Genetics Biomedical Research Center, Osaka University Medical School

KOJI UCHIDA (ch.68,74) Laboratory of Food and Biodynamics, Nagoya University School of Agricultural Sciences

KOZO UTSUMI (ch.9) Department of Biological Chemistry, Faculty of Agriculture, Yamaguchi University

TOSHIHIKO UTSUMI (ch.9) Department of Biological Chemistry, Faculty of Agriculture, Yamaguchi University

YOSHIHIRO YAMAMOTO (ch.37,38) Research Center for Advanced Science and Technology, University of Tokyo

SATOSHI YAMASHITA (ch.37,38) Research Center for Advanced Science and Technology, University of Tokyo

JUNJI YODOI (ch.13) Department of Biological Responses Institute for Virus Research, Kyoto University

MASAYUKI YOKODE (ch.40,82) Department of Geriatric Medicine, Kyoto University Faculty of Medicine

SHUUJI YONEI (ch.89) Graduate School of Science, Kyoto University
TADASHI YOSHIDA (ch.30) Department of Biochemistry, Yamagata University School of Medicine
TOSHIKAZU YOSHIKAWA (ch.78) 1st Department of Medicine, Kyoto Prefectual University of Medicine
QIU-MEI ZHANG (ch.89) Graduate School of Science, Kyoto University

Part 1

Measurement

1

Oxygen uptake, and spectrophotometric measurement of superoxide, hydrogen peroxide and hydroxyl radicals

J. M. C. GUTTERIDGE and G. J. QUINLAN

Oxygen uptake

Introduction

The uptake or evolution of oxygen during a reaction can be monitored by use of a Clark-type electrode. The method is particularly useful when turbidity or interfering chromogens preclude the use of spectrophotometry (1).

Protocol

Electrode calibration

1. Add deionized water (2 mL) to the temperature-controlled reaction chamber with magnetic stirring, and apply 0.7 V across the electrodes.
2. An output of 0.1 to 10 mV is used with a chart recorder set to a value of 100%.
3. Leave electrode to settle for 5 min.
4. Add a few crystals of sodium dithionate to remove O_2.
5. This value represents 0% O_2. The electrode is now calibrated from 0 to 100% oxygen.
6. Apply test samples to chamber after carefully washing each time.

Calculation

The concentration of oxygen can be calculated by use of the information that air-saturated water at 30 °C contains 0.460 μmol O_2.

References

1. Trudgill, P. W. (1985). Oxygen Consumption. In *Cell CRC Handbook of Methods for Oxygen Radical Research*, (ed. R. A. Greenwald), pp. 329–42. CRC Press, Boca Raton, USA.

Spectrophotometric measurement of superoxide anion radicals

Introduction

First applied to neutrophil studies in 1968, the nitroblue tetrazolium (NBT) test provides a simple assay for $O_2\cdot^-$ production. NBT is reduced by $O_2\cdot^-$ to a blue formazan product which is measured at 620 nm.

Spectrophotometric assay techniques are covered in full elsewhere (1, 2). Pulse radiolysis techniques are also described elsewhere (3). See also Chapter 34.

References

1. Rice-Evans, C. A., Diplock, A. T., and Symons, M. C. R. (1991). *Techniques in Free Radical Research*, pp. 85–91. Elsevier, Amsterdam.
2. Greenwald, R. A. (ed.) (1995). *CRC Handbook of Methods for Oxygen Radical Research*, pp. 121–32. CRC Press, Boca Raton, USA.
3. Punchard, N. A. and Kelly, F. J. (eds) (1996). *Free Radicals: A Practical Approach*, pp. 11–24. IRL Press, Oxford.

Spectrophotometric measurement of hydrogen peroxide

Introduction

Any system producing $O_2\cdot^-$ will also produce H_2O_2 by the dismutation reaction. H_2O_2 is an uncharged molecule that can readily enter cells as a water look-alike. Most assays utilize a peroxidase with coupled oxidation of a detector molecule.

H_2O_2 can also be determined by measuring the loss of scopoletin fluorescence (460 nm), an assay covered in detail elsewhere (1) or on the basis of its reaction with cytochrome-*c* peroxidase to form a stable complex with absorbance at 419 nm (A_{419}) (1)

In this section we describe a method using phenol red to measure hydrogen peroxide in breath condensates (2).

Protocol

1. Dilute stock hydrogen peroxide (30% *v*/*v*) 1:1000 with Chelex resin-treated distilled water.

2. Calculate the concentration of H_2O_2 by measuring absorbance at 240 nm using a molar absorption coefficient of 43.6 M^{-1} cm^{-1}.
3. Prepare buffered phenol red–peroxidase reagent containing NaCl (40 mM), potassium phosphate (pH 7.0, 10 mM), phenol red (0.1 g L^{-1}), and horseradish peroxidase (HRPO; 8.5 units mL^{-1}), adding phenol red and HRPO to buffer just before use.
4. To 1.0 mL phenol red reagent add 0.5 mL breath condensate. Hydrogen peroxide standards are similarly treated.
5. Mix and stand for 5 min at 25 °C.
6. Add NaOH (1 M, 10 μL) to give a pH of 12.5 and a stable purple colour.
7. Measure absorbance at 610 nm.

Calculation

Include a blank containing H_2O_2-free water and calculate values from an H_2O_2 standard curve.

References

1. Rice-Evans, C. A., Diplock, A. T., and Symons, M. C. R. (1991). *Techniques in Free Radical Research*, pp. 91–6. Elsevier, Amsterdam.
2. Pick, E. and Keisari, Y. (1980). A Simple colorimetric method for the measurement of hydrogen peroxide produced by cells in culture. *J. Immunol. Methods*, **38**, 161–70.

Spectrophotometric measurement of hydroxyl radical damage

Introduction

The hydroxyl radical is much too reactive to be measured directly by simple spectrophotometric techniques. Evidence characteristic of its damage is, therefore, usually sought by use of a detector molecule for such damage. Here we describe the use of the sugar 2-deoxy-D-ribose.

Protocol

1. Into clean glass tubes place 2-deoxy-D-ribose (5 mM, 0.2 mL) and sodium phosphate buffer (0.024 M containing 0.15 M NaCl, pH 7.4, 0.2 mL).
2. Add ·OH generating system, i.e.:

Protocol *Continued*

[A] 0.1 mL of ferrous salt (freshly prepared at pH < 6.0) **or**

[B] ferric-EDTA + ascorbate **or**

[C] hypoxanthine + xanthine oxidase + ferric salt.

3. Incubate reactants for 30 min at 37 °C.
4. Add thiobarbituric acid (1% *w/v*, 0.5 mL) and trichloroacetic acid (2.8% *w/v*, 0.5 mL).
5. Heat tube contents at 100 °C for 10 min.
6. Cool and read absorbance at 532 nm or fluorescence at 553 nm after excitation at 532 nm.

Calculation

The assessment of ·OH damage can be used to test the effectiveness of scavengers and inhibitors, and to determine second-order rate constants for reaction of ·OH with various scavengers.

Comments

·OH-generating systems [A] and [B] are not inhibited by SOD, whereas [C] is. All three ·OH-generating systems should be inhibited by catalase or other H_2O_2-removing enzymes. If solutions are turbid extract chromogen into butan-l-ol.

References

1. Gutteridge, J. M. C. (1981). Thiobarbituric acid reactivity following iron-dependent free-radical damage to amino acids and carbohydrates. *FEBS Lett.*, **128**, 343–6.
2. Halliwell, B. and Gutteridge, J. M. C. (1984). Formation of a thiobarbituric acid-reactive substance from deoxyribose in the presence of iron salt. The role of superoxide and hydroxyl radicals. *FEBS Lett.*, **128**, 347–52.
3. Gutteridge, J. M. C. (1987). Ferrous-salt-promoted damage to deoxyribose and benzoate. The increased effectiveness of hydroxyl-radical scavengers in the presence of EDTA. *Biochem. J.*, **243**, 709–14.

2

Measurements of nitric oxide and peroxynitrite

KOHJI ICHIMORI

Introduction

Although there is much interest in NO and peroxynitrite, their measurement in biological media is difficult because of their short lifetimes and low concentrations. Here, the use of an NO-sensitive electrode is first described for direct real-time measurement of NO in endothelial cells. A method for detecting peroxynitrite produced by the macrophage is also given. Brief references to other methods are included in the comments sections.

NO measurement with NO-sensitive electrodes (1)[1]

Calibration[2]

Protocol

1. Prepare KH solution (HEPES-buffered modified Krebs–Henseleit solution), containing NaCl (137 mM), KCl (1.1 mM), $CaCl_2.2H_2O$ (0.18 mM), $MgCl_2.6H_2O$ (0.1 mM), $NaHCO_3$ (4.2 mM), glucose (5.6 mM), and HEPES[1] (5 mM). After addition of optional reagents adjust pH to 7.4 with NaOH (0.1 M).
2. Immerse NO electrode in KH solution (20 mL) and read basal current after stabilization.
3. Immerse NO electrode in freshly prepared SNAP solution[2] (1 mM, 20 mL) and read NO current after stabilization (waiting for *ca* 5 min).
4. Repeat steps 2 and 3 three times and calculate mean NO-dependent current[3].

Although calibration of the electrode can be performed with authentic NO solution, this calibration is not practical because it is difficult and troublesome to maintain the buffer solution anaerobic; and the concentration of

Protocol *Continued*

authentic NO solution should be determined by other methods, e.g. the oxyhaemoglobin method (2).

[1] Although any buffer which keeps cells intact can be used instead of the KH solution, buffers with a high protein content might cause a change in sensitivity during measurements because of protein adhesion to the electrode surface. Ionic strength should be also kept constant during a series of experiments because changes in ionic strength cause transient currents.
[2] SNAP is *S*-nitroso-*N*-acetyl-D,L-penicillamine (Mr = 220.25, Alexis). After SNAP is dissolved in KH solution, pH should be adjusted to 7.4.
[3] The current is largely dependent on each individual electrode. Because SNAP in aqueous media slowly liberates NO, which is consumed by the reaction with dissolved oxygen, freshly prepared SNAP solution (1 mM, 25 °C) maintains a steady-state NO concentration of 1.3 μM. There is a linear correlation between electrode current and SNAP concentration ranging from 1 μM to 1 mM. Do not use electrodes (Inter Medical) with a sensitivity below 500 pA mM^{-1} SNAP.

Comments

1. NO measurements can be performed with an NO monitor (NO-501, Inter Medical, Japan) with an NO-sensitive electrode (NOE-47, 200 mmΦ, Inter Medical, Japan), or with an isolated nitric oxide meter (ISO-NO Mark II, World Precision Instruments, USA) with an NO sensor (ISO-NOP 200, World Precision Instruments, USA). All experiments must be performed with electromagnetic shields, made of an iron mesh screen, to avoid external noise. Plastics which are easily charged with electricity can cause unexpected noise.
2. New electrodes (both working and counter electrodes from Inter Medical) should be immersed for one day in buffer solution before use. It takes more than 30 min to stabilize basal current after first connecting the electrode and turning on the NO monitor. The electrodes are stored with the tips immersed in the buffer.

The measurement of NO produced by endothelial cells[1]

Protocol

1. Perfuse bovine aortic endothelial cells confluently cultured on a cover slip (15 mmΦ), with L-arginine solution (L-arg, MW = 174.2, 50 μM).
2. Place NO electrode as close as possible to the cover slip without making contact. Start recording the basal level of electrode current and wait until the current has stabilized (*ca* 30 min).
3. Switch the perfusate to 1 mM ATP/50 mM L-arg solution[2] to stimulate endothelial cells. ATP (Mr = 551) should be stored at −20 °C.

4. Re-switch the perfusate to 50 μM L-arg solution[3] and confirm that the current returns to the basal level.

[1] Other stimulants such as acetylcholine (1 μM), bradykinin (100 nM), and A23187 (1 μM) can be used instead of ATP. Because some stimulants can cause an artefactual electrode current, a control experiment with stimulant should be performed without endothelial cells.
[2] NO concentration can be calculated from the current difference before and after stimulation if the electrode is calibrated. Use 0.1 mM N^G-monomethyl-L-arginine, monoacetate (MW = 262.4 Alexis) of L-arg and confirm that the current increase associated with the stimulation is abolished by this nitric oxide synthase inhibitor.

Comments

1. Use a perfusion system as illustrated in Fig. 1. It is important to use a hydrostatic pressure of *ca* 40 cm H_2O so that the flow is constant, because fluctuations of the flow rate cause noise on the electrode current. All perfusate should be prepared by dissolving reagents in KH solution and adjusting the pH to 7.4.

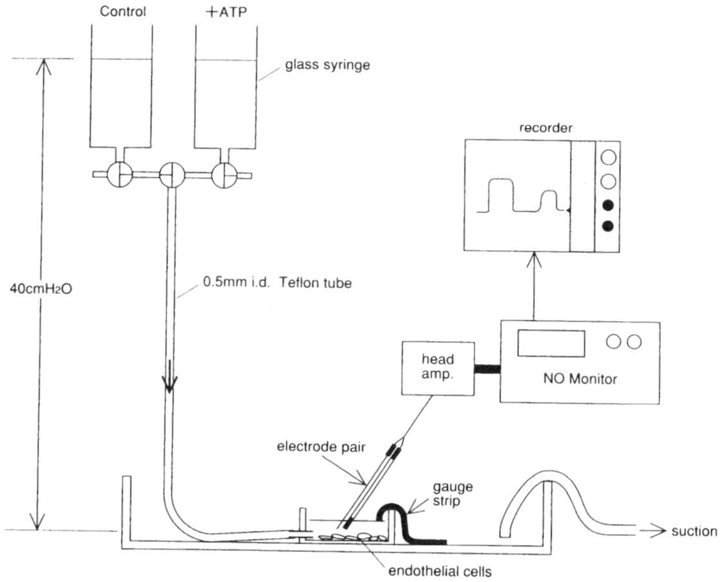

Figure 1. A perfusion system for measuring NO generated from endothelial cells.

Nitric oxide synthases (NOS) which produce NO in biological systems consist of three isoforms. Among these NOS-1 (neuronal type) and NOS-3 (endothelial type) transiently produce a relatively small amount of NO in response to external signalling molecules. To elucidate these transient NO responses, highly sensitive and real-time methods such as chemiluminescence methods (3, 4) or an electrochemical method (1) are required. On the other hand, spectrophotometric (5) and fluorimetric (6) methods are convenient for determining total NO production from NOS-2, which persistently produces a relatively large amount of NO. These methods can be applied to assay NO_2^- and NO_3^-, the end products of NO, accumulated in incubation media for 6 to 48 h.

Quantitation of peroxynitrite produced by macrophages (7)[1]

Protocol

1. Prepare Hank's balanced salt solution (HBSS) by dissolving HBSS powder (Gibco, 450–1201) and $NaHCO_3$ (0.35 g L^{-1}) in doubly distilled water (pH 7.4).
2. Prepare a stock solution of phorbol-12-myristate-12 acetate (PMA, Sigma; 1 mg mL^{-1}) in DMSO. Divide into 100-mL aliquots, and store below −80 °C (in the deep freeze).
3. Incubate macrophages[1] (8×10^6 cells mL^{-1}) in HBSS containing PMA (400 ng mL^{-1}), 4-hydroxyphenylacetic acid (4-HPA, MW 152.15; 1 mM) and Cu,Zn-superoxide dismutase (SOD; 0.1 mg mL^{-1}) for 4 h at 37 °C.
4. Centrifuge the solution (1000 rpm for 10 min) to remove macrophages.
5. Acidify supernatant with H_3PO_4 (10%) and add acetonitrile (final concentration 20% *v*/*v*).
6. Pass through a 0.45-μm membrane filter.
7. Apply the sample to a 4.6 mm × 150 mm C_{18} reversed-phase HPLC column equilibrated with 20:80 (*v*/*v*) acetonitrile–phosphate buffer (pH 3.2, 10%).
8. Elute the column[2] with a linear gradient of 20 to 60% acetonitrile over 10 min at a flow rate of 1 mL min^{-1}.
9. Monitor UV absorption at 360 nm to assay 4-hydroxy-3-nitrophenyl-acetic acid (NO_2–HPA)[3].

[1] In the original paper (7), macrophages were collected from the alveoli of Sprague–Dawley rats by bronchoalveolar lavage. PMA stimulates superoxide production by macrophages through NADPH oxidase. Lipopolysaccharide (10 ng mL^{-1}) and/or interferon-γ (10 units mL^{-1}) can be added 6 h before PMA stimulation to induce nitric oxide synthase-2.
[2] Because the retention time of NO_2–HPA depends on the HPLC equipment and/or column used, determine the optimum conditions for separation of HPA and NO_2–HPA. The duration of the gradient or the initial percentage of DMSO can be changed to obtain the optimum conditions. Prepare a standard curve by analysis of known amounts of NO_2–HPA.
[3] In the HBSS solution, 8% of produced peroxynitrite can be converted to NO_2–HPA. If 40 mM HEPES is used instead of HBSS, the recovery rate decreases to 3.4%. To estimate the precise amount of peroxynitrite production, determine the recovery rate first by using a known amount of peroxynitrite (9) under the same conditions without stimulated cells.

Comments

1. Peroxynitrite nitrates phenol derivatives in the presence of redox-active metal complexes such as Fe^{3+}– EDTA (1 mM) and Cu,Zn-SOD (0.1g mL^{-1}). Because relatively large amounts of Fe^{3+}– EDTA have toxic side-effects and promote reactions of reactive oxygen species, Cu,Zn-SOD is used as the copper catalyst although it scavenges superoxide and might reduce peroxynitrite generation. In this method, peroxynitrite is assayed as a nitrated phenol derivative (NO_2–HPA) and separated by HPLC. It has also been reported that a manganese porphyrin complex has negligible SOD activity, and is thus an excellent catalyst for this nitration reaction (8).

   Peroxynitrite formation *in vivo* can also be estimated by measuring nitrotyrosine by means of an HPLC (10) or immunohistochemical (11) method. Because there are several routes for nitrotyrosine formation (for example, nitrite + peroxidase + tyrosine), nitrotyrosine itself is not such a specific indicator for peroxynitrite, although the efficiency of peroxynitrite at producing nitrotyrosine is high. Thus, control experiments with NOS inhibitors and scavengers of reactive oxygen species should be performed to determine the origin of nitrotyrosine.

References

1. Ishimori, K., Ishida, H., Fukahori, M., Nakazawa, H., and Murakami, E. (1994). Practical nitric oxide measurement employing a nitric oxide-sensitive electrode. *Rev. Sci. Instrum.*, **65**, 2714–18.
2. Murphy, M. E. and Noark, E. (1994). Nitric oxide assay using haemoglobin method. *Methods Enzymol.*, **233**, 240–50.
3. Brien, J. F., McLaughlin, B. E., Nakatsu, K., and Marks, G. S. (1991). Quantitation of nitric oxide formation from nitrovasodilator drugs by chemiluminescence analysis of headspace gas. *J. Pharmacol. Method.*, **25**, 19–27.
4. Kikuchi, K., Nagano, T., Hayakawa, H., Hirata, Y., and Hirobe, M. (1993). Real-time measurement of nitric oxide produced ex vivo by luminol–H_2O_2 chemiluminescence method. *J. Biol. Chem.*, **268**, 23106–10.
5. Green, L. C., Wangner, D. A., Glogowski, J., Skipper, P. L., Wishnok, J. S., and Tannenbaum, S. R. (1992). Analysis of nitrate, nitrite, and [^{15}N]nitrate in biological fluids. *Anal. Biochem.*, **126**, 131–8.
6. Misko, T. P., Schilling, R. J., Salvemini, D., Moore, W. M., and Currie, M. G. (1993). A fluorimetric assay for the measurement of nitrite in biological samples. *Anal Biochem.*, **214**, 11–16.
7. Ischiropoulos, H., Zhu, L., and Beckman, J. S. (1992). Peroxy formation from macrophage-derived nitric oxide. *Arch. Biochem. Biophys.*, **298**, 446–51.
8. Groves, J. T. and Mrla, S. (1995). *J. Am. Chem. Soc.*, **117**, 9578.
9. Beckman, J. S., Beckman, T. W., Chen, J., Marshall, P. A., and Freeman, B. A. (1990). Apparent hydroxyl radical production by peroxynitrite: implications for

endothelial injury from nitric oxide and superoxide. *Proc. Natl. Acad. Sci. USA*, **87**, 1620–4.

10. Kaur, H. and Halliwell, B. (1994). Nitrotyrosine in serum and synovial fluid from rheumatoid patients. *FEBS Lett.*, **350**, 9–12.
11. Ye, Y. Z., Strong, M., Huang, Z. Q., and Beckman, J. S. (1996). Antibodies that recognize nitrotyrosine. *Methods Enzymol.*, **269**, 210–19.

3

Chemiluminescence

MINORU NAKANO and YASUHIRO KAMBAYASHI

Introduction

Inflammatory cells such as neutrophils, eosinophils, and macrophages produce a multitude of reactive oxygen species when they are activated. Of the reactive oxygen species, superoxide anion ($O_2\cdot^-$), a precursor of several oxidants such as hydrogen peroxide (H_2O_2), hydroxy radical, singlet oxygen ($^1O_2$), and hypochlorite (HOCl), can be formed in, and released from, each of three cell lines. On the other hand, $^1O_2$ and hypohalides can be generated by myeloperoxidase (MPO)-mediated reactions in the neutrophil phagosome at pH 4.5, and by the eosinophil peroxidase-mediated reaction in eosinophils at physiological pH. It is, therefore, important to detect and measure the generation of $O_2\cdot^-$ and $^1O_2$ in small amounts in such biological systems. Because production of highly bactericidal agents such as $^1O_2$ and HOCl is a result of myeloperoxidase activity, simple and sensitive methods to measure this enzyme activity are required. In this section, we describe the measurement of $O_2\cdot^-$, $^1O_2$ generation, and myeloperoxidase activity, in leukocytes by chemiluminescence using cypridina luciferin analogues as chemiluminescence probes.

$O_2\cdot^-$ production by stimulated leukocytes (1–3)

When granulocytes or macrophages are stimulated, they produce and release $O_2\cdot^-$. The $O_2\cdot^-$ released from these cells reacts efficiently with cypridina luciferin analogue (CLA, 2-methyl-6-phenyl-3,7-dihydroimidazo[1,2-*a*]pyrazin-3-one; MCLA, 2-methyl-6-(*p*-methoxyphenyl)-3,7-dihydroimidazo[1,2-*a*]pyrazin-3-one) to produce an excited carbonyl which emits light in the visible region; i.e., maximum emission at 380 nm for CLA and at 457 nm for MCLA.

Protocol

1. Dissolve powdered Hanks' balanced salt solution (HBSS) for tissue culture (Nissui Pharmaceutical Co. Ltd, Tokyo) in distilled water (9.8 g L^{-1}) and adjust to pH 7.4 by use of aqueous Na_2HPO_4 solution (280 mOsm).

Protocol *Continued*

2. Prepare granulocytes or macrophages by standard methods.
3. Dissolve MCLA ($\varepsilon_{430\ nm}$ = 9600 $M^{-1}\ cm^{-1}$) or CLA ($\varepsilon_{410\ nm}$ = 8900 $M^{-1}\ cm^{-1}$) in distilled water at a concentration of 40 μM.
4. Dilute granulocytes or macrophages with HBSS at 5×10^4 cells for MCLA or 4×10^5 cells for CLA.
5. Add MCLA or CLA (50 μL, final concentration: 1 μM)[1].
6. Preincubate at 37 °C for 3 min.
7. Add opsonized zymosan (OZ, in water, 20 mg mL^{-1}, 80 μL), formyl–methionyl–leucyl–phenylalanine (fMLP in Ca^{2+}- and Mg^{2+}-free HBSS/DMSO = 1:1, *v/v*, 200 μM, 10 μL), or 4β-phorbol myristate acetate (PMA in Ca^{2+}- and Mg^{2+}-free HBSS/DMSO = 1:1, *v/v*, 2 μg mL^{-1}, 20 μL). (Total volume 2.0 mL).
8. Measure chemiluminescence.[2]

[1] Although CLA and MCLA (Tokyo Kasei) are stable in pure water, they are likely to be auto-oxidized in HBSS and phosphate buffer and emit weak light (baseline chemiluminescence). On addition of CLA or MCLA to non-stimulated granulocytes or macrophages, chemiluminescence intensity is almost the same as the baseline chemiluminescence. When granulocytes are stimulated by OZ, some of the granule myeloperoxidase is released into the extracellular fluid and reacts with H_2O_2 in the reaction mixture to oxidize MCLA (or CLA), emitting light. Therefore, MPO-mediated oxidations interfere with the measurement of $O_2\cdot^-$; such interference does not, however, occur with less than 4×10^5 granulocytes/2 mL.

[2] Liquid scintillation counters, photon counters, or similar luminescence detectors can be used for measuring chemiluminescence (CL). A luminescence detector equipped with an agitator is required to avoid granulocyte sedimentation during chemiluminescence measurement. The Luminescence Reader (Aloka, Type: BLR101, 102, 301) is suitable for this experiment. The chemiluminescence intensity (CLI; counts min^{-1}), obtained by subtracting MCLA (CLA)-dependent CLI of non-stimulated cells from that of stimulated cells, reflects the rate of $O_2\cdot^-$ generated from stimulated cells. The maximum rate of $O_2\cdot^-$ production and release from granulocytes can be calculated by the following two methods.

1) Maximum MCLA (CLA)-dependent CLI in systems containing a fixed concentration (43 μM) of hypoxanthine and different concentrations of xanthine oxidase (XO; 50–400 units, Sigma) by subtracting baseline CLI plotted against XO units. Using this calibration line, the maximum rate of $O_2\cdot^-$ generated in the experimental system can be expressed as xanthine oxidase units (2, 3). Maximum CLI = 433 and 2000 XO units for CLA and MCLA, respectively (2).

2) Superoxide anion can reduce the oxidized form of cytochrome-*c* (5). The incubation mixture is essentially the same as the MCLA system, except that 40 μM cytochrome-*c* is used instead of MCLA in a total volume of 3.0 mL. Reference controls contain all components in the reaction mixture and 0.5 μM SOD. The reaction is initiated by addition of a stimulant and measurement of the increase in the difference between absorbance at 550 nm and that at 540 nm, or the increase of absorbance at 550 nm ($\varepsilon_{(550-540\ nm)}$ = 19.1 $mM^{-1}\ cm^{-1}$ or $\varepsilon_{550\ nm}$ = 21.1 $mM^{-1}\ cm^{-1}$) (4, 5), is recorded under gentle agitation in a spectrophotometer. The initial rate of cytochrome-*c* reduction should correspond to the maximum CLI with subtraction of baseline CLI (6, 7). A calibration curve can be prepared by plotting maximum CLI against initial rate of cytochrome-*c* reduction.

MPO in granulocytes ($^1O_2$ measurements) (8)

When peripheral leukocytes derived from bone marrow are separated the neutrophils contain abundant granule myeloperoxidase as a host enzyme. Under acidic conditions, myeloperoxidase (MPO) catalyses the production of HOBr and HOCl in the presence of Br^- and Cl^-, respectively (reaction 1). The HOBr or HOCl can efficiently react with H_2O_2 to generate $^1O_2$ (reaction 2).

$$H_2O_2 + HBr\,(Cl) \xrightarrow[pH4.5 \sim 5.0]{MPO} HOBr\,(Cl) + H_2O_2 \qquad (1)$$

$$HOBr\,(Cl) + H_2O_2 \longrightarrow {}^1O_2 + H_2O + HBr\,(Cl) \qquad (2)$$

MCLA reacts not only with $O_2\cdot^-$ but also with $^1O_2$ to emit light with maximum intensity at 457 nm. The addition of 0.5 μM SOD to the system, in which both $O_2\cdot^-$ and $^1O_2$ are generated, completely quenches the $O_2\cdot^-$-dependent luminescence leaving only the $^1O_2$-dependent luminescence.

Protocol

1. Mixing human neutrophils and hexadecyltrimethylammonium bromide (0.02% *w*/*v*) in potassium phosphate buffer (pH 6.0, 50 mM) for 10 min.
2. Sonicate the mixture on ice for 30 s (0.5-s bursts at 0.5-s intervals) at an output level of 30 W (Branson Sonifier 250, Branson Ultrasonics Corporation, Danbury, CT, USA).
3. Centrifuge at 40 000 g for 20 min at 4°C (The supernatant is used as the sample.)
4. Preincubate a standard solution containing MCLA (in water, 200 μM, 100 μL), SOD (in water, 50 μM, 20 μL), KBr (in water, 50 μM, 20 μL), desferrioxamine (in water, 2 mM, 20 μL), H_2O_2 (40 mM, 25 μL), and acetate buffer (0.2 M, 1 mL) (total volume 1.95 mL).
4. Measure chemiluminescence until baseline is constant.
5. Inject samples rapidly (50 μL)[1] by means of a microsyringe.
6. Measure chemiluminescence.[2]

[1] The measurement should be performed rapidly after suitable dilution of the neutrophil extract. Granulocytes infiltrated into tissue can also be measured by this method.

[2] Use of a chemiluminescence detector equipped with temperature controller and agitator is desirable. The reaction is performed in a chemiluminescence reader (Aloka, BLR-301) at 25°C. At the time at which the chemiluminescence intensity of the reaction mixture (without sample) becomes constant, the sample is taken with a microsyringe and injected into the incubation mixture to start the reaction. The amount of MPO in granulocytes is calculated using purified MPO
1) MCLA-dependent maximum chemiluminescence intensity (MCLI, counts min^{-1}) induced by purified MPO is measured. The calibration curve (C.1) can be obtained by plotting MCLI against the amount of MPO used.
2) The same experiment is performed using neutrophil extract. MCLA-dependent MCLI measured is plotted against the number of granulocytes used for extraction (C.2).
3) The amount of MPO in granulocytes, and granulocyte number are calculated by use of calibration curves C.1 and C.2, respectively.

References

1. Nakano, M., Sugioka, K., Ushijima, Y., and Goto, T. (1986). Chemiluminescence probe with Cypridina luciferin analog, 2-methyl-6-phenyl-3,7-dihydroimidazo [1,2-*a*]pyrazin-3-one, for estimating the ability of human granulocytes to generate $O_2{\cdot}^-$. *Anal. Biochem.*, **159**, 363–9.
2. Nishida, A., Kimura, H., Nakano, M., and Goto, T. (1989). A sensitive and specific chemiluminescence method for estimating the ability of human granulocytes and monocytes to generate $O_2{\cdot}^-$. *Clin. Chim. Acta*, **179**, 177–82.
3. Sugioka, K., Nakano, M., Kurashige, S., Akuzawa, Y., and Goto, T. (1986). A chemiluminescent probe with a Cypridina luciferin analog, 2-methyl-6-phenyl-3,7-dihydroimidazo[1,2-*a*]pyrazin-3-one, specific and sensitive for $O_2{\cdot}^-$ production in phagocytizing macrophages. *FEBS Lett.*, **197**, 27–30.
4. Babior, B. M., Kipnes, R. S., and Curnutte, J. T. (1973). Biological defence mechanisms. The production by leukocytes of superoxide, a potential bactericidal agent. *J. Clin. Invest.*, **52**, 741–4.
5. McCord, J. M. and Fridovich, I. (1969). Superoxide dismutase. An enzymic function for erythrocuprein (hemocuprein). *J. Biol. Chem.*, **244**, 6049–55.
6. Koga, S., Nakano, M., and Tero-Kubota, S. (1992). Generation of superoxide during the enzymatic action of tyrosinase. *Arch. Biochem. Biophys.*, **292**, 570–5.
7. Hayakawa, H., Sato, A., Yagi, T., Uchiyama, H., Ide, K., and Nakano., M. (1995). Superoxide generation by alveolar macrophages from aged rats: improvement by *in vitro* treatment with IFN-γ. *Mech. Ageing Dev.*, **80**, 199–211.
8. Uehara, K., Hori, K., Nakano, M., and Koga, S. (1991). Highly sensitive chemiluminescence method for determining myeloperoxidase in human polymorphonuclear leukocytes. *Anal. Biochem.*, **199**, 191–5.

4

ESR Spin trapping

HIROAKI KOSAKA

Introduction

Election spin resonance (ESR) spectroscopy is a technique that can be applied to the detection and measurement of free radicals, because it detects the presence of unpaired electrons. An unpaired electron has a spin of either + or − and behaves as a small magnet. ESR spectroscopy is highly sensitive and can detect radicals at levels as low as 10^{-10} M, if they are sufficiently long-lived to be measured. For very unstable radicals in biological systems, however, alternative approaches, are required. One such example is spin-trapping, in which a highly reactive radical reacts with a trapping reagent producing a long–lived nitroxide radical, for example, that can be detected and measured (1).

Protocol

1. Set ESR spectrometer as follows:

   incident microwave power, 5–20 mW;
   modulation frequency, 100 kHz;
   modulation amplitude 0.3–1.0 Gauss;
   response time 0.5–1.0 s;
   sweep rate 12.5 Gauss min^{-1}.
2. Add the following spin trapping agent to the reaction mixture:

   5,5-dimethyl-1-pyrroline *N*-oxide (DMPO);
   N-*t*-butyl-α-phenylnitrone (PBN);
   α-4-Pyridyl 1-oxide *N*-*tert*-butyl nitrone (4-POBN).
3. Add the agent to start the reaction.
4. Place the reaction solution in an ESR quartz flat cell.
5. Place the flat cell in the cavity of the ESR spectrometer.
6. Adjust frequency control to get Q dip at the centre of display.
7. Adjust the flat cell to obtain a maximum Q dip.
8. Adjust phase and iris.

Protocol *Continued*

9. Start the measurement, sequential ESR scans are then recorded.
10. Analyse the ESR signal of spin-trapping adduct.
11. To determine the spin concentration, integrate doubly the ESR signal and calculate the spin concentration by comparing the signal with that of a standard Tempol (2,2,6,6-tetramethyl-4-hydroxypiperidine-1-oxyl) solution. The concentration of Tempol can be determined optically from the extinction coefficient at 240 nm of 1440 M^{-1} cm^{-1}.

Spin-trapping of NO with HbCO in the peritoneal cavity

It takes a long time to displace oxygen from Hb. Because CO readily replaces oxygen in Hb and NO readily replaces CO in Hb, HbCO can be used as a trapping agent for NO. This method can be applied under low pO_2 (2).

Protocol

1. Treat male Wistar rats (*ca* 100 g) intraperitoneally with nitric oxide synthetase-inducing agent, i.e. *E. coli* LPS.
2. Stir Hb solution in a closed flask and introduce CO gas (100%). After conversion of Hb to HbCO, remove excess CO gas in the HbCO solution by purging with nitrogen gas.
3. After induction of nitric oxide synthase, inject the HbCO solution into the peritoneal cavity.
4. A few hours later, aspirate peritoneal fluid by syringe, transfer to a quartz ESR tube through a long needle, and freeze immediately in liquid nitrogen. The syringe, long needle and the ESR tube should be prepared oxygen-free by means of flowing nitrogen.
5. Set the ESR spectrometer as follows:

   incident microwave power, 5–20 mW;
   modulation frequency, 100 kHz;
   modulation amplitude, 2–5 Gauss;
   response time, 0.5–1.0 s;
   sweep rate, 125 Gauss min^{-1};
   temperature 77–150 K.

   For stability and sensitivity, low-temperature operation is recommended. The cavity must be flushed with dry nitrogen gas during measurement.
6. Start the measurement after adjustment of the ESR spectrometer.

7. To see a three-line hyperfine structure characteristic of HbNO, the peritoneal fluid should be added in oxygen-free inositol hexaphosphate solution.

Measurement of HbNO in circulating blood

HbNO in circulating blood can be detected, probably because of the low pO_2 of the venous circulation, and there is a spectral transition between A and V (3, 4).

Protocol

1. Treat rats with LPS or cytokines.
2. Collect blood (0.4 mL) from a vein or artery, transfer to an ESR tube through a long needle, and freeze immediately in liquid nitrogen.
3. Measure the ESR spectrum.
4. Determine HbNO concentration on the basis of double integration of the first derivative ESR spectrum, because of the spectral diversity of HbNO.
5. Prepare HbNO standard anaerobically with NO gas (*via* HbCO). Briefly, oxyHb is converted to HbCO. Then introduction of nitric oxide gas converts HbCO to HbNO.

References

1. Shimizu, S., Eguchi, Y., Kosaka, H., Kamiike, W., Matsuda, H., and Tsujimoto, Y. (1995). Prevention of hypoxia-induced cell death by Bcl-2 and Bcl-xL. *Nature*, **374**, 811–13.
2. Kosaka, H. and Shiga, T. (1996). Detection of nitric oxide by ESR using haemoglobin. In *Methods in Nitric Oxide Research*, (ed. M. Feelisch and J. S. Stamler), pp. 373–81. Wiley and Sons.
3. Kosaka, H., Sawai, Y., Sakaguchi, H., Kumura, E., Harada, N., Watanabe, M., and Shiga, T. (1994). ESR spectral transition by arteriovenous cycle in nitric oxide haemoglobin of cytokine-treated rats. *Am. J. Physiol.*, **266**, C1400-5.
4. Kosaka, H. and Seiyama, A. (1997). Elevation of oxygen release by nitroglycerine without increase in blood flow in hepatic microcirculation. *Nature Medicine*, **3**(4), 456–9.

5

Nitrosothiol detection by HPLC coupled with flow reactors of Hg^{2+} and Griess reagent

TAKAAKI AKAIKE and KATSUHISA INOUE

Introduction

NO-related intermediates, including NO^{+}-like species, can participate in nitrosating additions to nucleophilic centres of biological molecules. Sulfhydryl-containing molecules such as glutathione are particularly susceptible to nitrosation and form nitrosothiol adducts (nitrosothiols; RS-NOs). It seems that these adducts in biological systems play an important role in NO-mediated signalling cascades such as the downregulation of *N*-methyl-D-aspartate receptor, and the regulation of transcriptional factors (1); they might also be involved in non-adrenergic and non-cholinergic neuronal responses (2). It is, therefore, essential to identify nitrosothiols specifically and to quantify them in biological systems.

In this section we describe a sensitive and specific HPLC method coupled with Hg^{2+} and Griess reagent for nanomolar quantification of a wide range of RS-NOs including low-molecular weight RS-NOs such as nitrosoglutathione (GS-NO) and nitrosocysteine (Cys-NO), and also *S*-nitrosoproteins, in particular *S*-nitrosoalbumin and *S*-nitrosohaemoglobin (3).

Preparation of *S*-nitrosoproteins

Protocol

Reagents

Bovine serum albumin (BSA) (Nacalai Tesque, Osaka, Japan), dithiothreitol (DTT; Wako Pure Chemicals, Osaka, Japan), isoamyl nitrite (Wako), ethylenediaminetetraacetic acid (EDTA; Dojindo Laboratories, Kumamoto, Japan), diethylenetriaminepentaacetic acid (DTPA; Dojindo), and 5,5′-dithiobis[2-nitrobenzoic acid] (DTNB; Nacalai).

Procedure

Reduction of BSA[1]

1. Add DTT (100 mM, 10 μL) to BSA (1 mM, 1 mL) in sodium phosphate buffer (pH 7.0, 100 mM).
2. Incubate for 30 min at 37 °C.
3. Apply to a Sephadex G-100 (Pharmacia, Uppsala, Sweden) column equilibrated with sodium phosphate buffer (pH 7.0, 10 mM) containing EDTA (1 mM; buffer A), and elute with buffer A.
4. Collect the fractions containing BSA and concentrate the fractions by ultrafiltration[2].

Nitrosation of the reduced BSA

1. Add isoamyl nitrite (100 mM, 10 μL) to reduced BSA (0.1 mM, 1 mL) in sodium phosphate buffer (pH 7.8, 100 mM) containing DTPA (0.5 mM).
2. Incubate for 30 min at 37 °C.
3. Apply to a Sephadex G-25 (Pharmacia) column equilibrated with buffer A, and elute with buffer A.
4. Collect the fractions containing *S*-NO-BSA and concentrate the fractions by ultrafiltration.

[1] Any type of protein containing sulfhydryl groups other than BSA can be nitrosated after reduction to form free sulfhydryl (–SH), which becomes reactive and accessible to NO^+ or nitrosation reagent, i.e., isoamyl nitrite.
[2] The amount of free SH moiety per molecule of the protein should be quantified by the DTNB method; details of the assay are described elsewhere (4).

Quantification of nitrosothiols

Protocol

Reagents

Nitrosoproteins (e.g., *S*-NO-BSA) prepared above, GS-NO (Dojindo), Cys-NO (Dojindo), $HgCl_2$, and Griess reagent (naphthylethylenediamine; sulfanilamide (Wako)).

Principle

RS-NOs of different molecular size separated by HPLC are converted to NO^+ or NO_2^-, in a flow reactor system connected serially to the HPLC (Fig. 1), by rapid and quantitative metal-catalysed reaction with Hg^{2+}. As

Protocol *Continued*

reported by Saville, Hg^{2+} decomposes the nitrosothiols stoichiometrically to nitrite according to the equations (5, 6):

$$RS\text{–}NO + Hg^{2+} \longleftrightarrow [RS\text{–}(Hg)NO]^{2+}$$
$$[RS\text{–}(Hg)NO]^{2+} + H_2O \rightarrow RSHg^{+} + NO_2^{-} + 2H^{+}$$

Peak detection is based on the colorimetric assay of nitrite using Griess reagent, which forms a diazo dye having strong absorbance at 540 nm under acidic conditions; the extinction coefficient is 53 000 M^{-1} cm^{-1} (7).

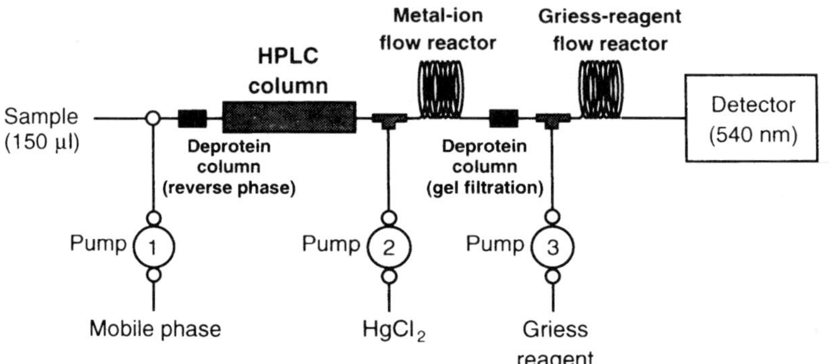

Figure 1. Flow diagram of the HPLC-flow reactor system for measurement of different RS-NOs by use of Hg^{2+} and Griess reagent flow reactors.

HPLC conditions

The injection volume is 150 µL

Column

1. Separation: a C_{18}-reversed-phase column (4.6 × 250 mm; TSKgel ODS-80Ts; Tosoh, Tokyo) for low molecular weight RS-NOs; a gel filtration column (8 × 300 mm; Diol-120; YMC, Kyoto) for the nitrosoproteins.
2. Deproteinating column: small columns (3 × 10 mm) packed with C_{18}-based resin are placed just before the separation column and just after Hg^{2+}-flow reactor coil, in the reversed-phase and gel filtration systems, respectively[1].

Mobile phase

1. Pump 1 for HPLC: sodium acetate buffer (pH 5.5, 10 mM) containing DTPA[2] (0.5 mM) and methanol (0–7%), for low molecular weight RS-NOs (reversed-phase HPLC); sodium acetate buffer (pH 5.5, 10 mM) containing DTPA (0.5 mM) and NaCl (150 mM) for the nitrosoprotein (gel filtration HPLC).
2. Pump 2 for Hg^{2+}-flow reactor[3]: $HgCl_2$[2] (1.75 mM) in sodium acetate buffer (pH 5.5, 10 mM).

3. Pump 3 for Griess reagent flow reactor[3]: naphthylethylenediamine (0.1%) in H_2O; sulfanilamide (1.0%) + phosphoric acid (2.0%) in H_2O[4].

Flow rate:

0.55 mL min^{-1} (pump 1), 0.2 mL min^{-1} (pump 2), 0.24 mL min^{-1} (pump 3)

Detector:

visible detector (Eicom, Kyoto) operated at 540 nm.

[1] The use of a deproteinating column is critical, in particular for the determination of nitrosoproteins, because the protein, irrespective of nitrosation, often gives somewhat non-specific absorbance with the visible detector.

[2] DTPA added to the mobile phase could prevent decomposition of RS-NOs eluted from the HPLC column, but does not interfere with the reaction of Hg^{2+} with RS-NOs, because the concentration of $HgCl_2$ (911 mM) is much higher than that of DTPA (348 mM) in the Hg^{2+} flow reactor system.

[3] The flow reactor coils must be long enough to enable complete RS-NOs decomposition and azo compound formation with the Griess reagent. In addition, the temperature of the flow reactors is maintained at 40°C by means of a thermostat. One pump can be saved for the measurement of the low molecular mass RS-NOs (e.g., GS-NO and Cys-NO) by providing a mixture of Hg^{2+} solution and Griess reagent from a single pump. The Hg^{2+} and Griess reagent flow reactors must, however, be placed separately and serially by using two pumps for the nitrosoprotein assay, because the capacity of the deproteinating column diminishes markedly under acidic elution conditions if the column is eluted with the acidic Griess solution.

[4] Griess reagent can be stored for as long as two months, if 0.2% naphthylethylenediamine in H_2O and 2.0% sulfanilamide plus 4.0% phosphoric acid in H_2O are prepared separately. These solutions are mixed together just before the measurement of RS-NOs.

Typical elution profiles of GS-NO, Cys-NO and S-NO-BSA are shown in Fig. 2. Fig. 3 illustrates the correlation between the peak area of GS-NO calculated

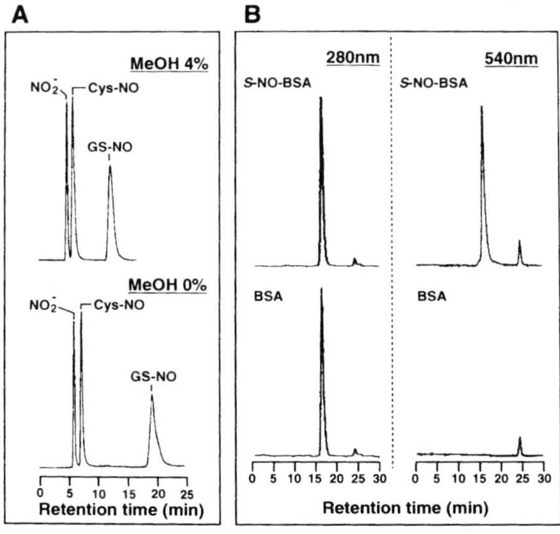

Figure 2. Elution profiles of different RS-NOs for the HPLC/Hg^{2+} flow reactor system. A: Low molecular weight RS-NOs (reversed-phase HPLC). B: High molecular weight RS-NOs (gel filtration HPLC).

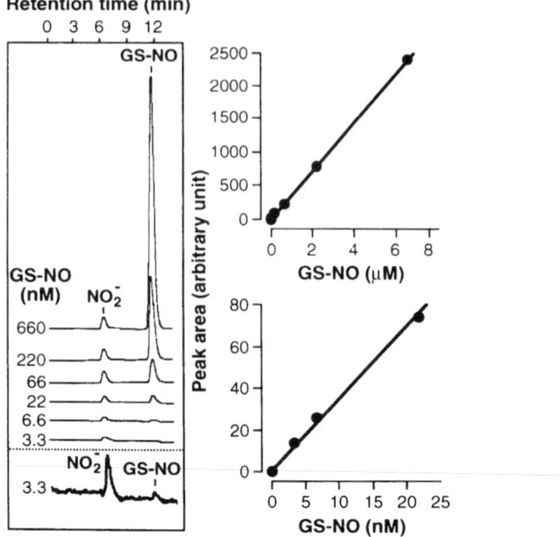

Figure 3. Quantitation of GS-NO by means of the HPLC-flow reactor system. The peak area of GS-NO correlates well with GS-NO concentrations ranging from 3.3 nM to 6.6 mM ($r^2 = 1.00$).

by the integrator and different concentrations examined by use of the HPLC flow reactor. The detection limit for GS-NO was found to be >3 nM.

References

1. Hausladen, A., Privalle, C. T., Keng, T., DeAngelo, J., and Stamler, J. S. (1996). Nitrosative stress: activation of the transcription factor OxyR. *Cell*, **86**, 719–29.
2. Rand, M. J. and Li, C. G. (1995). Discrimination by the NO-trapping agent, carboxy-PTIO, between NO and the nitrergic transmitter but not between NO and EDRF. *Br. J. Pharmacol.*, **116**, 1906–10.
3. Akaike, T., Inoue, K., Okamoto, T., Nishino, H., Otagiri, M., Fujii, S., and Maeda H. (1997). Nanomolar quantification of various nitrosothiols by high performance liquid chromatography coupled with flow reactors of metals and Griess reagent. *J. Biochem.*, **122**, 459–66.
4. Jenke, D. R. and Brown, D. S. (1987). Determination of cysteine in pharmaceuticals via liquid chromatography with post-column derivatization. *Anal. Chem.*, **59**, 1509–12.
5. Saville, B. (1958). A scheme for the colorimetric determination of microgram amounts of thiols. *Analyst*, **83**, 670–2.
6. Williams, D. H. L. (1996). *S*-Nitrosothiols and role of metal irons in decomposition to nitric oxide. *Methods Enzymol.*, **268**, 299–308.
7. Wishnok, J. S., Glogowski, J. A., and Tannenbaum, S. R. (1996). Quantitation of nitrate, nitrite, and nitrosating agents. *Methods Enzymol.*, **268**, 130–42.

6

NO-generating system

JUNICHI FUJII

Introduction

Nitric oxide (NO) produced by nitric oxide synthases (NOS) *in vivo* plays multiple roles in the body. There are three NOS isozymes whose tissue-distribution and regulation are variable. Effects of NO produced by NOS, which is expressed constitutively or induced by various stimuli, can be examined *in vitro* and *in vivo*. Precise control of NO production by NOS *in vivo* or in culture is, however, not possible. Because NO is such a simple molecule and a fairly stable radical, NO gas and NO donors are available for various experiments. If high concentrations are required, NO gas can be used by simply flushing target cells or tissues. Various NO donors, such as nitrosocysteine, nitrosoglutathione, *S*-nitroso-*N*-acetyl-D,L-penicillamine (SNAP), 3-morpholinosyndonimine (SIN-1) (produces both NO and superoxide), which spontaneously release NO species, are also commercially available.

Protocol

NO generation by peritoneal macrophages

1. Collect macrophages from the peritoneal cavity with Hank's balanced salt solution (without calcium and magnesium but with heparin (10 units mL^{-1})).
2. Centrifuge at 250 g for 10 min at 4 °C.
3. Isolate the supernatant by aspiration.
4. Add an appropriate volume of supplemented Eagle's minimum essential medium (SMEM) without phenol red and supplemented with sodium bicarbonate (2.0 g L^{-1}); sodium pyruvate (110 mg L^{-1}); glucose (3.5 g L^{-1}); L-glutamine (584 mg L^{-1}); penicillin (5000 units L^{-1}); streptomycin (50 µg L^{-1}); Hepes (15 mM); and 10% heat-inactivated foetal calf serum
5. Repeat twice.
6. Adjust number of macrophages to 1×10^6 mL^{-1}.

Protocol *Continued*

7. Incubate macrophages to adhere at 37 °C for more than 1 h.
8. Isolate the supernatant by aspiration.
9. Add an appropriate volume of SMEM containing lipopolysaccharide + interferon-γ.
10. After several hours the macrophages produce nitrite and nitrate.

NO gas

1. Displace O_2 in pipework with flowing inert gas.
2. Introduce NO gas for reaction with reactant. (A mixture of 10% NO in N_2 gas is commercially available.)
3. After the reaction NO must be displaced with flowing inert gas.

NO donors

Various NO donors are available and applicable for *in vivo* and *in vitro* experiments. They generally release reactive nitrogen species spontaneously when dissolved in culture medium or administered to animals.

Results

The results are illustrated in Fig. 1.

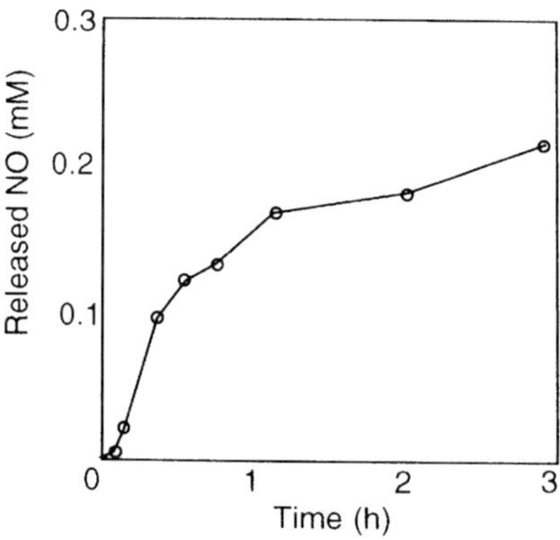

Figure 1. Time course of NO released from 0.5 mM SNAP.

Comments

NO generation by peritoneal macrophages.

A mouse-derived cell line such as RAW264.7 (commercially available from ATCC) is also often used.

Bloodletting of animals should be conducted to avoid contamination with blood cells. After injection of Hank's balanced salt solution into the peritoneal cavity, the abdomen should be massaged vigorously for *ca* 1 min to free adhered macrophages. Take care not to aspirate the internal organs or fatty tissue when the intraperitoneal fluid is collected. It is better to use young rats.

In this system, NO produced can be reacted with oxygen and spontaneously converted to NO_2, N_2O_3, N_2O_4, and finally NO_2^- and NO_3^-. NO_2^- and NO_3^- can be easily measured by use of an automated Griess method (1). It is important to know whether or not the sample dose contains inhibitors of the reduction of NO_3^- to NO_2^- (2, 3).

NO gas

Stainless steel gas regulators should be used on NO gas tanks to avoid leaks resulting from corrosion. High concentrations of NO react with oxygen producing NO_2 and the solution becomes acidic. To avoid this, sample solutions should be freed from oxygen by purging with inactive gas (N_2, Ar, He).

Because NO and NO_2 are harmful, all experiments should be conducted in a fume cupboard. An outlet pipe from equipment should be vented into the fume cupboard.

The concentration of NO can be measured by use of a thermal energy analyser.

NO donors

Because each NO donor releases specific NO species with different time courses and different efficiencies, special attention should be paid to distinguish whether results arise as a result of NO species or as a side-effect of the donor molecule.

References

1. Kosaka, H., Wishnok, J. S., Miwa, M., Leaf, C. D., and Tannenbaum, S.R. (1989). Nitrosation by stimulated macrophages. Inhibitors, enhancers and substrates. *Carcinogenesis*, **10**, 563–6.
2. Granger, D. L., Taintor, R. R., Boockvar, K. S., and Hibbs, J. B. Jr. (1996). Measurement of nitrate and nitrite in biological samples using nitrate reductase and Griess reaction. *Methods Enzymol.*, **268**, 142–51.
3. Kumura, E., Kosaka, H., Shiga, T., Yoshimine, T., and Hayakawa, T. (1994). Elevation of plasma nitric oxide end products during focal cerebral ischaemia and reperfusion in the rat. *J. Cereb. Blood. Flow. Metab.*, **14**, 487–91.

7

Superoxide-generating system

KEIICHIRO SUZUKI

Introduction

There are several chemical ways of generating superoxide anions ($O_2\cdot^-$); in the biochemical and medical research fields an enzymatic method using xanthine oxidase is widely used (Fig. 1).

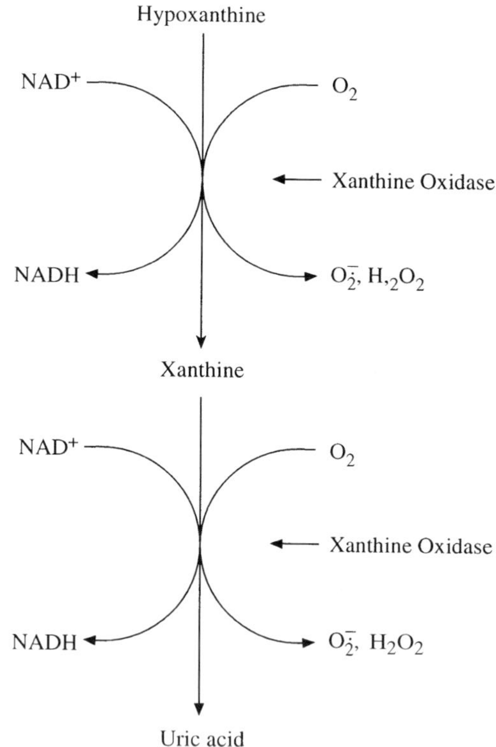

Figure 1. Generation of superoxide. ($O_2\cdot^-$) and hydrogen peroxide from hypoxanthine and xanthine oxidase.

Protocol

Reagents

Hypoxanthine solution (1 mM) and xanthine oxidase.

Procedure

1. Take hypoxanthine solution (1 mM). For cell culture, use phosphate-buffered saline or foetal calf serum-free medium instead of usual culture medium.
2. Add xanthine oxidase (0.01~100 munits mL^{-1}).
3. Incubate for an appropriate time determined experimentally.
4. For cell culture, remove the hypoxanthine solution by aspiration and add the culture medium. The cells are harvested when ready.

Comments

If xanthine oxidase is used H_2O_2 and $O_2\cdot^-$ will be produced. If transition metal ions or their redox active complexes are also present hydroxyl radical will be produced. To remove H_2O_2 add catalase to the reaction.

8

ROS-generating systems in mitochondria, microsomes and peroxisomes

DONGCHON KANG and KOICHIRO TAKESHIGE

Preparation of each fraction

Bovine heart submitochondrial particles (SMP)

Protocol

Keep the sample temperature below 4 °C whenever possible.

1. Put fresh bovine heart obtained from a slaughter house in a vinyl bag and keep on ice during transfer.
2. Remove fat and endothelium, cut into 2–3-cm cubes and keep on ice in a vinyl bag until mincing.
3. Mince with an electric food cutter for *ca* 1 min and keep on ice in a vinyl bag. It is possible to stop experiments by freezing the mince in a vinyl bag at this step. For quick freezing and thawing, the mince should be spread thinly in the bag. Usually, 1–2 kg of the mince is used for the preparation of SMP.
4. Suspend in ice-cold 0.25 M sucrose–0.185% K_2HPO_4 (3 L/kg mince).
5. Adjust pH to *ca* 7.5 quickly by adding 6 M KOH (8–10 mL kg^{-1}).
6. Homogenize with a food mixer at maximum power for 2 min. The homogenates should be kept on ice until centrifugation.
7. Centrifuge at 2000 g for 10 min (Hitachi RPR 12–2 rotor, 4600 rpm).
8. Decant the supernatant and filter through four layers of gauze.
9. Centrifuge the supernatant at 11 000 g for 40 min (Hitachi RPR 12–2 rotor, 10 600 rpm).
10. Suspend the pellet in a small volume of sucrose (0.25 M, 40–80 mL) and homogenize with a Potter–Elvehjem type homogenizer (3–5 strokes). The homogenizer should be kept in the ice–water bath during homogenization.

11. Dilute the homogenate with sucrose (0.25 M, 400–800 mL).
12. Centrifuge at 11 000 g for 40 min.
13. Repeat procedures 10–12 twice more.
14. Suspend in sucrose (0.25 M, 10–20 mL) containing EDTA (pH 7.5, 2 mM, ~30 mg mL^{-1}).
15. Freeze at −20 °C for at least 3 days.
16. Thaw quickly in flowing water.
17. Sonicate 10 mL of the mitochondrial sample six times for 45 s at 1-min intervals (Branson sonifier, cell disrupter 200, output 3.5–5.0 with a flat tip). The sample should be kept in the ice–water bath during and between sonications.
18. Centrifuge at 12 000 g for 15 min (Hitachi RPR20 rotor, 13 000 rpm).
19. Centrifuge the supernatant at 77 000 g for 1 h (Hitachi RP30 rotor, 30 000 rpm).
20. Suspend the pellet in a small volume of sucrose (0.25 M) containing EDTA (2 mM) and homogenize with a Potter–Elvehjem type homogenizer (3–5 strokes).
21. Repeat procedures 18–20 twice more.
22. Finally suspend and homogenize in several millilitres sucrose (0.25 M) containing Hepes (5 mM), NaOH (5 mM), and EDTA (0.1 mM) (pH 7.5).
23. Store at −20 °C.

Comments

1–2 g mitochondria are obtained from 1 kg mince. The recovery of submitochondrial particles from the mitochondrial fraction is 20–30%.

Rat liver microsomes

Protocol

1. Perfuse rat liver with cold saline.
2. Cut into pieces with scissors and suspend in 8 mL/g wet weight of sucrose (0.25 M), Hepes (5 mM), NaOH (5 mM), and EDTA (0.1 mM) at pH 7.4.
3. Homogenize with a Potter–Elvehjem type homogenizer (3–5 strokes).
4. Centrifuge at 7000 g for 10 min (Hitachi RPR20 rotor, 10 000 rpm).
5. Centrifuge the supernatant at 77 000 g for 1 h (Hitachi RP30, 30 000 rpm).
6. Suspend the pellet in sucrose (0.25 M), Hepes (5 mM), NaOH (5 mM), and EDTA (0.1 mM) at pH 7.4.

Protocol *Continued*

7. Homogenize with a Potter–Elvehjem type homogenizer (3–5 strokes).
8. Centrifuge at 77 000 g for 1 h.
9. Repeat procedures 6–8 twice more.
10. Suspend and homogenize in a small volume of sucrose (0.25 M), Hepes (5 mM), NaOH (5 mM), and EDTA (0.1 mM) at pH 7.4.
11. Store at −20 °C.

Comments

Approximately 50 mg microsomes are obtained from 10 g wet weight liver.

Peroxisomes of rat liver

Protocol

1. Inject clofibrate (200 mg kg^{-1}) subcutaneously once a day consecutively for two weeks.
2. Perfuse rat liver with cold saline.
3. Cut into pieces with scissors and suspend in 4 mL/g of wet weight sucrose (0.25 M) containing ethanol (0.1% *v/v*). (Ethanol is added to prevent inactivation of catalase.)
4. Homogenize with a Potter–Elvehjem type homogenizer (3–5 strokes).
5. Centrifuge at 1000 g for 10 min (Hitachi RP20, 3700 rpm).
6. Homogenize the pellet in 2 vol. of the same buffer.
7. Centrifuge at 1000 g for 10 min.
8. Combine the supernatant from procedures 5 and 7.
9. Centrifuge the combined supernatant at 20 000 g for 15 min (Hitachi RP20, 16 600 rpm).
10. Suspend the pellet in the same buffer and homogenize.
11. Centrifuge at 20 000 g for 15 min.
12. Repeat procedures 10 and 11.
13. Suspend and homogenize in a small volume of sucrose (0.25 M) containing ethanol (0.1% *v/v*).
14. Store at −20 °C.

Comments

50–100 mg peroxisomes are obtained from 10 g wet weight liver.

Measurement of the production of superoxide anion ($O_2\cdot^-$)

$O_2\cdot^-$ is produced by complex I and ubisemiquinone in mitochondria, and by cytochrome P-450 in microsomes.

Mitochondria (1)

Epinephrine (adrenaline) is reduced to adrenochrome by $O_2\cdot^-$.

Protocol

1. Prepare reaction mixture: Hepes–NaOH (pH 7.5, 100 mM) containing sucrose (0.5 M), (500 μL; final concentration 50 mM/0.25 M); water (450 μL); rotenone in ethanol (100 μM, 10 μL; (final concentration 1 μM); epinephrine (200 mM, 5 μL; final concentration 1 mM); bovine heart SMP (20 mg mL^{-1}, 25 μL; final concentration 0.5 mg mL^{-1}); total volume 990 μL.
2. Preincubate at 37 °C for 5 min.
3. Record the baseline at 485–575 nm (Hitachi spectrophotometer 557) for at least 1 min.
4. Start the reaction by adding NADH (20 mM, 10 μL) in Hepes–NaOH (pH 7.5, 50 mM) or in sodium succinate (pH 7.0, 2 M) (final concentrations 0.2 mM or 20 mM, respectively).
5. Add 10 μL 1 mg mL^{-1} superoxide dismutase (SOD).

Results and calculations

The SOD inhibitable reduction of epinephrine is calculated as the reduction by $O_2\cdot^-$ (Fig. 1). The absorbance coefficient for adrenochrome is 2.96 mM^{-1} cm^{-1}.

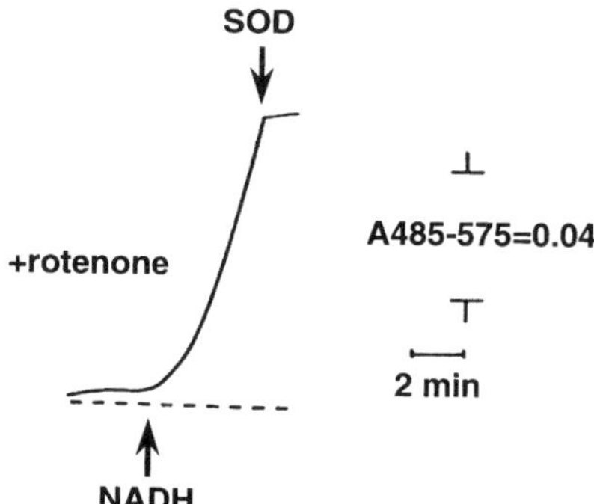

Figure 1. $O_2\cdot^-$ production by bovine heart SMP.

Reference

1. Takeshige, K. and Minakami, S. (1979). NADH- and NADPH–dependent formation of superoxide anions by bovine heart submitochondrial particles and NADH–ubiquinone reductase preparation. *Biochem. J.*, **180**, 129–35.

Microsomes

Preparation of acetylated cytochrome-*c*

The reduction of cytochrome-*c* by endogenous electron-transfer systems in microsomes is reduced to below 10% by acetylation of cytochrome-*c*.

Protocol

All procedures should be performed at temperatures below 4 °C.

1. Dilute saturated sodium acetate (2.5 mL) with distilled water (2.5 mL).
2. Add cytochrome-*c* (horse heart, Sigma; 50 mg) with stirring.
3. Add acetic anhydride (64 μL) slowly.
4. Allow the reaction to continue for 60 min.
5. Dialyse twice for 12 h in 2 L water at 4 °C.
6. Determine the concentration and store at −20 °C.

Measure the increase in absorbance at 550–540 nm in Hepes–NaOH (pH 7.5, 50 mM) after reduction by addition of a small amount of powdered $Na_2S_2O_4$. The millimolar absorption coefficient is 19.1 mM^{-1} cm^{-1}. Acetylated cytochrome-*c* is stable to several freeze thaw cycles.

Measurements

Protocol

1. Reaction mixture: Hepes–NaOH (pH 7.7, 100 mM, 500 μL; final concentration 50 mM); water (280 μL); acetylated cytochrome-*c* (0.3 mM, 200 μL; final concentration 60 μM); rat liver microsomes (2 mg mL^{-1}, 10 μL; final concentration 20 μg mL^{-1}); total volume 990 μL.
2. Preincubate at 37 °C for 5 min.
3. Record the baseline for *ca* 1 min at 550–540 nm (Hitachi spectrophotometer 557).
4. Start the reaction by adding NADPH (20 mM, 10 μL) in Hepes–NaOH (pH 7.7, 50 mM) (0.2 mM final).
5. Add SOD (1 mg mL^{-1}, 10 μL).

Results and calculations

The SOD-inhibitable part is calculated as the reduction of acetylated cytochrome-*c* by $O_2{\cdot}^-$ (Fig. 2).

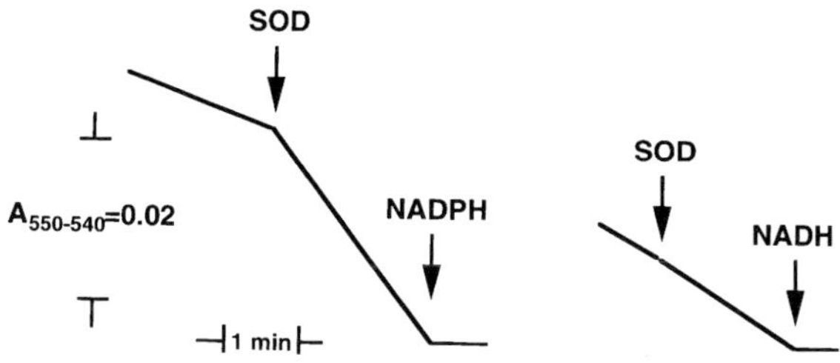

Figure 2. $O_2{\cdot}^-$ production by rat liver microsomes.

Measurement of the production of hydrogen peroxide (H_2O_2)

H_2O_2 results from the dismutation of $O_2{\cdot}^-$ in mitochondria and microsomes. In peroxisomes H_2O_2 is produced by acyl coenzyme A, an oxidase which is involved in the beta-oxidation of fatty acids.

Mitochondria–scopoletin method

Protocol

1. Reaction mixture: Tris-MOPS (pH 7.4, 60 mM) containing mannitol (0.46 M) and sucrose (0.14 M), 1000 μL (final concentration 30 mM, 0.23 M, and 0.07 M); water (740 μL); horseradish peroxidase (RZ > 3; Sigma; 100 μM, 20 μL; final concentration 1 μM); SMP (200 μL, final concentration 2 mg mL^{-1}); antimycin A in ethanol (200 μM, 10 μL; final concentration 1 μM); scopoletin in ethanol (200 μM, 10 μL; final concentration 1 μM); total volume 2980 μL.
2. Preincubate at 37 °C for 5 min.
3. Record baseline fluorescence (λ_{ex} 365 nm, λ_{em} 450 nm).
4. Start the reaction by adding sodium succinate (1 M 20 μL; 10 mM final) and record the decrease in the fluorescence.

Microsomes–catalase method

Protocol

1. Reaction mixture: Tris-HCl (pH 7.5, 100 mM) containing KCl (0.3 M) and $MgCl_2$ (20 mM, 3.75 mL); water (2.13 mL); trisodium isocitrate (80 mM, 750 μL; final concentration 8 mM); isocitrate dehydrogenase (Oriental; 100 units mL^{-1}, 25 μL; total amount 0.3 units); catalase (Boehringer–Mannheim; 13×10^5 units mL^{-1}, 5 μL; total amount 867 units); methanol (15 μL; final concentration 3 mM); rat liver microsomes (750 μL, 15 mg mL^{-1}; final concentration 1.5 mg mL^{-1}); total volume 7.425 mL.
2. Preincubate at room temperature for 5 min.
3. Start the reaction by adding NADPH (40 mM, 75 μL; 0.4 mM final).
4. Take aliquots (500 μL) every 2 min and add to ice-cold trichloroacetic acid (TCA; 15%, 500 μL) in an Eppendorf tube.
5. Centrifuge at 6000 g for 10 min (Tomy TMA-2, 10 000 rpm).
6. Mix the supernatant (750 μL) with Nash reagent (750 μL)[1].
7. Heat at 58 °C for 10 min.
8. Cool to room temperature.
9. Measure absorbance at 412 nm.

[1] To prepare Nash reagent dissolve ammonium acetate (150 g) and acetyl acetone (2 mL) in water (~900 mL), adjust the pH to 6.0–6.2 with acetic acid (3–5 mL), and finally dilute to 1 L. This reagent is stable at room temperature for 2 weeks.

Peroxisomes–catalase method

Protocol

1. Reaction mixture: Tris-HCl (pH 7.5, 100 mM) containing KCl (0.3 M, 1.0 mL; final concentrations 50 mM and 0.15 M); water (896 μL); methanol (4 μL; final concentration 3 mM); rat liver peroxisomes (20 mg mL^{-1}, 50 μL; final concentration 0.5 mg mL^{-1}); total volume 1.95 mL.
2. Preincubate at room temperature for 5 min.
3. Start the reaction by adding palmitoyl-CoA (10 mM, 50 μL; dissolve before use, 0.2 mM final).
4. Take aliquots (500 μL) at 0, 1, and 2 min and add to ice-cold TCA (15%, 500 μL).
5. Proceed as for microsomes.

Lipid peroxidation

Ferric salts are reduced to the ferrous state by the mitochondrial or microsomal electron transport systems, and ferrous salts are more active in redox reactions leading to lipid peroxidation. Malondialdehyde (MDA), a degradation product resulting from lipid peroxidation reactions, is a widely used quantitative marker for lipid peroxidation.

Mitochondria MDA quantitation by HPLC

Preparation of standard MDA

Protocol

1. Dissolve 1,1,3,3-tetraethoxypropane (Sigma; 0.1 mmol) in HCl (10 mM, 10 mL).
2. Heat at 50 °C for 1 h.
3. Dilute the sample (20 μL) to 10 mL with water and store at 4 °C. The millimolar absorption coefficient at 267 nm is 31.8 mM^{-1} cm^{-1}.

Lipid peroxidation reaction and quantitation of MDA

Protocol

1. Reaction mixture: Hepes–NaOH (pH 7.4, 100 mM, 250 μL; final concentration 50 mM); water (205 μL); bovine heart SMP (10 mg mL^{-1}, 15 μL; final concentration 0.3 mg mL^{-1}); ADP (100 mM, 10 μL; final concentration 2 mM); $FeCl_3$ (freshly prepared; 10 mM, 10 μL; final concentration 0.2 mM); rotenone in ethanol (100 μM, 5 μL; final concentration 1 μM); total volume, 495 μL.
2. Preincubate at room temperature for 5 min.
3. Start the reaction by adding NAD(P)H (10 mM, 5 μL; 0.1 mM final).
4. Stop the reaction by adding acetonitrile (1 mL).
5. Leave to stand at room temperature for more than 5 min.
6. Centrifuge at 6000 g for 10 min (Tomy TMA-2, 10 000 rpm).
7. Use the supernatant (50–100 μL) for HPLC analysis on a 4.6 mm i.d. × 150 mm Chemcopac Spherisorb-NH_2 column with 2:8 (*v*/*v*) Tris-HCl (pH 7.4, 30 mM)–acetonitrile as mobile phase at 2.0 mL min^{-1}; detection wavelength 267 nm.

Results and calculation

The amount of MDA in the sample is calculated by comparing its peak area with that from standard MDA (Fig. 3).

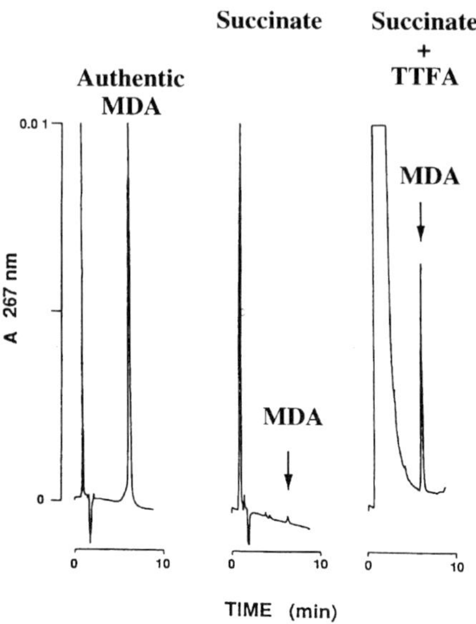

Figure 3. Lipid peroxidation in SMP (2).

Comments

When succinate is used as an electron donor, rotenone is replaced by 2-theonyltrifluoroacetone (TTFA, 1 mM final). The buffer solution used to suspend the SMP should be replaced by Hepes–NaOH (pH 7.4, 10 mM) if MDA is measured by the following TBA method.

Reference

2. Eto, Y., Kang, D., Hasegawa, E., Takeshige, K., Minakami, S. (1992). Succinate-dependent lipid peroxidation and its prevention by reduced ubiquinone in beef heart submitochondrial particles. *Arch. Biochem. Biophys.*, **295**, 101–6.

Microsomes–thiobarbituric acid (TBA) method

Protocol

1. Reaction mixture: Hepes–NaOH (pH 7.4, 100 mM, 250 μL; final concentration 50 mM); water (220 μL); ADP (100 mM, 10 μL; final concentration 2 mM); $FeCl_3$ (freshly prepared; 10 mM, 10 μL; final concentration 0.2 mM); rat liver microsomes (20 mg mL^{-1}) in KCl (0.15 M, 5 μL; final concentration 0.2 mg mL^{-1}); total volume 495 μL.

2. Preincubate at room temperature for 5 min.
3. Start the reaction by adding NADPH (20 mM, 5 μL; 0.2 mM final).
4. Stop the reaction by adding ice-cold TCA (100%, 0.5 mL).
5. Add butylhydroxytoluene (10 mM, 5 μL) in ethanol to prevent non-enzymic lipid peroxidation in subsequent steps.
6. Stand on ice for at least 10 min.
7. Centrifuge at 6000 g for 10 min (Tomy TMA-2, 10 000 rpm).
8. Add supernatant (0.5 mL) to TBA[1] (0.375%, 0.5 mL)[1].
9. Heat at 80 °C for 15 min.
10. Cool to room temperature.
11. Measure absorbance at 535 nm.

[1] Dissolve TBA by heating or by adding NaOH. If the latter is used its pH should be below 3.0.

Comments

The reaction of TBA is not specific for MDA. The absorbance values are, therefore, used for the quantitation of lipid peroxidation and expressed as TBA-reactive substances (TBARS).

9

Neutrophils and priming. Isolation of neutrophils/assay of $O_2\cdot^-$ generation by cytochrome-*c* reduction

TOSHIHIKO UTSUMI and KOZO UTSUMI

Introduction

Human neutrophils play critical roles in host defence against microorganisms and in inflammatory responses. Because neutrophils in the peripheral blood of healthy individuals are not primed, stimulation-dependent responses *in vitro* are weak. Cytokines such as TNF-α and G-CSF induce a 'primed state' in neutrophils characterized by an increased capacity to produce $O_2\cdot^-$, adherence to endothelial cells, migration, lysosomal enzyme release, and cytocidal activity. The enhancement of neutrophil responses in this fashion has been termed 'priming'.

Here, we describe procedures for the preparation of human peripheral neutrophils and for the analysis of cytokine-induced priming of neutrophils monitored by measuring $O_2\cdot^-$ generation.

Protocol

Isolation of neutrophils

Method I: Isolation of human neutrophils by Ficoll–Paque solution (Fig. 1)
(When the use of heparin is unfavourable, Method II should be used.)

1. Sample human peripheral blood (20 mL; heparinized with heparin (500–1000 units mL^{-1} Noboheparin, 0.4 mL) in a 50-mL plastic syringe).
2. Mix well and change the needle to 18 gauge (18G × 3.5).
3. Aspirate dextran[1] (2% *w*/*v*, 13 mL) and mix well by inverting the syringe.
4. Leave at room temperature for 20 min for sedimentation of RBCs.

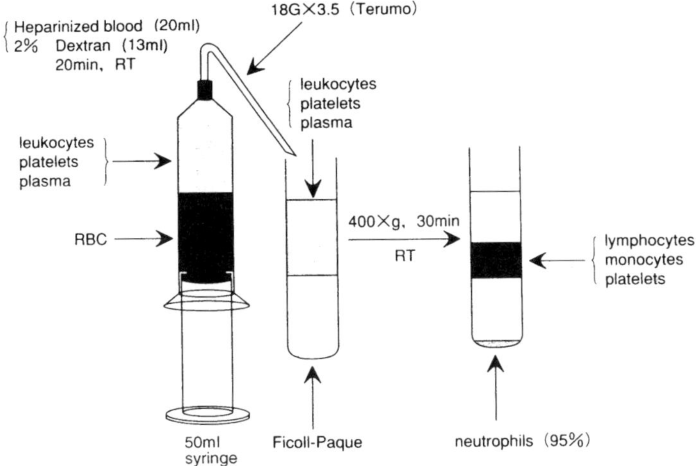

Figure 1. Isolation of human neutrophils by use of Ficoll–Paque solution.

5. Bend the syringe needle and dispense the upper layer into a 50-mL plastic centrifuge tube.
6. Carefully layer Ficoll–Paque solution[2] (5 mL) at the bottom of the centrifuge tube.
7. Centrifuge at 1500 rpm (400 g) for 30 min at room temperature.
8. Remove the upper layer by aspiration.
   [I] If many RBCs contaminate the cells (precipitate), add NaCl (0.033 M, 50 mL), mix well with a plastic pipette, then divide cell suspension into two tubes.
   [II] After 30 s add NaCl (0.27 M, 25 mL) to each tube and mix.
   [III] Spin at 900 rpm (150 g) for 5 min at room temperature.
   [IV] Remove the upper layer by aspiration.
9. Add ice-cold Ca^{2+}-free KRP[3] (50 mL) and re-suspend cells (as precipitate).
10. Centrifuge at 900 rpm (150 g) for 5 min at room temperature.
11. Remove the upper layer by aspiration and re-suspend cells (as precipitate) in ice-cold Ca^{2+}-free KRP (1×10^8 cells mL^{-1}). More than 95% of the obtained cells are neutrophils.

[1] To prepare dextran (2% *w/v*) solution dextran T500 (Pharmacia; 10 g) and NaCl (4.5 g) are dissolved in distilled water, diluted to 500 mL, sterilized and stored at 4 °C.
[2] Ficoll–Paque (Pharmacia) or mono-poly resolving medium (M-PRM; Flow Laboratories) or 3:7 mixture of sodium metrizate (32.8%, *w/v*; Daiichi Pharmaceutical Co. Ltd.) and Ficoll 400 (8%, *w/v*; Pharmacia).
[3] Ca^{2+}-free Krebs–Ringer phosphate(Ca^{2+}-free KRP, pH 7.4) contains phosphate buffer (pH 7.4, 0.1 M, 21 vol.), NaCl (0.9% *w/v*; 0.15 M, 100 vol.), KCl (0.154 M, 4 vol.), and $MgCl_2$ (0.154 M, 1 vol.). Mix just before use and keep at 4 °C

Protocol *Continued*

Method II: Isolation of neutrophils using mono-poly resolving medium (M-PRM) (Fig. 2)

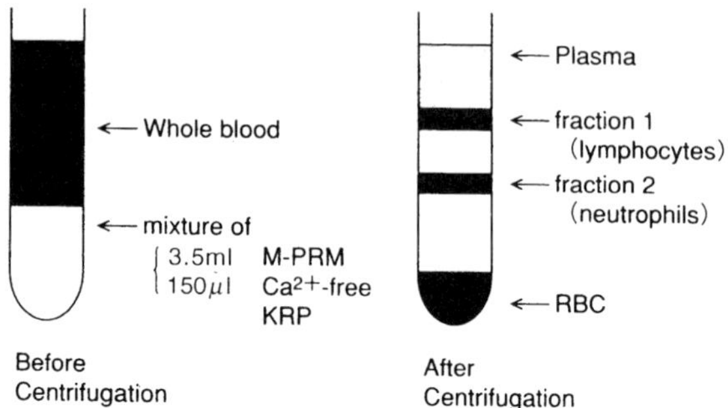

Figure 2. Isolation of neutrophils by use of mono-poly resolving medium.

This method is for the rapid preparation of fresh neutrophils.

1. Aspirate human peripheral blood (20 mL) with a 50-mL plastic syringe containing sodium citrate solution (3.8% *w*/*v*, 2 mL) and mix well.
2. Carefully layer blood (5 mL) on a mixture of M-PRM solution[1] (3.5 mL) and Ca^{2+}-free KRP (150 μL) in a 10 mL plastic centrifuge tube.
3. Centrifuge at 1500 rpm (400 g) for 30 min at room temperature.
4. Remove fraction 1 by Pasteur pipette.
5. Transfer fraction 2 by plastic pipette to a 50-mL plastic centrifuge tube containing Ca^{2+}-free KRP (10 mL).
6. Centrifuge at 1500 rpm (400 g) for 5 min at room temperature.
7. Wash twice with Ca^{2+}-free KRP (1200 rpm, 250 g, 5 min, room temperature).
8. Re-suspend in ice cold Ca^{2+}-free KRP (1×10^8 cells mL^{-1}). If many RBCs contaminate the cells, treat with hypotonic solution as described in Method I.

[1] Mono-poly resolving medium (M-PRM, sg 1.114) or a solution of sodium metrizate (13.8% *w*/*v*) and dextran T500 (8% *w*/*v*) prepared just before use as a mixture of sodium metrizate (32.8%, *w*/*v*, 30 mL) and dextran T500 (5.9 g) in 71.3 mL.

Assay of neutrophil superoxide generation

Cytochrome-*c* reduction method, using a dual beam spectrophotometer equipped with a thermostatically controlled cuvette holder and a magnetic stirrer.

1. Set the dual beam spectrophotometer[1] and recorder (37 °C, stirring, absorbance $A_{550-540\ nm}$).

2. Add Ca^{2+}-free KRP (2 mL) containing glucose (10 mM) and cytochrome-*c*[2] (20 μM) to the cuvette (cytochrome-*c* (2 mM, 20 μL) and glucose (1 M, 20 μL) are added to Ca^{2+}-free KRP (1.96 mL)).
3. Add neutrophils (1×10^8 cells mL^{-1}, 10 μL; final concentration 5×10^5 cells mL^{-1}).
4. After 2 min add TNF-α (2000 units mL^{-1}, 10 μL; final concentration 10 units mL^{-1}) or G-CSF (5 μg mL^{-1}, 20 μL; final concentration 50 ng mL^{-1}) or other stimuli.
5. After 10 min add FMLP[3], PMA[4] or other stimuli (10 μL; final concentrations of FMLP and PMA 5×10^{-8} and 5×10^{-10} M, respectively) and record absorbance $A_{550—540\ nm}$ continuously.

[1] When a dual-beam spectrophotometer is not available, a double-beam spectrophotometer can be used. In this case, after centrifugation (150 g) of the reaction mixture at the appropriate time, the absorbance (550 nm) of the supernatant should be measured.
[2] To minimize the involvement of non-specific reductions by the mitochondrial electron transport chain, acetylated cytochrome-*c* (3) should be used. The procedure for the preparation of acetylated cytochrome-*c* is described in Chapter 2 of this book. Store 2 mM (26 mg mL^{-1}) cytochrome-*c* solution at −20 °C.
[3] Formyl–methionyl–leucyl–phenylalanine (FMLP) 10^{-5} M in ethanol (stock solution 2 mM; 886.2 μg mL^{-1}) in DMSO or ethanol, store at −80 °C)
[4] Phorbol myristate acetate (PMA) 10^{-7} M in ethanol (stock solution 10^{-4} M, 23.2 μg mL^{-1}) in DMSO or ethanol, store at −80 °C)
[5] Store 3 mg mL^{-1} superoxide dismutase (SOD) solution at −80 °C.

Results and calculations

The amount of $O_2^{\cdot -}$ generated was calculated from $\Delta A_{550—540\ nm}$ min^{-1} and a millimolar absorption coefficient of 19.1 mM^{-1} cm^{-1} according to the equation:

$$\text{Cytochrome-}c\text{ reduction (nmol min}^{-1}/10^6\text{cells mL}^{-1}) = (\Delta A_{550—540\ nm}\ \text{min}^{-1}) \times 1/19.1 \times 1000$$

Because cytochrome-*c* can be reduced by other radical species, a control experiment using superoxide dismutase (SOD[5], final concentration of 15 μg mL^{-1}) should be performed to confirm that the reduction is dependent on $O_2^{\cdot -}$.

Human peripheral neutrophils from healthy individuals are not primed and, hence, only a low level of $O_2^{\cdot -}$ generation was induced by the treatment of FMLP (1.25×10^{-8} M), as shown by a dotted line in Fig. 3. This weak $O_2^{\cdot -}$generation was not affected by subsequent treatment with TNF-α (10 units mL^{-1}). In contrast, when neutrophils were first treated with TNF-α (10 units mL^{-1}), they underwent priming and the rate of $O_2^{\cdot -}$generation induced by the treatment with FMLP (1.25×10^{-8} M) was remarkably increased, as shown by a solid line in Fig. 3 (1, 2).

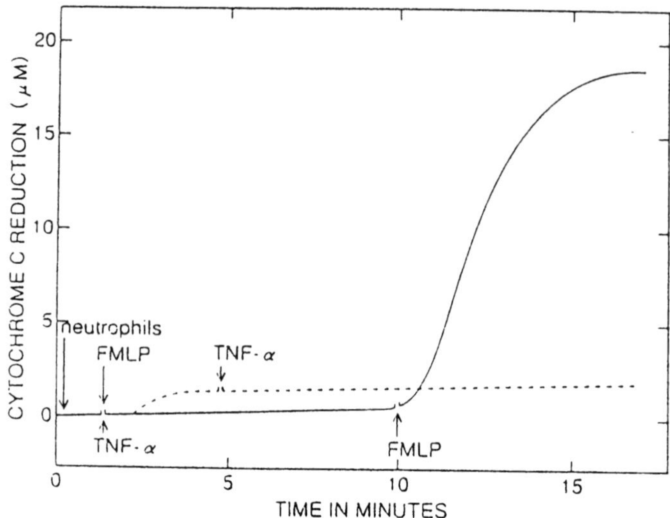

Figure 3. Enhancement of FMLP-dependent $O_2^{\cdot -}$ generation by TNF-α. $O_2^{\cdot -}$ generation was assayed by reduction of cytochrome-*c* as described in the text. $O_2^{\cdot -}$ generation was triggered by adding FMLP (1.25×10^{-8} M); TNF-α (10 units mL^{-1}) was added before or after addition of FMLP.

References

1. Utsumi, T., Klostergaard, J., Akimatu, K., Edashige, K., Sato, E., and Utsumi, K. (1992). Modulation of TNF-α-priming and stimulation-dependent superoxide generation in human neutrophils by protein kinase inhibitors. *Arch. Biochem. Biophys.*, **294**, 271–8.
2. Akimaru, K., Utsumi, T., Sato, E., Klostergaard, J., Inoue, M., and Utsumi, K. (1992). Role of tyrosyl phosphorylation in neutrophil priming by tumor necrosis factor-α and granulocyte colony stimulating factor. *Arch. Biochem. Biophys.*, **298**, 703–9.
3. Takeshige, K. and Minakami, S. (1979). NADH- and NADPH-dependent formation of superoxide anions by bovine heart submitochondrial particles and NADH-ubiquinone reductase preparation. *Biochem. J.*, **180**, 129–35.

10

Bioassay for antioxidant activity: determination of scavenging capacity of antioxidants against lipid peroxyl radicals

YOICHI MIYAMOTO, TAKAAKI AKAIKE, and HIROSHI MAEDA

Introduction

Much attention has been paid to the beneficial effect of antioxidants in the modulation of oxidative stress; it is thus important to the investigate the different antioxidants. Several techniques including the TBA method, chemiluminescence analysis, and ESR spectroscopy can be applied to the evaluation of antioxidant activity. Most methods, however, have disadvantages, such as need for expensive equipment, and for some the chemical reactions employed are poorly understood. We have previously discovered the potent bactericidal action of alkyl (lipid) peroxyl radicals (ROO·) generated by haem-iron-catalysed decomposition of organic hydroperoxides (1). This can be used as a novel bioassay for antioxidant activity (2). In this section, we describe a simple bioassay which measures the scavenging capacity of different substances when they protect *Staphylococcus aureus* against the *t*-butyl hydroperoxyl radical (*t*-BuOO·) generated by reacting *t*-butyl hydroperoxide (*t*-BuOOH) and haemoglobin (2).

Bactericidal assay system comprising *t*-BuOOH, methaemoglobin and *S. aureus*

Protocol

Reagents

1. Brain–heart infusion broth (Eiken Chemical Co., Tokyo, Japan).
2. *t*-Butyl hydroperoxide (200 mM; prepare immediately before use).

Protocol *Continued*

3. Methaemoglobin (1 mg mL^{-1}).
4. Mannitol–phenol red broth[1].
5. 2 × Mannitol–phenol red broth.

Procedure

1. Inoculate brain–heart infusion broth (5 mL) with *S. aureus* 209P and culture overnight at 37 °C.
2. Wash the bacteria three times with sodium phosphate (0.01 M)-buffered saline (0.15 M) (PBS; pH 7.2) by centrifugation.
3. Re-suspend the bacteria at the concentration of 1×10^7 colony forming units (CFU) mL^{-1} in PBS (pH 7.2)
4. Prepare a mixture of bacterial suspension[2] (1×10^7 CFU mL^{-1}, 100 μL; final concentration 1×10^6 CFU mL^{-1}), antioxidant substance[3] (100 μL), methaemoglobin[4] (1 mg mL^{-1}, 100 μL; final concentration 100 μg mL^{-1}), and PBS (pH 7.2, 600 μL); the total volume of the mixture is 900 μL.
5. Add *t*-BuOOH (200 mM, 100 μL) to this solution (final concentration; 20 mM) and mix the reaction solution immediately[5] to start the radical-generating reaction[6].
6. Incubate the reaction mixture for 30 min at 37 °C. *t*-BuOO· generated by the reaction has a strong bactericidal effect on *S. aureus*; this is inhibited by the antioxidant added to the reaction mixture (Fig. 1).

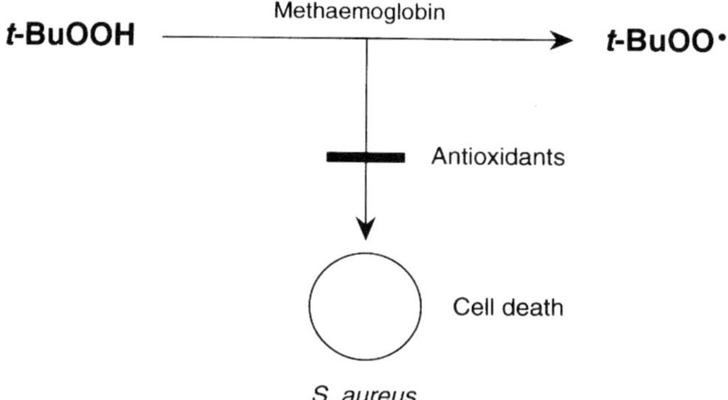

Figure 1. The principle of the bioassay of antioxidant activity.

7. Stop the reaction by adding 2 × mannitol–phenol red broth (1 mL).

[1] To make 1 L mannitol–phenol red broth mix beef extract (3 g), peptone (10 g), sodium chloride (5 g), D-mannitol (10 g), and phenol red (0.035 g) and adjust the pH to 7.4 ± 0.1.
[2] A suspension of *S. aureus* 209P with an absorbance of 0.1 at 610 nm usually contains $1\text{–}1.5 \times 10^7$ CFU bacteria mL^{-1}.

[3] Hydrophobic antioxidant substances are dissolved in dimethyl sulfoxide (DMSO). Up to 20% DMSO does not interfere with the assay.

[4] In the author's laboratory methaemoglobin is prepared by reaction of sodium nitrite with haemoglobin isolated from human blood (1). Methaemoglobin is also available from Sigma (St. Louis, MO, USA).

[5] Because the reaction of *t*-BuOOH with haemoglobin is very fast, the reaction mixture must be vortex stirred immediately after the addition of *t*-BuOOH solution.

[6] Preparation of a reference control without haemoglobin (without generation of *t*-BuOO·) is recommended in each assay.

Measurement of the survival of the bacteria using multiwell plates (Fig. 2)

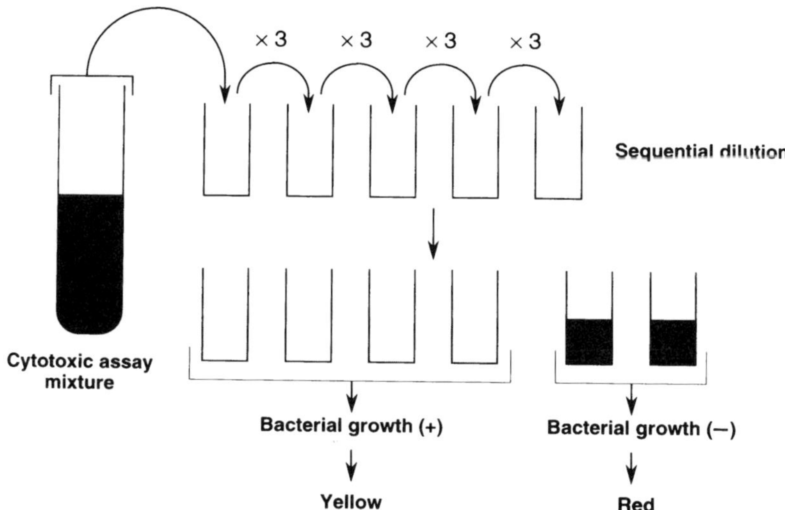

Figure 2. Schematic illustration of the method used to determine the viability of the bacteria by means of a 96-well plate.

Protocol

1. Dispense mannitol–phenol red broth (160 μL) into each well of a 96-well plate.
2. Add the bactericidal reaction mixture described above to the wells at the left end.
3. Make threefold dilutions serially from the left to the right along the plate[1].
4. Incubate the plate overnight at 37 °C.

Protocol *Continued*

5. Because the colour of the reaction mixture containing the viable bacteria changes from red to yellow because of the growth of the bacteria after cultivation, survival of bacteria can be detected macroscopically[2]. The bacterial growth in each well is quantitated colorimetrically by subtracting the absorbance at 655 nm (turbidity) from that at 570 nm (red colour) by use of an automatic multiwell plate reader (Model 450, Bio-Rad Laboratories, Richmond, CA, USA).
6. The antioxidant capacity of a sample is expressed in terms of the minimum inhibitory concentration (MIC) needed to inhibit *t*-BuOO·-induced bactericidal action[3]. The antioxidant activity of α-tocopherol is shown in Fig. 3[4].

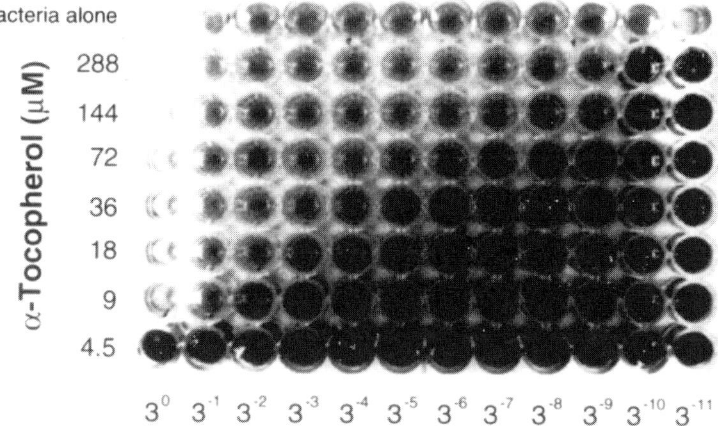

Figure 3. Antioxidant activity of α-tocopherol assessed by the method presented in this chapter. Determination of the survival of bacteria using a 96-well plate.

[1] The reaction mixture is diluted by thorough pipetting with tips which are changed after each dilution.

[2] The colour of the medium which was prepared as a control reaction mixture without production of *t*-BuOO· (the reaction mixture without haemoglobin) changes to yellow from the 1st to the 11th or 12th well, depending on the growth of the bacteria. The medium containing the reaction mixture, in which *t*-BuOO· is scavenged by the antioxidant substance, also changes colour because of the growth of the surviving bacteria. In contrast, the colour of the medium remains red when the bacteria were completely killed by *t*-BuOO·, or when the medium does not contain bacteria initially (Fig. 3).

[3] Probucol has highly reproducible antioxidant activity in this assay system and it is sufficiently soluble in DMSO. We therefore recommend probucol as a standard antioxidant substance in each assay.

[4] α-Tocopherol solution is prepared by mixing 1:13 (*v/v*) α-tocopherol with 20% egg yolk lecithin in ethanol. The ethanol is removed by evaporation and the resulting mixture is diluted with PBS.

References

1. Akaike, T., Sato, K., Ijiri, S., Miyamoto, Y., Kohno, M., Ando, M., and Maeda, H. (1992). Bactericidal activity of alkyl peroxyl radicals generated by haem-iron-catalysed decomposition of organic peroxides. *Arch. Biochem. Biophys.*, **294**, 55–63.
2. Akaike, T., Ijiri, S., Sato, K., Katsuki, T., and Maeda, H. (1995). Determination of peroxyl radical-scavenging activity in food by using bactericidal action of alkyl peroxyl radical. *J. Agric. Food Chem.*, **43**, 1864–1870.

11

Iron-binding and iron-oxidizing antioxidant activities

J. M. C. GUTTERIDGE and G. J. QUINLAN

Plasma iron-binding antioxidant activity

Protection against lipid-derived radicals

Introduction

Human plasma contains the iron-binding protein transferrin and an iron-oxidizing protein called caeruloplasmin. These two proteins, representing less than 4% of total plasma proteins, are responsible for over 90% of the inhibitory properties of plasma towards iron-driven free-radical reactions.

Protocol

1. Into clean glass tubes place bovine brain phospholipid (5 mg mL^{-1}, 0.2 mL) as a liposomal suspension (from Sigma or prepared as in *Anal. Biochem.*, **82** (1997) 76); phosphate buffer (0.2 mL, 0.1 M, pH 7.4; Na salt); plasma (0.01 mL; or standard of apotransferrin 3 g L^{-1}) and ascorbic acid (7.5 mM, 0.02 mL; freshly prepared).
2. Incubate at 37 °C for 20 min.
3. Add HCl (25% *v/v*, 0.5 mL) and thiobarbituric acid (1% *w/v*, 0.5 mL)
4. Heat for 15 min at 100 °C
5. Cool and add butan-1-ol (1.5 mL)
6. Vortex mix for 2 min.
7. Centrifuge at 6000 rpm for 5 min.
8. Read upper organic (butan-1-ol) phase at A 532 nm.

Calculation

1. Blank (no plasma, no ascorbate)
2. Test (all reagents + plasma)
3. Control (all reagents + apotransferrin)
4. 100% peroxidation standard (all reagents, no plasma)

5. Subtract blank value
6. The percentage inhibition by the plasma sample is derived from the formula:

   (Test reading / 100% Peroxidation standard reading) $\times$ 100 = Value χ
7. Subtract χ from 100 to give the percentage inhibition.

Comments

If contaminating iron present in reagents is insufficient to enable iron-binding, add ferric chloride (1 mM, 10 μL) to the phosphate buffer (100 mL). Iron overload (100% iron-saturated transferrin) plasma will not inhibit lipid peroxidation, it will stimulate. Normal plasma values are 86.4 $\pm$ 1.14% inhibition.

Reference

Gutteridge, J. M. C. and Quinlan, G. J. (1992). Antioxidant protection against organic and inorganic oxygen radicals by normal human plasma: the important primary role for iron-binding and iron-oxidizing proteins. *Biochim. Biophys. Acta*, **1159**, 248–54.

Protection against hydroxyl radicals

Protocol

1. Into new clean glass tubes place 2-deoxyribose (10 mM, 0.2 mL); sodium phosphate buffer (pH 7.4, 0.1 M, 0.2 mL); plasma (or apotransferrin standard 3 g L^{-1}; 0.01 mL); freshly prepared ascorbic acid (7.5 mM, 0.02 mL); and hydrogen peroxide (8.8 mM, 0.02 mL).
2. Incubate at 37 °C for 60 min.
3. Add trichloroacetic acid (2.8% *w/v*, 0.5 mL) and thiobarbituric acid (1% *w/v*, 0.5 mL)
4. Heat for 15 min at 100 °C.
5. Cool, and add butan-1-ol (1.5 mL) and vortex mix for 2 min.
6. Centrifuge at 6000 rpm for 5 min.
7. Read upper organic (butan-1-ol) phase at A 532 nm

Calculation

1. Blank (no plasma, no ascorbate)
2. Test (all reagents + plasma)
3. Control (all reagents + apotransferrin)
4. 100% Oxidative damage standard (all reagents, no plasma)

5. Subtract blank value
6. The percentage inhibition by the plasma sample is derived from the formula:

   $$(\text{Test reading} / 100\% \text{ Oxidative damage}) \times 100 = \text{Value } \chi$$

7. Subtract χ from 100 to give the percentage inhibition.

Comments

A small amount of iron might have to be added to the buffer to achieve the required sensitivity (see previous protocol). Normal plasma values are 96.0 ± 0.57% inhibition.

Reference

Gutteridge, J. M. C. and Quinlan, G. J. (1992). Antioxidant protection against organic and inorganic oxygen radicals by normal human plasma: the important primary role for iron-binding and iron-oxidizing proteins. *Biochim. Biophys. Acta*, **1159**, 248–54.

Plasma iron-oxidizing antioxidant activity

Protection against lipid-derived radicals

Introduction

Human plasma contains an iron-oxidizing (ferroxidase) protein called caeruloplasmin, a copper-containing protein that catalyses the oxidation of ferrous ions to the ferric state; the latter are considerably less reactive in most free-radical reactions.

Protocol

1. Into new clean glass tubes place bovine brain phospholipid (5 mg mL^{-1}, 0.2 mL); sodium phosphate buffer (pH 6.5, 0.1 M, 0.2 mL); plasma (or caeruloplasmin standard 300 mg L^{-1}; 0.01 mL); ferric chloride (1 mM, 0.02 mL); and freshly prepared ascorbic acid (0.125 mM, 0.03 mL).
2. Incubate for 20 min at 37 °C.
3. Add HCl (25% *v/v*, 0.5 mL) and thiobarbituric acid (1% *w/v*, 0.5 mL).
4. Heat at 100 °C for 15 min.
5. Cool, add butan-1-ol (1.5 mL) and vortex mix for 2 min.
6. Centrifuge at 6000 rpm for 5 min.
7. Read upper organic (butan-1-ol) phase at A 532 nm.

Calculation

1. Blank (no plasma, no ascorbate).
2. Test (all reagents + plasma).

3. Control (all reagents + caeruloplasmin).
4. 100% peroxidation standard (all reagents, no plasma).
5. Subtract blank value.
6. The percentage inhibition by the plasma sample is derived from the formula:

   (Test reading / 100% Peroxidation standard reading) $\times$ 100 = Value χ
7. Subtract χ from 100 to give the percentage inhibition.

Comments

Do not prepare solutions of ferric chloride in Chelex resin-treated water, or in alkaline solution, as it will form insoluble polyhydrate complexes. Normal plasma values are 64.0 $\pm$ 2.9% inhibition.

Reference

Gutteridge, J. M. C. and Quinlan, G. J. (1992). Antioxidant protection against organic and inorganic oxygen radicals by normal human plasma: the important primary role for iron-binding and iron-oxidizing proteins. *Biochim. Biophys. Acta*, **1159**, 248–54.

Protection against hydroxyl radicals

Protocol

1. Into new clean glass tubes place 2-deoxyribose (10 mM, 0.20 mL); sodium phosphate buffer (pH 6.5, 0.1 M, 0.2 mL); plasma (or caeruloplasmin 300 mg L^{-1}; 0.01 mL); ferric chloride (1 mM, 0.02 mL); freshly prepared ascorbic acid (0.125 mM, 0.03 mL); and hydrogen peroxide (8.8 mM, 0.02 mL)
2. Incubate for 60 min at 37 °C.
3. Add trichloroacetic acid (0.5 mL, 2.8% *w/v*) and thiobarbituric acid (1% *w/v*, 0.5 mL).
4. Heat at 100 °C for 15 min.
5. Cool, add butan-1-ol (1.5 mL) and vortex mix for 2 min.
6. Centrifuge at 6000 rpm for 5 min.
7. Measure absorbance of upper organic (butan-1-ol) phase at 532 nm.

Calculation

1. Blank (no plasma, no ascorbate).
2. Test (all reagents + plasma).
3. Control (all reagents + caeruloplasmin).

4. 100% oxidative damage standard (all reagents, no plasma).
5. Subtract blank value.
6. The percentage inhibition by the plasma sample is derived from the formula: (Test reading / 100% Peroxidation standard reading) × 100 = Value χ
7. Subtract χ from 100 to give the percentage inhibition.

Comments

Normal plasma values are 73.8 ± 1.3% inhibition.

Reference

Gutteridge, J. M. C. and Quinlan, G. J. (1992). Antioxidant protection against organic and inorganic oxygen radicals by normal human plasma: the important primary role for iron-binding and iron-oxidizing proteins. *Biochim. Biophys. Acta*, **1159**, 248–54.

Plasma scavenging antioxidant activity

Protection against lipid-derived radicals

Introduction

Human plasma contains many low molecular mass molecules that can scavenge alkoxyl and peroxyl (organic) radicals. These are secondary antioxidants the activity of which will be seen in an iron-driven reaction only when the iron-binding and iron-oxidizing activities are removed.

Protocol

1. Into new clean glass tubes place bovine brain phospholipid (5 mg mL^{-1}, 0.2 mL); phosphate buffer (pH 7.4, 0.1 M (Na salt), 0.2 mL); plasma (or BHT; 0.5 mM, 0.01 mL); ferric chloride (1 mM, 0.02 mL); and ascorbate (7.5 mM, 0.02 mL; prepare freshly)
2. Incubate for 20 min at 37 °C.
3. Add HCl (25% *v*/*v*, 0.5 mL) and thiobarbituric acid (1% *w*/*v*, 0.5 mL).
4. Heat at 100 °C for 15 min.
5. Cool, add butan-1-ol (1.5 mL) and vortex mix for 2 min.
6. Centrifuge at 6000 rpm for 5 min.
7. Read upper organic (butan-1-ol) phase at A 532 nm.

Calculation

1. Subtract blank values (see iron-binding and iron-oxidizing assays).
2. The percentage inhibition by the plasma sample is derived from the formula:

(Test reading / 100% Peroxidation standard reading) × 100 = Value χ

3. Subtract χ from 100 to give the percentage inhibition.

Comments

BHT, propyl gallate, or other chain-breaking antioxidants can be used as internal controls. Iron is added to saturate the plasma transferrin, and a high ascorbate concentration is present to inhibit the ferroxidase activity of caeruloplasmin. Normal plasma values are 16.8 ± 2.9% stimulation, because redox-active molecules in plasma stimulate iron cycling.

Reference

Gutteridge, J. M. C. and Quinlan, Gj. (1992). Antioxidant protection against organic and inorganic oxygen radicals by normal human plasma: the important primary role for iron-binding and iron-oxidizing proteins. *Biochim. Biophys. Acta*, **1159**, 248–54.

Protection against hydroxyl radicals

Introduction

Almost any high or low molecular mass molecule present in plasma will be attacked extremely rapidly by hydroxyl radicals.

Protocol

1. Into new clean glass tubes place 2-deoxyribose (10 mM, 0.20 mL); sodium phosphate buffer (pH 7.4, 0.1 M, 0.2 mL); plasma (0.01 mL); ferric chloride (1 mM, 0.1 mL); ascorbic acid (7.5 mM, 0.02 mL); and hydrogen peroxide (8.8 mM, 0.02 mL).
2. Incubate at 37 °C for 60 min.
3. Add trichloroacetic acid (2.8% *w/v*, 0.5 mL) and thiobarbituric acid (1% *w/v*, 0.5 mL)
4. Heat for 15 min at 100 °C.
5. Cool, and add butan-1-ol (1.5 mL) and vortex mix for 2 min.
6. Centrifuge at 6000 rpm for 5 min.
7. Measure absorbance of upper organic (butan-1-ol) phase at 532 nm

Calculation

1. The percentage inhibition by the plasma sample is derived from the formula:

   (Test reading / 100% Peroxidation standard reading) × 100 = Value χ

2. Subtract χ from 100 to give the percentage inhibition.

Comments

Controls are set up as previously indicated, and the iron-binding and iron-oxidizing activities eliminated. Normal plasma values are 67.1 ± 3.9% inhibition.

Reference

Gutteridge, J. M. C. and Quinlan, G. J. (1992). Antioxidant protection against organic and inorganic oxygen radicals by normal human plasma: the important primary role for iron-binding and iron-oxidizing proteins. *Biochim. Biophys. Acta*, **1159**, 248–54.

12

Glutathione (oxidized and reduced form): assay mixture and measurement

TAKAHIKO KONDO and FUMI SAWADA

Introduction

Enzymatic recycling is described for determining the concentration of glutathione (GSH), of which there is a reduced form (γ-glutamylcysteinyl glycine, GSH) and an oxidized form (GSSG). Most of the glutathione present in cells is in the reduced form, with a small percent as GSSG. Because glutathione dissolves in acid, biological samples should first be treated with protein-precipitating agents to remove proteins. We use trichloroacetic acid (TCA) for precipitating proteins and diethyl ether for removing them. The principle of the method for determining glutathione utilizes the absorbance of *p*-nitrophenol formed from 5,5-dithiobis 2-nitrobenzoic acid (DTNB) with a maximum absorbance (ε_{412} = 13 600) at 412 nm.

Protocol

Estimation of total glutathione

Preparation of materials from red blood cells[1]

1. Prepare red blood cells from whole blood (2 mL) by passage through a column containing α-cellulose and microcrystalline cellulose (1:1, *w/w*; Sigma; 2 mL) previously equilibrated with isotonic saline.
2. Suspend cells in ice-cold isotonic saline at an approximate haematocrit of 10%.
3. Estimate the concentration of haemoglobin or level of haematocrit by use of an aliquot of the sample.
4. Treat with 1 vol. TCA (10% *v/v*) at 0 °C. Mix well using a vortex mixer.
5. Centrifuge at 1000 g for 15 min.
6. Treat supernatant (TCA extract; 0.5 mL) with diethyl ether (approx. 3 mL) previously cooled at 0 °C, and mix the supernatant using a vortex mixer.

Protocol *Continued*

7. Evaporate diethyl ether with N_2 gas.
8. Wash with diethyl ether and evaporate it with N_2 gas 5 times to remove excess TCA.

Preparation of materials from cultured cells.(2)

1. Remove cells (2×10^6) from culture medium, and suspend in isotonic saline (1 μL).
2. Add ice-cold TCA (10%, 1 μL).
3. Homogenize cells by means of a Potter-tyme homogenizer, and treat with TCA as described above.

Preparation of materials from rat kidney

1. Dialyse kidney tissues with ice-cold phosphate buffer (pH 7.4, 0.1 M) to effect removal of red blood cells.
2. Suspend the tissue (1 g) in ice-cold phosphate buffer (1 mL) and suspend in TCA solution (10%, 1 mL) on ice.
3. Homogenize and centrifuge as described above.

DTNB–glutathione reductase recycling method

Titration of NADPH

1. Dissolve NADPH (2 mg) in $NaHCO_3$ (5%, 1 mL).
2. Read absorbance at 340 nm (A_0) of blank (0.85 mL H_2O and 0.1 mL 1 M Tris-HCl, pH 8.0).
3. Read absorbance at 340 nm (A_1) after the addition of the NADPH solution (50 mL).
4. The millimolar concentration of NADPH = $(A_1 - A_0)/0.311$.

Standard curve[1]

1. Prepare GSH standard solution (5 mM) by dissolving GSH (15.37 mg) in phosphate buffer (10 mM, 10 mL)[2].
2. Dilute GSH standard solution[3] (5 mM) with phosphate buffer (0.1 M) (100-fold to give 50 μM GSH, 200-fold to give 25 μM GSH, 500-fold to give 10 μM GSH, and 1000-fold to give 5 μM GSH).

Assay system[4]

1. Prepare mixtures of phosphate buffer (0.1 M, 500 μL), NADPH (4 mM, 50 μL), and glutathione reductase solution (6 units mL^{-1}, μL). For assay add sample or standard (100 μL) and for blank add water (100 μL).
2. Incubate for 5 min at 37 °C.

3. Add DTNB[5] (50 μL).

4. Measure absorbance at 412 nm at 37 °C[6].

[1] When estimating the total glutathione concentration of many samples using the enzymatic method, it is necessary to examine the standard curve at regular intervals because of changes in the activity of NADPH and glutathione reductase. Some batches of glutathione reductase are difficult to dissolve.

[2] Prepare sodium phosphate buffer (pH 7.5, 0.1 M) containing EDTA (5 mM). Adjust the pH to 7.5 by adding 5 vols 0.2 M Na_2HPO_4 to 1 vol. 0.2 M NaH_2PO_4. Readjust the pH to neutrality with 3.7 mL 0.27 M Na-EDTA per 100 mL buffer and dilute twofold. Store at −4 °C. Sodium phosphate buffer (pH 7.5, 10 mM) containing EDTA (0.5 mM) is prepared by tenfold dilution of 0.1 M buffer immediately before use.

[3] To prepare GSH standard solution, mix NADPH (4 mM; prepared by dissolving NADPH (5.3 mg) in $NaHCO_3$ (5%) and adjusting the volume to 1 mL) with glutathione reductase (prepared by diluting glutathione reductase (120 units mL^{-1}; Behringer) twentyfold with sodium phosphate buffer (10 mM).

[4] When estimating total glutathione concentration by use of this enzymatic recycling method, dilute the sample to be in the range 5–100 mM GSH, because the GSH concentration of cells is high.

[5] Dissolve DTNB (20 mg) in sodium phosphate buffer (0.1 M, 5 mL). This can be stored in a dark cool place for several months.

Results and calculation

The GSH concentration can be calculated from the standard curve. The concentration of total glutathione or GSSG is corrected for the dilution factor of the precipitating reagent and *N*-ethylmaleimide added. A typical standard curve for GSH is shown in Fig. 1.

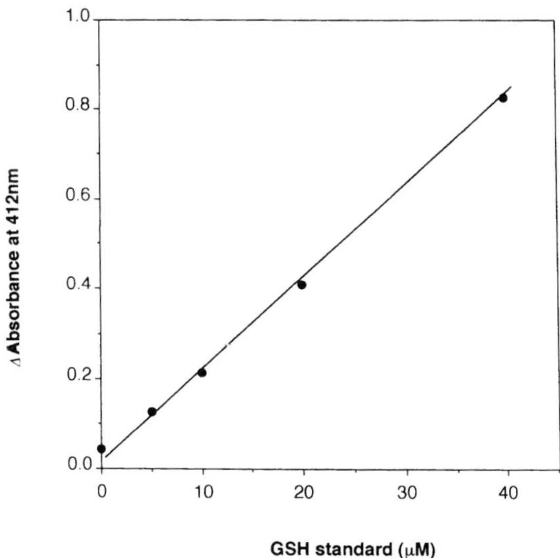

Figure 1. Standard curve of GSH concentration.

The concentrations of total glutathione in some typical cells are: human erythrocytes, 6.57 ± 1.04 μmol g^{-1} haemoglobin; human cancer cells, 9.4–119 nmol/10^6 cells; mouse vascular endothelial cells, 11.7 ± 1.3 nmol/10^6 cells.

Preparation of samples for estimation of GSSG

Protocol

1. Centrifuge the sample[1] after treatment with TCA.
2. Add *N*-ethylmaleimide (0.25 mM) to the TCA extract.
3. Estimate the concentration of GSSG as described above for the GSH concentration.

[1] Sufficient sample must be used when measuring GSSG because the concentration of GSSG in cells is much lower than that of GSH. More than 10^6 cultured cells are required for estimating GSSG.

Calculation

Calculate the GSSG concentration by use of the standard curve for GSH—1 mol GSSG corresponds to 2 mol GSH.

References

1. Beutler, E. (1986). *Red Cell Metabolism: A Manual of Biochemical Methods*, p. 1888. Grune and Stratton, Orlando, USA.
2. Anderson, M. E. (1985). Determination of glutathione and glutathione disulfide in biological samples. *Methods Enzymol.*, **113**, 548–70.

13

Measurement of proteinaceous plasma antioxidants, low molecular weight antioxidants, and redox related compounds

YUKI KITAOKA and JUNJI YODOI

ADF/Thioredoxin: purification of thioredoxin reductase/thioredoxin assay (insulin reducing assay)

Introduction

Adult T cell leukaemia-derived factor (ADF), originally defined as an inducer of interleukin 2 receptor/α chain (IL-2R/p55) of human T-lymphotropic virus type I (HTLV-I) positive T cells, is a human analogue of the redox-active coenzyme thioredoxin of *Escherichia coli*. ADF/thioredoxin contains a redox-active dithiol (–Cys–Gly–Pro–Cys–) and has a variety of biological activities as a hydrogen donor, including dithiol-dependent reducing activity and radical scavenging activity.

This section describes a thioredoxin assay (insulin reducing assay), an enzymatic means of measurement of ADF/thioredoxin based on its dithiol-dependent insulin-reducing activity. The method uses NADPH and thioredoxin reductase purified from human placenta. In the insulin reducing assay, a proton from NADPH is transferred to thioredoxin reductase, which donates the proton to ADF/thioredoxin. ADF/thioredoxin reduces the insulin to separate the A- and B-chains. The rate of conversion of NADPH to $NADP^+$ is determined by measuring the decrease min^{-1} of the absorbance at 340 nm; this is then used to determine ADF/thioredoxin activity (Fig. 1).

An enzyme-linked immunosorbent assay (ELISA) system for ADF/thioredoxin has been developed, as has an assay for determination of its concentration as a protein using two monoclonal antibodies against ADF/thioredoxin.

Protocol

Purification of thioredoxin reductase (1,2)

Stock solutions of buffers A and B should be prepared immediately before use from Tris-HCl (pH 7.4, 1 M) and EDTA (pH 8.0, 500 mM), both stored at room temperature.

1. Homogenize human placenta (200 g) with Tris-HCl (pH 7.4, 50 mM, 200 mL)–2 mM EDTA (buffer A) after removal of the cord, membrane, and major vessels.
2. Centrifuge at 2000 g for 30 min at 4°C and collect supernatant.
3. Adjust pH to 5.0 by addition of acetic acid (1 M).
4. Centrifuge at 2000 g for 30 min at 4°C and collect supernatant.
5. Adjust pH to 7.4 by addition of aqueous ammonia (1 M).
6. Centrifuge at 2000 g for 30 min at 4°C and collect supernatant.
7. Heat to 60°C in a water bath, and then cool immediately on ice to room temperature.
8. Centrifuge at 2000 g for 30 min at 4°C and collect supernatant.
9. Add powdered ammonium sulfate to give 85% saturation, and continue stirring at 4°C overnight.
10. Centrifuge at 2000 g for 30 min at 4°C collect precipitate.
11. Dissolve precipitate in buffer A (200 mL) and dialyse twice, overnight, at 4°C, against 10 L buffer A.
12. Centrifuge at 2000 g for 30 min at 4°C and collect supernatant.
13. Filter through a 0.45-μm pore ultrafiltration membrane (Falcon)
14. Apply to DEAE-sepharose CL-6B (Pharmacia; 500 mL) in a column equilibrated with buffer A, and wash the column with buffer A (1000 mL).
15. Elute with NaCl linear gradient (0–300 mM in buffer A).
16. Collect active fractions determined by the insulin reducing assay, and dialyse these against buffer A.
17. Apply to 2′5′-ADP-Sepharose 4B (Pharmacia; 25 mL) in a column equilibrated with buffer A and wash the column with buffer A (50 mL).
18. Elute with $NADP^+$ (Kojin; 10 mM, 10 mL) in buffer A.
19. Apply to Superose 12 (Pharmacia; 25 mL) in a column and perform gel-filtration with buffer A as mobile phase.
20. Collect the active fractions.
21. Check the efficiency of purification by SDS–PAGE and measure the amount of protein by means of a protein-assay kit (Bio-Rad).
22. Dilute the purified thioredoxin reductase to 200 $\mu g\ mL^{-1}$ with buffer A and store at −80°C

Thioredoxin assay (insulin reducing assay) (2,3)

Insulin (Sigma) stock solution (10 mg mL^{-1}) should be prepared before starting the assay and stored at −80 °C.

1. Mix assay mixture (90 μL) with NADPH (Kojin; 0.27 mM) in Tris-HCl (pH 7.5, 0.1 M)– EDTA (2 mM) (buffer B) and place in a 200 μL microcuvette in a spectrophotometer (Beckman DU-650, or compatible).
2. Add insulin (10 mg mL^{-1}, 10 μL; Sigma).
3. Add thioredoxin reductase from human placenta (200 μg mL^{-1}, 10 μL).
4. Add sample (10 μL).
5. Place microcuvette in spectrophotometer and record the rate of conversion of NADPH to $NADP^+$ by monitoring the decrease in absorbance at 340 nm min^{-1} at 25 °C.

Results and calculations

Calculate the thioredoxin activity as μmol NADPH oxidized min^{-1} from the formula:

$$(\Delta A_{340}\ min^{-1} \times 0.12 \times 10^2)/6.2\ (units\ mL^{-1})$$

The change in absorbance at 340 nm represents NADPH consumption in the insulin-reducing assay (2). A schematic diagram of the cascade of proton transfer in the insulin-reducing assay is shown below the graph in Fig. 1. The concentration of each molecule was 1 μM. The closed circles represent recombinant ADF (rADF)/thioredoxin, the open circles mutated (redox-inactive) ADF/thioredoxin, the closed squares cytochrome-*c* and the closed triangles glutathione (reduced form).

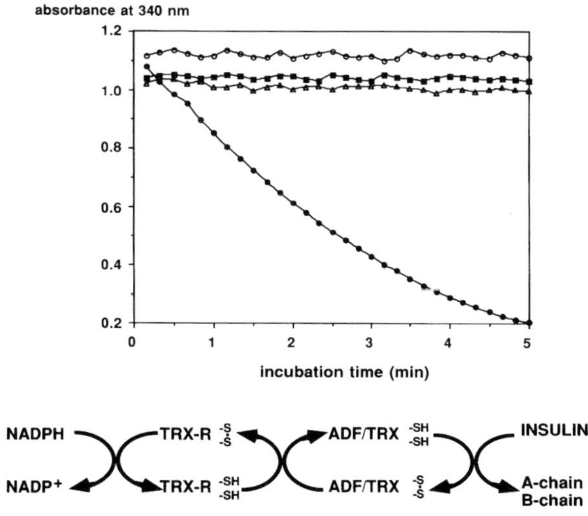

Figure 1. The principles of the assay for thioredoxin activity.

Comments

1. For measurement of thioredoxin reductase activity, 1 mg mL^{-1} rADF/thioredoxin is used instead of the 200 μg mL^{-1} thioredoxin reductase used in the insulin reducing assay.
2. The insulin reducing assay should be standardized by use of rADF/thioredoxin, and the assay mixture without thioredoxin reductase or insulin used for background readings.
3. Microplate-readers capable of measuring absorbance at 340 nm in the UV range, with a constant temperature function, have recently become available. With such equipment 96 samples can be measured simultaneously, although it is necessary to compensate for the length of the light path for each type of microtiter plate used.
4. Human samples are known to contain several redox-active proteins with ADF/thioredoxin moieties, e.g. glutaredoxin, protein disulfide isomerase, trypsin inhibitor, luteinizing hormone, and follicular stimulating hormone. To differentiate between the contributions to thioredoxin activity made by these proteins and by ADF/thioredoxin, anti-rADF/thioredoxin monoclonal antibody (anti-ADF/thioredoxin mAb) conjugated affinity column-depleted samples obtained from an identical source should be used for analysis.
5. rADF/thioredoxin and two anti-ADF/thioredoxin mAbs are not currently commercially available, but are distributed by the 'Society of Redox and Signal Information (President: Professor Junji Yodoi, Department of Biological Responses, Institute for Virus Research, Kyoto University)' to collaborators.

References

1. Luthmann, M. and Holmgren, A. (1982). Rat liver thioredoxin and reductase: purification and characterization. *Biochemistry*, **21**, 6628.
2. Kitaoka, Y., Sorachi, K., Nakamura, H., Masutani, H., Mitsui, A., Kobayashi, F., Mori, T., and Yodoi, J. (1994). Detection of adult T cell leukaemia-derived factor (ADF)/human thioredoxin in human serum. *Immunology Letters*, **41**, 155.
3. Kitaoka, Y., Sachi, Y., Mori, T., and Yodoi, J. (1994). Measurement of ADF/Thioredoxin in human serum and its clinical significance. *Japanese Journal of Clinical Pathology*, **42**, 853.

14

Vitamin E

HIROSHI TAMAI

Introduction

Extraction of vitamin E from tissue usually involves saponification with ethanolic potassium hydroxide. Direct extraction without saponification, however, is used for plasma. HPLC equipped with a spectrofluorimeter (λ_{ex} 298 nm, λ_{em} 325 nm) or with electrochemical detection have enabled measurement of small amounts of tocopherols in biological samples (1).

Plasma or cell suspensions

Protocol

1. Place sample (0.5 mL) in a tube.
2. Add phosphate buffer (pH 7.4, 150 mM, 0.5 mL) containing EDTA (0.27 mM).
3. Add pyrogallol in ethanol (6% *w/v*, 1 mL).
4. Add a solution of tocol (2-methyl 2-phytyl 6-chromanol; 2 μg) in ethanol (1 mL) as internal standard.
5. Shake.
6. Preincubate for 2 min at 70 °C.
7. Add KOH (60%, 0.2 mL).
8. Incubate for 30 min at 70 °C.
9. Add *n*-hexane (5 mL) and distilled water (2.5 mL).
10. Shake.
11. Centrifuge for 5 min at 3000 rpm.
12. Take the *n*-hexane layer (4 mL) and evaporate under N_2 gas.
13. Add ethanol (100 mL).
14. Analyse by HPLC.

Protocol *Continued*

HPLC was performed with an Irica Instruments (Kyoto, Japan) Σ-871 chromatograph equipped with a 250 mm × 4 mm i.d. Irica RP-18 column and electrochemical detection (ECD). The mobile phase was 100:2:7 (*v*/*v*/*w*) methanol–H_2O–$NaClO_4$ at 1 mL min^{-1}. Typical results are illustrated in Fig. 1.

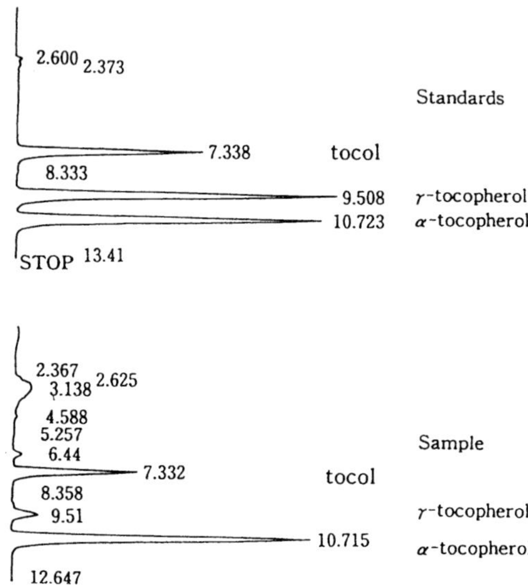

Figure 1. Typical chromatograms obtained from standards and a sample (plasma).

Reference

1. Tamai, H., Manago, M., Yokota, K., Kitagawa, M., and Mino, M. (1988). Determination of α-tocopherol in buccal mucosal cells using an electrochemical detector. *Intern. J. Nutr. Res.*, **58**, 202–7.

15

Vitamin C

TOHRU OGIHARA

Introduction

Although several indirect methods are available for measurement of ascorbic acid (AA), the specificity is generally poor. In addition, for accurate evaluation of vitamin C status, dehydroascorbic acid (DHAA), the oxidation product of AA should also be measured simultaneously. Although direct measurement of DHAA is possible, the indirect method (1) described uses highly sensitive HPLC–ECD equipment based on the method reported by Iriyama *et al.* (2). With this method, total AA is estimated after the reduction of DHAA to AA and DHAA is then calculated from the difference between total AA and free AA (reduced form) originally present in the sample.

Protocol

Sample preparation

1. Mix plasma (25 μL) with metaphosphoric acid (5%, 225 μL) containing desferrioxamine mesylate (Sigma; 20 mM)
2. Centrifuge at 3000 g for 10 min.
3. Isolate supernatant for measurement of reduced and total AA.

Measurement of reduced AA

1. Mix supernatant (75 μL), distilled water (50 μL), and metaphosphoric acid (50%, 25 μL).
2. Centrifuge at 3000 g for 10 min.
3. Isolate supernatant (5 μL) for direct HPLC injection.

Measurement of total AA

1. Mix supernatant (75 μL), dithiothreitol (DTT; 10 mM, 25 μL), and K_2HPO_4 (40 mM, 25 μL; to maintain the pH of the mixture at 6.8).
2. React for 20 min at room temperature in the dark.
3. Add metaphosphoric acid (50%, 25 μL).

Protocol *Continued*

4. Centrifuge at 3000 g for 10 min.
5. Isolate supernatant (5 μL) for direct HPLC injection.

HPLC was performed with an Irika Instruments (Kyoto, Japan) Σ-871 chromatograph equipped with a 250 mm × 4 mm i.d. Irica RP-18T (ODS) column and Irika Amperometric S-875 electrochemical detector with glossy carbon working electrode set at 700 mV relative to Ag/AgCl. The mobile phase was 0.2 M KH_2PO_4–H_3PO_4 (pH 3.0) containing 50 μM EDTA.

The amount of DHAA is total AA minus reduced AA.

Comments

Although the ratio of DHAA to AA is reported to be constant for whole blood stored at 4 °C for 6 h, AA in plasma or serum is highly labile and readily oxidized to DHAA and subsequently to diketogulonic acid. To prevent oxidation of AA in plasma samples, add desferrioxamine. In the presence of 20 mM desferrioxamine, more than 98% of the AA content is preserved when plasma is stored at 4 °C in the dark for 6 h. Dithiothreitol is used as the reductant for DHAA, because it is the most suitable at a neutral pH (4). The recovery of DHAA added to plasma is 98.9% in the presence of 10 mM dithiothreitol at pH 6.8. This reaction was complete within 20 min at 25 °C in the dark.

References

1. Ogihara, T., Kim, H. S., Hirano, K., Imanishi, M., Ogihara, H., Tamai, H., Okamoto, R., and Mino, M. (1998). Oxidation products of uric acid and ascorbic acid in preterm infants with chronic lung disease. *Biol. Neonate*, **73**, 24–33.
2. Iriyama, K., Yoshiyra, M., Iwamoto, T., and Ozaki, Y. (1984). Simultaneous determination of uric and ascorbic acids in human serum by reversed-phase high-performance liquid chromatography with electrochemical detection. *Anal. Biochem.*, **141**, 238–42.
3. Morgalis, S. A. and Davis, T. P. (1988). Stabilization of ascorbic acid in human plasma, and its liquid chromatographic measurement. *Clin. Chem.*, **34**, 2217–23.
4. Okamura, M. (1980). An improved method for determination of L-ascorbic acid and L-dehydroascorbic acid in blood plasma. *Clin. Chim. Acta*, **103**, 259–68.

16

CoQ: Extraction of CoQ homologues from animal tissues / HPLC

TATSUYA MATSURA

Introduction

Coenzyme Q (CoQ) plays an important role in ATP synthesis as an electron-carrying component of the mitochondrial respiratory chain. In recent years, increasing attention has focused on the reduced form of CoQ homologues (CoQ_nH_2) as antioxidants.

During lipid peroxidation in disease states it is important to detect the decrease in endogenous antioxidants and to measure the increase in lipid peroxides. Measurement of CoQ homologues in animal tissues by high-performance liquid chromatography (HPLC) is described in this chapter.

Protocol

Extraction procedure[1]

1. Animal tissue (approx. 300 mg) is stored in liquid nitrogen.
2. Add ice-cold distilled water (8 vol.).
3. Homogenize tissues by means of a Polytron under a stream of nitrogen[2] for 20 s at 4°C. A sample of the homogenate is used for protein assay.
4. Place homogenate (1 mL) in individual 10-mL test tubes with a screw cap[8].
5. Add HPLC-grade ethanol (2 mL) to each tube.
6. Extract each tube with HPLC-grade *n*-hexane (3 × 5 mL, shaking vigorously for 10 min during each extraction). After each extraction centrifuge at 750 g for 5 min, remove the upper *n*–hexane layer carefully[3] and pour into 60-mL dark-brown centrifuge tubes with stoppers, previously filled with nitrogen gas.
7. Evaporate the combined *n*–hexane layers (15 mL) under a stream of nitrogen[2].

Protocol *Continued*

8. Redissolve the residue in ice-cold ethanol (0.5–1.0 mL) and pass the solution through a 0.45-μm filter.

9. Analyse by HPLC[4].

HPLC was performed with a JASCO 880-PU pump and an injector. Compounds were separated on a 250 mm × 4.6 mm i.d., 7 μm particle, Chemcosorb ODS-H column. The mobile phase was 0.7% (*w/v*) $NaClO_4.H_2O$ in 700:300:1 HPLC-grade ethanol–HPLC-grade methanol–70% $HClO_4$ at a flow rate of 1.2 mL min^{-1}. The injection volume was 10 μL. A JASCO 870-UV detector (operated at 275 nm) was used for determination of oxidized CoQ_n[5] and a JASCO 840-EC electrochemical detector (ECD)[6,7] (+700 mV[8] relative to Ag^+/AgCl) for CoQ_nH_2. The chart speed was 2 cm min^{-1}

Preparation of CoQ homologue standards

Standard solutions of CoQ_9 and CoQ_{10}
Chromatographically pure CoQ_9 and CoQ_{10} from Eisai (Tokyo, Japan) were dissolved in HPLC-grade ethanol at a concentration of 550 μg mL^{-1} and diluted to 5.5 μg mL^{-1} with ethanol. When stored in dark-brown vials at −80 °C these standard solutions are stable for approx. 1 year.

Standard solutions of CoQ_9H_2 and $CoQ_{10}H_2$
Standard solutions of CoQ_9 and CoQ_{10} were reduced with $NaBH_4$. CoQ_n stock solution (200 μL) was vortex-mixed with a mixture of $NaBH_4$ (0.3%, 40 μL) in water and 160 μL HPLC-grade ethanol. CoQ_nH_2 standards should be freshly prepared before use.

[1] We usually perform extraction of CoQ_n and CoQ_nH_2 and HPLC analysis within one day to prevent oxidation of CoQ_nH_2.
[2] Sample preparation should be performed under a stream of nitrogen gas to prevent oxidation of CoQ_nH_2 in the samples; this is especially important when the tissues are homogenized and when the combined *n*-hexane extracts are evaporated.
[3] To collect as much as possible of the *n*-hexane layer we use a 5-mL transfer pipette attached to a piece of silicone rubber tubing, the other end of which is connected to a 5-mL syringe.
[4] In this HPLC system, α-tocopherol can be measured simultaneously by ECD.
[5] The concentrations of CoQ_n and CoQ_nH_2 in the sample were determined by comparing the peak areas in the HPLC chromatogram with standard plots of peak area against CoQ_n and CoQ_nH_2 concentrations. The detection limits were 0.7 pmol for CoQ_9 and CoQ_{10}, and 0.5 pmol for CoQ_9H_2 and CoQ_{10} H_2. The extraction recoveries of CoQ_9, CoQ_{10}, CoQ_9H_2 and $CoQ_{10}H_2$ from rat liver were >95%.
[6] The ECD is connected to the outlet of the UV detector.
[7] ECD is sensitive to electrical noise and the baseline on the ECD chromatogram is not always stable. This seems to be because of dirt in the HPLC system, including the column, and because of mechanical and electrical noise. The working electrode should be cleaned when the sensitivity of the electrochemical detector falls.
[8] In some animal species, and tissues, an unknown peak coelutes with the CoQ_9H_2 peak on the chromatogram if the applied potential is +700 mV. If this happens, selectivity is improved by use of a setting near +600 mV. It has previously been reported that electrochemical detection can be performed with an applied potential of +500 mV (1).

Results

Typical UV and EC chromatograms of CoQ homologues from the liver of a three-week-old rabbit are shown in Fig. 1.

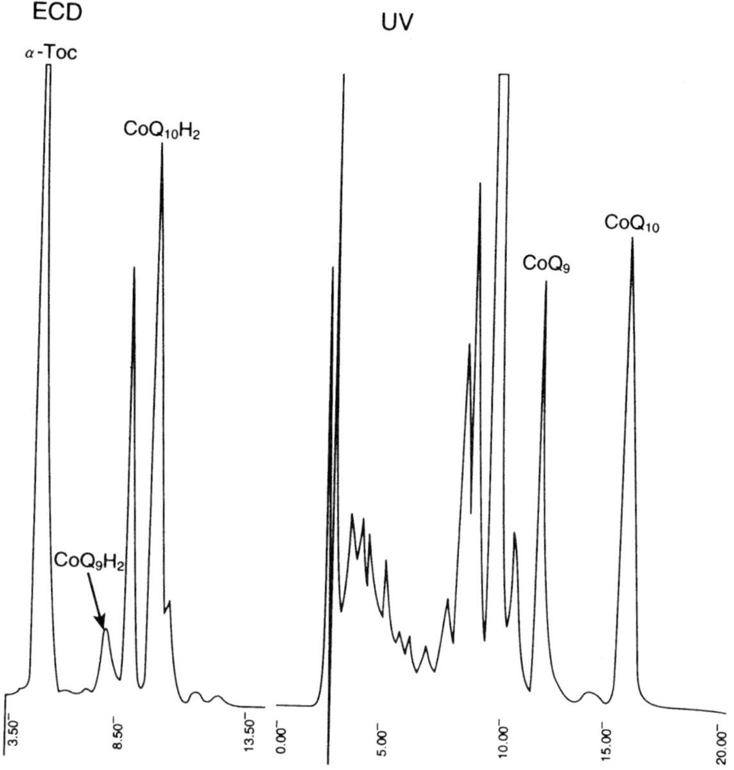

Figure 1. Typical HPLC chromatogram of liver CoQ derivatives.

Reference

1. Lang, J. K., Gohil, K., and Packer, L. (1986). Simultaneous determination of tocopherols, ubiquinols, and ubiquinones in blood, plasma, tissue homogenates, and subcellular fractions. *Anal. Biochem.*, **157**, 106–16.

17

Vitamin A and carotenoids

HIROSHI TAMAI

Introduction

Retinol and carotenoids are extracted from plasma with ethanol containing butylated hydroxytoluene (BHT). Retinol is determined by HPLC with spectrofluorimetric detection (λ_{ex} 340 nm; λ_{em} 460 nm). Carotenoids are also determined by HPLC with UV detection ($A_{453\ nm}$) or with an electrochemical detector (ECD). Fig. 1 shows a typical chromatogram from human plasma carotenoids. (1)

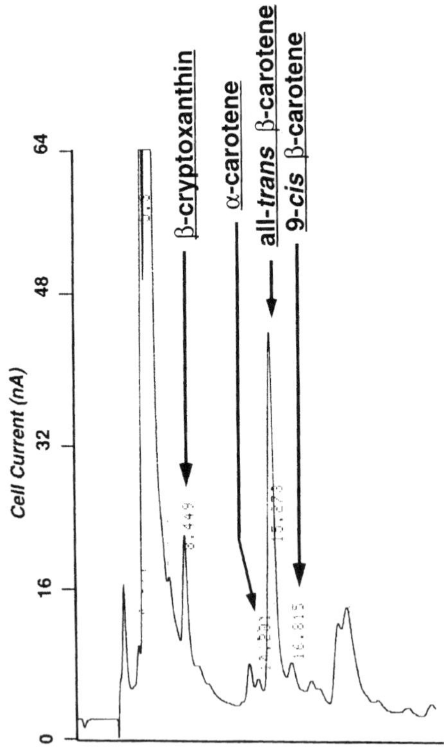

Figure 1. Typical HPLC chromatogram of carotenoids from plasma.

Protocol

1. Add BHT in ethanol (0.015%, 1 mL) to plasma (0.2 mL) and shake.
2. Add *n*-hexane (5 mL) and shake.
3. Centrifuge for 10 min at 3000 rpm.
4. Evaporate upper layer (4 mL) under N_2 gas.
5. Add ethanol (100 mL) and analyse by HPLC.

HPLC was performed with an Irica Instruments (Kyoto, Japan) Σ-871 chromatograph equipped with a 250 mm × 4 mm i.d. Irica RP-18 column and fluorimetric detection (λ_{ex} 340 nm; λ_{em} 460 nm). The mobile phase was 95:5 (*v*/*v*) ethanol–H_2O at a flow rate of 0.7 mL min^{-1}. The external standard was retinol prepared from retinyl palmitate

Comment

ECD is highly sensitive and is used when small amounts of biological sample are available.

Reference

1. Murata, T., Tamai, H., Morinobu, T., Manago, M., Takenaka, A., Takenaka, H., and Mino, M. (1992). Determination of β-carotene in plasma, blood cells and buccal mucosa by electrochemical detection. *Lipids*, **27**, 840–3.

18

Determination of flavonoids (catechins) by HPLC–ECD

ISAO TOMITA and MITSUAKI SANO

Introduction

Flavonoids are ubiquitous in the plant kingdom and are very efficient antioxidants. High-performance liquid chromatography (HPLC) with electrochemical detection (ECD) is a sensitive, simple and selective method for determination of flavonoids. Fig. 1 shows typical HPLC chromatograms and the legend gives details of the method of preparation of HPLC samples of flavonoids (catechin) from the plasma of rats given green-tea extracts. The major flavonoids (catechins) in green-tea extracts are the four epicatechins: (–)-epicatechin, EC; (–)-epicatechin gallate, ECG; (–)-epigallocatechin, EGC; and (–)-epigallocatechin gallate, EGCG. They were separated on a ODS C_{18}

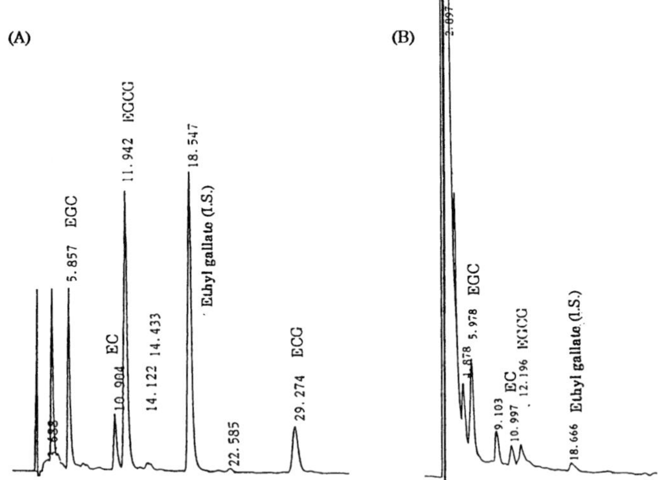

Figure 1. HPLC–ECD chromatogram of flavonoids (catechins). Sample A was prepared by adding a mixture of tea catechins (50 mg mL^{-1}) to rat plasma. Sample B was a plasma sample (200 μL) prepared from blood obtained 1 h after oral administration of the catechin mixture (100 mg kg^{-1}). The catechin mixture contained EC 5.81%, GC 1.44%, ECG 12.51%, EGC 17.57% and EGCG 53.90%.

reversed-phase column by isocratic elution with 85:15 0.1% phosphoric acid–acetonitrile solution containing 0.1 mM Na_2EDTA (1).

Protocol

1. Add ethyl gallate (0.1–50 μg mL^{-1}, 50 μL; internal standard) to plasma (200 μL).
2. Add metaphosphoric acid solution (30% *w/v*, 200 μL).
3. Mix for 1 min.
4. Sonicate at 20 kHz for 30 s.
5. Leave to stand in an ice bath for 10 min.
6. Centrifuge at 3000 rpm for 10 min.
7. Dilute with HPLC mobile phase and pass through 0.45-μm filter.
8. Take 10–20 μL of sample for analysis by HPLC with electrochemical detection (ECD)

HPLC was performed with a Shimadzu (Kyoto, Japan) LC-10AT pump and a Shiseido (Tokyo Japan) Nanospace SI-1/2005 ECD with the applied voltage set at 600 mV. Compounds were separated on a 4.6 mm i.d. × 250 mm TSK gel ODS80Ts reversed-phase column (Tosoh, Tokyo, Japan) maintained at 30 °C in a Shimadzu CTO-10 AC column oven. The mobile phase was 85:15 0.1 M phosphoric acid–acetonitrile containing 0.1 mM Na_2EDTA at a flow rate of 1.0 mL min^{-1}. The limits of detection for epicatechins are approx. 1 ng mL^{-1} (signal-to-noise ratio, $S/N = 3$) in plasma or bile.

Comments

Aqueous solution of flavonoids (catechins) are readily oxidized and polymerized under alkaline conditions, but are fairly stable under acidic conditions. Oxidation is accelerated in the presence of metal ions such as Fe^{2+}, Fe^{3+}, and Cu^{2+}. Plasma samples for HPLC analysis should be stored in the presence of an antioxidant and metal chelator. Epicatechins in rat or human plasma are relatively stable when mixed 1:50 (*v/v*) with Vit.C–EDTA solution (0.4 M NaH_2PO_4 buffer containing 20% vitamin C and 0.1% Na_2EDTA, pH 3.6) and stored at −80 °C (2).

Free epicatechins are detected in plasma and bile 1–2 h after oral administration of tea catechin or green tea powder to rats or man (3–5). Epicatechins are also present in the conjugated form (e.g. glucuronide and sulfate) in plasma and bile, and hence, whole catechins can be detected after pretreatment with β-glucuronidase (Sigma G-7896) and sulfatase (Sigma S-9754) (2). When EGCG, a major green tea catechin, is incubated with rat plasma or bile at 37 °C, three small peaks arising from three dimers can be detected in both fluids by following the disappearance of EGCG (5, 6).

References

1. Umegaki, K., Esashi, T., Tezuka, M., Ono, A., Sano, M., and Tomita, I. (1996). Determination of tea catechins in food by HPLC with an electrochemical detector. *J. Food Hygiene Soc. Jpn.* (in Japanese), **37**, 77–82.
2. Lee, M. J., Wang, Z.-Y., Li, H., Chen, L., Sun, Y., Gobbo. S., Balentine, D. A., and Yang C. S. (1995). Analysis of plasma and urinary tea polyphenols in human subjects. *Cancer Epidemiology Biomarkers and Prevention*, **4**, 393–399.
3. Unno, T., Kondo, K., Itakura, H., and Takeo, T. (1996). Analysis of (–)-epigallocatechin gallate in human serum obtained after ingesting green tea. *Biosci. Biotech. Biochem.* **60**, 2066–2068.
4. Okushio, K., Matumoto, N., Kohri, T., Suzuki, M., Nanjo, F., and Hara, Y. (1996). Absorption of tea catechins into rat portal vein. *Biol. Pharm. Bull.*, **19**, 326–329.
5. Tomita, I., Sano, M., Sasaki, K., and Miyase, T. (1998). Tea catechin (EGCG) and its metabolites as bioantioxidants, ACS Symposium Series, No. 701, Oxford Univ. Press, pp. 209–216.
6. Yoshino, K., Suzuki, M., Sasaki, K., Miyase, T., and Sano, M. (1999). Formation of antioxidants from (-)-epigallocatechin gallate in mild alkaline fluids, such as authentic intestinal juice and mouse plasma. *J. Nutr. Biochem.* **10**, 223–229

19

Catalase

NAOHISA KAWAMURA

Introduction

The enzyme catalase, widely distributed in animal tissues, consists of four identical subunits. The molecular weight is approximately 240 000. The major active component of the enzyme, ferri protoporphyrin (haematin), which occurs once in each subunit, is responsible for catalysing the dismutation $2H_2O_2 \rightarrow O_2 + 2H_2O$

Catalase activity can be measured by monitoring the decrease in absorbance of H_2O_2 at 240 nm as it is removed, by monitoring the consumption of potassium permanganate by H_2O_2 (1, 2), and by monitoring the production of oxygen from H_2O_2 by means of an oxygen electrode. This chapter reports a simpler and more accurate method with UV detection for measurement of catalase activity in blood cells and tissue samples.

Protocol

Peripheral blood samples

1. Wash heparinized peripheral blood three times with saline.
2. Obtain the packed RBC fraction.
3. Dilute 1:20 with 2-mercaptoethanol–EDTA stabilizing solution.
4. Dilute 1:100 and haemolyse with 2% ethanol.

Tissue samples (e.g. liver, kidney)

1. Homogenize fresh tissue samples with 20 vol. RIPA buffer.
2. Centrifuge at 12 000 g for 30 min at 4 °C
3. Isolate the supernatant (tissue sample).
4. Perform protein assay (BCA, Lowry, etc.).

Measurement of catalase activity by monitoring absorbance ($A_{240\ nm}$) to follow H_2O_2 destruction

1. Mix Tris-HCl (1 M), EDTA (pH 8.0, 5 mM, 50 μL), H_2O_2 (10 mM, 900 μL) and deionized water (30 μL). For blanks replace the H_2O_2 with an equal volume of deionized water

Protocol *Continued*

2. Incubate at 37 °C for 10 min.
3. Add sample (1:2000 haemolysate or tissue samples; 20 μL).
4. Continuously measure absorbance at 240 nm.

Calculation

$$\text{Catalase unit (Int. units/g Hb or protein)} = (\Delta\text{absorbance}/0.071/2) \times (V_C/V_X) \times (100/\text{Hb or protein})$$

where Δabsorbance is the change in absorbance at 240 nm in 1 min, 0.071 is the absorbance of 1 mM catalase, and $V_C/V_X = 50$ (the dilution level in the cuvette).

Comments

At room temperature, catalase dissociates into a 120 K subunit and loses its activity. Catalase diluted 1:20 is frozen for storage. The sample is used soon after defrosting and adjusted to 1:100 for each measurement.

The substrate H_2O_2 is measured at a low level (approximately 10 mM). The substrate is also adjusted before use, because of the potential instability of H_2O_2.

The surfactant Triton-X, which has UV absorbance, cannot be used at high concentrations. Digitonin and cholic sodium can be used, however.

References

1. Beutler, E. (1975). *Red Cell Metabolism. A Manual of Biochemical Methods*, (2nd edn), pp. 89–90. Grune and Stratton, Orlando, USA.
2. Aebi, H. (1974). *Catalase, Methods of Enzymatic Analysis*, Vol. II, pp. 673–83. Academic Press; New York.

20

Glutathione peroxidase: coupled enzyme assay

KAZUHIKO TAKAHASHI

Introduction

Glutathione peroxidase (GPx) is a cytosolic selenocysteine-containing enzyme that catalyses the reduction of many hydroperoxides (ROOH: hydrogen peroxide, fatty acid hydroperoxides and phospholipid hydroperoxides) in the presence of glutathione (GSH).

$$2GSH + H_2O_2 \xrightarrow{GPx} GSSG + 2\,H_2O$$
$$2GSH + ROOH \xrightarrow[GPx]{} GSSG + ROH + H_2O$$

t-Butyl hydroperoxide is used as a substrate and the rate of formation of GSSG is measured by coupling it to the glutathione reductase (GR) reaction (1).

$$GSSG + NADPH + H^+ \xrightarrow[GR]{} 2GSH + NADP^+$$

The oxidation of NADPH is followed at 340 nm.

Protocol

1. Set up a double-beam spectrophotometer equipped with recorder and thermostatted semi-micro cuvettes (1 mL).
2. To both sample and reference cuvettes add Tris-HCl (1 M), containing EDTA (5 mM) (pH 8.0, 100 μL), GSH (0.1 M, 20 μL)[1], GR (10 units mL^{-1}, 100 μL)[1], NADPH (2 mM, 100 μL)[1], and the enzyme-containing sample (10 μL). In addition 660 and 670 μL water are added to sample and reference cuvettes, respectively.
3. Preincubate at 37 °C for 2 min.
4. Add *t*-butyl hydroperoxide (7 mM, 10 μL) to the sample cuvette only.
5. Measure the decrease in absorption at 340 nm.

[1] These reagents should be freshly prepared and kept on ice before use.

Calculation

The decreasing absorption at 340 nm is calculated from the slope of the graph obtained. The blank rate measured in the absence of the enzyme fraction should be subtracted from the rate measured in the presence of enzyme. The molar absorption coefficient for NADPH (6.22 $\text{mM}^{-1}\ \text{cm}^{-1}$) is used to calculate the activity. One mol NADPH is oxidized for each mol *t*-butyl hydroperoxide reduced.

Comments

t-Butyl hydroperoxide is the most suitable peroxide substrate for assay of GPx in erythrocytes and plasma samples. As tissues contain non-selenium-dependent GPx (isozyme of glutathione S-transferase) which does not reduce H_2O_2 but reduces *t*-butyl hydroperoxide, H_2O_2 should be used with sodium phosphate buffer (0.1 M, pH 7.0) and NaN_3 (2 mM) at 25 °C and with H_2O_2 (1.5 mM) as substrate.

At least three forms of GPx are reported to exist and differ in many properties, including enzymatic nature. The classical cellular GPx (cGPx), found in the cytosol of various tissues and blood cells, reduces H_2O_2 and organic hydroperoxides. A second GPx, phospholipid hydroperoxide GPx (PH-GPx), reduces both phospholipid hydroperoxide and H_2O_2 (2). A third GPx in plasma, now called extracellular GPx (eGPx), is reported to reduce both H_2O_2 and phospholipid hydroperoxides (3). It is important to know that only cGPx activity can be detected in crude extracts by this assay.

Measurement of PH-GPx activity is described elsewhere (4).

References

1. Beutler, E. (1984). Glutathione peroxidase. In *Red Cell Metabolism*, pp. 74–6. Grune and Stratton, New York.
2. Roveri, A., Máiorino, M., Nishii, C., and Ursini, F. (1994). Purification and characterization of phospholipid hydroperoxide glutathione peroxidase from rat testis mitochondrial membranes. *Biochim Biophys Acta*, **1208**, 211–21.
3. Yamamoto, Y. and Takahashi, K. (1993).Glutathione peroxidase isolated from plasma reduces phospholipid hydroperoxide. *Arch Biochem Biophys.*, **305**, 541–5.
4. Saito, Y., Hayashi, T., Tanaka, A., Watanabe, Y., Suzuki, M., Saito, E., Takahashi, K. (1999). Selenoprotein P in human plasma as an extracellular phospholipid hydroperoxide glutathione peroxidase. *J. Biol. Chem.*, **274**, 2866–71.

21

Glutathione reductase

KENJI AKI

Introduction

Glutathione reductase (EC1.6.4.2) catalyses the reduction of GSSG by NADPH, producing 2 mol GSH from 1 mol GSSG. The enzyme contains FAD and redox-active disulfide at the active centre (1). The optimum pH is broad and centred at pH 7.6. The reaction is irreversible reduction of GSSG and its metabolic function is, therefore, to supply GSH. GSH is utilized in two types of reaction—it is the substrate for enzymes such as glutathione peroxidase and it is a non-enzymatic reducing equivalent in the cell. The enzyme is distributed widely in aerobic life forms and is located in the cytoplasm of cells.

Protocol

1. Prepare assay mixture containing 50 mM phosphate buffer[1], 1 mM EDTA[2], 0.1 mM NADPH[3], 1 mM GSSG[2], 0.1% bovine serum albumin (BSA)[4], pH 7.6. Control mixture, in which GSSG is not contained but other constituents are the same, should also be prepared.
2. Place 1 ml of the assay mixture into a cuvette (semi-micro cell, 1 cm light path).
3. Preincubate the mixture for 5 min at 25 or 30 °C.
4. Start reaction by addition of enzyme solution.
5. Monitor the time course of the change of absorbance at 340 nm.[5]
6. Read the absorbance decrease per min ($\Delta A/\Delta t$) in a linear region (initial velocity) of the recording.
7. Perform Steps (2)–(6) using the control mixture in place of the assay mixture except that pure enzyme are used.[6]
8. Calculate the enzyme activity (units mL^{-1} of the enzyme solution) from:

Activity $= (\Delta A/\Delta t_{(GSSG\ plus)} - \Delta A/\Delta t_{(GSSG\ minus)}) \times (1000 + V_{enz})/V_{enz}/6.22$

where V_{enz} is a volume of enzyme solution added into the mixture and should be less than 25 μl to avoid changes in the substrate concentrations.

[1] The values of K_m and V_{max} depend on ionic strength. At high ionic strength the K_m value is small and the value of V_{max} is large. Kinetic constants for the enzyme of human erythrocytes: K_m for NADPH is 13 μM in 0.3 M phosphate and 9.5 μM in 0.03 M phosphate; K_m for GSSG is 125 μM in 0.3 M phosphate and 19 μM in 0.03 M phosphate. V_{max} is 14300 min^{-1}/FAD in 0.3 M phosphate and 7700 min^{-1}/ FAD in 0.03 M phosphate (3).
[2] EDTA and GSSG solutions are adjusted to neural pH.
[3] NADH is a poor co-substrate.
[4] BSA is added into the reaction mixture to stabilize the enzyme. The effect of addition of BSA is dominant on the purified enzyme.
[5] Dual-wavelength spectrophotometry is useful for the turbid reaction mixture obtained (λ_1 = 340 nm, $\lambda_2 > 400$ nm).
[6] Because crude extracts of cells contain non-specific NADPH reductase activity, a control experiment in the absence of GSSG is required.

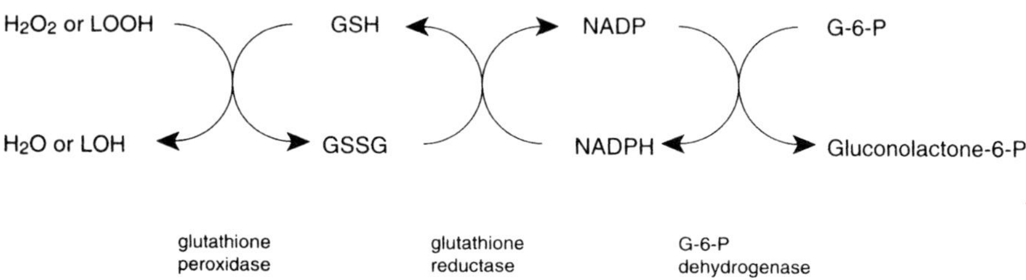

Figure 1. System for scavenging hydrogen peroxide or lipoperoxide.

Hydrogen peroxide and lipid peroxide are scavenged by a coupling reaction including glutathione reductase, as shown in Fig. 1.

References

1. Ghisla, S. and Massey, V. (1989). Mechanisms of flavoprotein-catalyzed reactions. *Eur. J. Biochem.* **181**, 1–17.
2. Williames, C. H. Jr, Zanetti, G., Arscott, L. D., and McAllister, J. K. (1967). Lipoamide dehydrogenase, glutathione reductase, thioredoxin reductase, and thioredoxin. *J. Biol. Chem.*, **242**, 5226–31.
3. Williames, C. H. Jr (1976). *Glutathione Reductase in the Enzymes*, Vol. 13, (ed. P. D. Boyer), pp. 129–42. Academic Press, New York.

22

Glutathione transferase

SHIGEKI TSUCHIDA

Introduction

The glutathione transferases (GSTs) are a family of multi-functional proteins which act as enzymes and also as binding proteins in detoxification processes (1–5). GSTs catalyse the nucleophilic attack of the sulfur atom of reduced glutathione by electrophilic groups in a second substrate. Most GSTs are located in the cytosol, and are grouped into five classes, Alpha, Mu, Pi, Theta and Sigma. Some forms of GST have selenium-independent glutathione peroxidase activity towards lipid hydroperoxides.

Protocol

Assay for activity towards 1-chloro-2,4-dinitrobenzene (CDNB)[1] (1)

1. Mix potassium phosphate (pH 6.5, 0.2 M, 0.5 mL), GSH[2] (10 mM, 0.1 mL), water (0.2 mL), enzyme[3]-containing sample (0.1 mL) and CDNB[4] (10 mM 0.1 mL) (total volume 1.0 mL).
2. Incubate at 25 °C.
3. Record increase of absorbance at 340 nm.

Assay for activity towards cumene hydroperoxide[5] (3, 4)

1. Mix sodium phosphate (pH 7.0, 0.25 M, 0.2 mL), EDTA (10 mM, 0.1 mL), NaN_3[6] (10 mM, 0.1 mL), water (0.1 mL), GSH (10 mM, 0.1 mL), glutathione reductase[7] (2 unit mL^{-1}, 0.1 mL), NADPH[8] (2 mM, 0.1 mL), cumene hydroperoxide[9] (15 mM, 0.1 mL) and enzyme-containing sample (0.1 mL) (total volume 1.0 mL).
2. Incubate at 25 °C.
3. Record decrease of absorbance at 340 nm.

[1] Glutathione conjugate of CDNB has maximum absorbance at 340 nm. CDNB is used for the assay because most forms of GST have activity towards this substrate.
[2] GSH solution is stable at 4 °C for 1 month.
[3] Tissue extract can be used as an enzyme source. The homogenizing buffer routinely used is Tris-HCl (10 mM, pH 7.8), containing NaCl (0.2 M), and the supernatant centrifuged at 105000 g

Protocol *Continued*

for 45 min. The enzyme solution is diluted with potassium phosphate (pH 6.5, 10 mM) to adjust the change in absorbance to approx. 0.05 min^{-1}.
[4] CDNB (102 mg) is dissolved in ethanol (20 mL) and then diluted to 50 mL with water. This solution is added to the assay mixture containing enzyme to start the reaction. CDNB solution in a capped bottle should be stored at 37 °C in the dark. This solution is stable for approx. 1 month.
[5] Some forms of GST have glutathione peroxidase activity towards lipid hydroperoxides, producing oxidized glutathione and alcohols. Unlike Se-dependent glutathione peroxidase, GST is not active towards hydrogen peroxide. This method measures the decrease in the level of NADPH, coupled to glutathione reductase for quantitating oxidized glutathione produced from cumene hydroperoxide and reduced glutathione. When a crude extract is used as an enzyme source, values determined by this method also reflect the activity of Se-dependent glutathione peroxidase. The contribution of GST to activity values can be evaluated by subtracting the activity with hydrogen peroxide from that with cumene hydroperoxide.
[6] Sodium azide is added to inhibit catalase activity.
[7] Glutathione reductase from yeast (250 units mL^{-1}; Oriental Yeast Co., Tokyo) is diluted 125-fold with sodium phosphate buffer (pH 7.0, 10 mM).
[8] NADPH (18.1 mg) is dissolved in $NaHCO_3$ solution (0.1% *w/v*, 10 mL).
[9] Cumene hydroperoxide (46 mg) is dissolved in water (20 mL). When hydrogen peroxide is used as substrate, hydrogen peroxide (2.5 mM, 0.1 mL) is added to the assay mixture in place of cumene hydroperoxide. H_2O_2 solution (30%, 0.5 mL) is added to water (50 mL) and this mixture (0.25 mL) is further diluted with water (8.5 mL) to prepare 2.5 mM H_2O_2 solution.

Results and calculations

For CDNB:

$$\text{Enzyme activity (units mL}^{-1}) = [(\Delta A_{enzyme} - \Delta A_{no\ enzyme}{}^{1}) \times (\text{dilution factor}) \times 10]/\varepsilon \times 10^{-3}$$

where ΔA is the change in absorption min^{-1} and ε is the molar absorption coefficient (9600 $M^{-1}cm^{-1}$ for CDNB). One unit is the amount of enzyme which catalyses the conjugation of 1 µmol CDNB min^{-1} at 25 °C. For cumene hydroperoxide, enzyme activity (units mL^{-1}) can be calculated by use of a similar formula except that ε is 6220 $M^{-1}cm^{-1}$.

Comments

1. The non-enzymatic change in absorbance is in the range 0.005–0.007 min^{-1} at 25 °C.

References

1. Habig, W. H., Pabst, M. J., and Jakoby, W. B. (1974). Glutathione S-transferases. The first enzymatic step in mercapturic acid formation. *J. Biol. Chem.*, **249**, 7130–9.
2. Keen, J. H. and Jakoby, W. B. (1978). Glutathione transferases. Catalysis of nucleophilic reactions of glutathione. *J. Biol. Chem.*, **249**, 5654–7.

3. Paglia, D. E. and Valentine, W. N. (1967). Studies on the quantitative and qualitative characterization of erythrocyte glutathione peroxidase. *J. Lab. Clin. Med.*, **70**, 158–69.
4. Lawrence, R. A. and Burk, R. F. (1976). Glutathione peroxidase activity in selenium-deficient rat liver. *Biochem. Biophys. Res. Commun.*, **71**, 952–8.
5. Hayes, J. D. and Pulford, D. J. (1995). The glutathione S-transferase supergene family: regulation of GST and the contribution of the isoenzymes to cancer chemoprotection and drug resistance. *Crit. Rev. Biochem. Mol. Biol.*, **30**, 445–600.

23

γ-Glutamyl transpeptidase

YOSHITAKA IKEDA

Introduction

γ-Glutamyl transpeptidase catalyses transfer of a γ-glutamyl group from glutathione or other γ-glutamyl compounds to amino acids or dipeptides, and also hydrolyses the γ-glutamyl bond. This enzyme, which is also known as γ-glutamyltransferase, γ-GTP, γ-GT, or GGT, plays an important role in glutathione metabolism because the reactions catalysed by the enzyme are the first steps in the degradation of glutathione. In the assay for γ-glutamyl transpeptidase activity, L-γ-glutamyl-*p*-nitroanilide and glycylglycine are most commonly used as donor and acceptor substrates, respectively. The transfer reaction using the *p*-nitroaniline derivative leads to the stoichiometric release of *p*-nitroaniline, which can be detected spectrophotometrically.

Protocol

Preparation of the assay mixture for γ-glutamyl transpeptidase

1. Completely dissolve L-γ-glutamyl-*p*-nitroanilide (0.5 mmol) in HCl (0.2 M, 10 mL).
2. Add double-distilled water (DDW; 350 mL).
3. Add glycylglycine (1.32 g) and dissolve.
4. Add Tris HCl (pH 9.0–9.5, 0.5 M, 100 mL).
5. Adjust pH to 8.0 with HCl.
6. Dilute volume to 500 mL with DDW.

Measurement of γ-glutamyl transpeptidase activity

1. Place the assay mixture (1 mL) into a cuvette (semi-micro cell, light path 1 cm).
2. Add the enzyme solution (5–20 μL) to start the reaction.
3. Record the increase in absorbance at 410 nm continuously on chart paper.
4. Read the absorbance increase per min ($\Delta A/\Delta t$) in a linear region (initial velocity) of the recording.

Calculation

Calculate the enzyme activity (units mL^{-1} of the original enzyme solution) from:

Activity = $(\Delta A/\Delta t)/\varepsilon$ (= 8.8) × 1000/volume of enzyme solution added (μL).

Comments

1. The composition of the substrate mixture is L-γ-glutamyl-*p*-nitroanilide (1 mM), glycylglycine (20 mM), and Tris-HCl (pH 8.0, 0.1 M).
2. The substrate mixture is stable for several months if stored at −20 °C.
3. In the assay, a spectrophotometer should be equipped with a cuvette holder which can be equilibrated at 37 °C. It is important that a cuvette and its contents (assay mixture) are warmed to 37 °C before assay.
4. To start the reaction a small volume of an enzyme solution (< 20 μL) is added to the substrate solution in the cuvette. The cuvette is then closed with Parafilm and the thumb, inverted gently but quickly two or three times to mix the contents, then placed in the holder and the recording is started.
5. If a linear portion is not obtained in the recording, because the activity is too high, it is necessary to dilute the enzyme solution.
6. The molar absorption coefficient (ε) for *p*-nitroaniline is 8.8 mM^{-1} cm^{-1}. Although addition of the enzyme solution to the assay mixture changes the volume and the concentrations of the substrates by a few percent, these changes are negligible in the calculation of enzyme activity.
7. L-γ-Glutamyl-3-carboxy-4-nitroanilide is also used as a donor substrate. This substrate is more convenient because of its greater solubility, but more expensive. The ε value for carboxynitroaniline is 9.5 mM^{-1} cm^{-1}.

References

1. Tate, S. S. and Meister, A. (1985). γ-Glutamyl transpeptidase from kidney. *Methods Enzymol.* **113**, 400–19.
2. Taniguchi, N. and Ikeda, Y. (1998). γ-Glutamyl transpeptidase: catalytic mechanism and gene expression. *Adv. Enzymol. RAMB*, **72**, 239–78.

24

γ-Glutamylcysteine synthetase: assay mixture and measurement

TAKAHIKO KONDO and FUMI SAWADA

Introduction

Glutathione (GSH) is synthesized by two enzymes, γ-glutamylcysteine synthetase (γ-GCS) and glutathione synthetase. γ-GCS, a rate-limiting enzyme of GSH synthesis, catalyses the reaction:

$$\text{L-glutamylcysteine} + \text{glycine} + \text{ATP} \rightarrow \text{GSH} + \text{ADP} + \text{Pi}.$$

Glutathione synthetase catalyses the reaction

$$\text{L-glutamate} + \text{L-cysteine} + \text{ATP} \rightarrow \text{L-}\gamma\text{-glutamylcysteine} + \text{ADP} + \text{Pi}.$$

Determination of the activity of γ-GCS is based on a method which utilizes the oxidation of NADH to NAD. The formation of ADP is measured by converting it to ATP in the pyruvate kinase reaction:

$$\text{PEP} + \text{ADP} \rightarrow \text{Pyruvate} + \text{ATP}$$

The pyruvate formed oxidizes NADH in the lactic dehydrogenase reaction:

$$\text{Pyruvate} + \text{NADH} + \text{H}^+ \rightarrow \text{Lactate} + \text{NAD}$$

The oxidation of NADH to NAD is measured at 340 nm.

Protocol

Preparation of Materials

Samples can be stored at −80 °C, but are rapidly inactivated by repeated freezing and thawing.

Preparation of materials from red blood cells[1]

1. Prepare red blood cells from whole blood (2 mL) by passing through a column packed with an equal volume of a 1:1 (*w/w*) cellulose–microcrystaline cellulose (Sigma) previously equilibrated with isotonic saline.

2. Collect cell pellets after centrifugation at 1000 g for 10 min and store at −80 °C before use.

Preparation of materials from cultured cells[(2)]

1. Remove cells (1×10^7) from culture medium and wash in isotonic saline (3×1 mL).
2. Collect cell pellets after centrifugation and store at −80 °C until use.

Preparation of materials from rat kidney

1. Dialyse kidney tissues against ice-cold 0.1 M phosphate buffer (KH_2PO_4/K_2HPO_4, pH 7.4) to effect removal of red blood cells.
2. Store tissue in gram portions at −80 °C before use.

Homogenization[1]

1. Thaw cells or tissues, suspend in ice-cold phosphate buffer (0.1 M, pH 7.4, 1 mL or 100 mL) and homogenize by means of a Potter–Elvehjem-type homogenizer.
2. Isolate the supernatant of the samples and use as material for tests.

Assay system

1. Prepare pre-mix solution by mixing Tris-HCl (pH 8.2, 1 M) containing EDTA (5 mM, 2.50 mL)[2], KCl (1 M, 1.67 mL)[3], $MgCl_2$ (1 M, 0.42 mL)[4], EDTA (0.5 M, 8 mL)[5], water (5.4 mL), and NADH (3 mg) (total volume 10 mL).
2. Confirm absorbance of this pre-mixture solution at 340 nm is near 2.0.
3. Prepare the assay system and blank by mixing pre-mix solution (600 μL), ATP (0.1 M, 100 μL)[6], phosphoenolpyruvate (PEP; 0.1 M, 100 μL)[7], L-glutamate (0.1 M, 50 μL)[8], and PK/LDH (10 μL)[9].
4. To the assay system only add L-α-aminobutyric acid (L-α-AB; 0.1 M 50 μL)[10].
5. To the blank only add water (50 μL).
6. Measure the background absorbance at 340 nm and 37 °C.
7. Add the sample (90 μL) to both assay and blank.
8. Measure the decrease in the concentration of NADH at 340 nm and 37 °C[11].

[1] The temperature of the sample should not exceed 0 °C during homogenization of the tissue samples or cultured cells. Two freeze and thaw cycles are sufficient to the lyse red blood cells.
[2] Tris-HCl (pH 8.2, 1 M) containing EDTA (5 mM, 2.50 mL) is prepared by dissolving Tris (12.11 g) and Na-EDTA (186 mg) in water, adjusting the pH to 8.2 with HCl (1 M) and diluting to 100 mL with water.
[3] KCl (1 M) is prepared by dissolving KCl (7.456 g) in water and diluting to 100 mL with water.
[4] $MgCl_2$ (1 M) is prepared by dissolving $MgCl_2$ (20.33 g) in water and diluting to 100 mL with water.

Protocol *Continued*

[5] EDTA (0.5 M) is prepared by dissolving EDTA (sodium salt; 1.86 g) in water, adjusting the pH to neutral with NaOH (1 M), and diluting to 100 mL with water.
[6] ATP (0.1 M) is prepared by dissolving ATP (disodium salt; 551.1 mg) in water, adjusting the pH to neutral with NaOH (1 M) and the total volume to 10 mL. The solution is divided into small amounts and stored at −20°C.
[7] Phosphoenolpyruvate (PEP; 0.1 M) is prepared by dissolving PEP (41.1 mg) in water, adjusting the pH to neutral with NaOH (1 M) and adjusting the volume to 2 mL on use.
[8] L-glutamate (0.1 M) is prepared by dissolving L-glutamate (14.7 mg) in water, adjusting the pH to neutral with NaOH (1 M) and adjusting the volume to 1 mL on use.
[9] PK/LDH solution is prepared by mixing pyruvate kinase (Pk, Boehringer; 100 μL), lactate dehydrogenase (LDH, Boehringer; 50 μL) and water (100 μL) and adjusting the volume on use.
[10] L-α-aminobutyric acid (L-α-AB; 0.1 M) is prepared by adjusting the pH of L-α-aminobutyric acid (20.6 mg) to neutral with NaOH (1 M) and adjusting the volume to 2 mL on use.
[11] The time taken for the decrease in the amount of NADH to become a straight line depends on the sample; if it is less than 5 min the determination should be performed promptly. Excess sample rapidly consumes NADH. In this circumstance diluted sample should be used. Many cells ($\sim 1 \times 10^7$) are, however, required for estimation of activity in culture cell lines.

Results and calculation

The activity is calculated by use of the equation:

$$(A_s - A_b \text{ min}^{-1}) \times 100/90 \times 1/6.22 = \text{units mL}^{-1}$$

where A_s is the change in the absorbance of the system min^{-1} and A_b is the change in the absorbance of the blank min^{-1}. The activity can be expressed represented in units (mg protein)$^{-1}$ if the result is adjusted for the volume of protein in the sample.

Typical activity in some cells is: human red blood cells, 0.11 ± 0.22 units g^{-1} haemoglobin; mouse vascular endothelial cells, 2.53 ± 0.35 milliunits/10^6 cells; human cancer cells, 1.4–7.1 milliunits/10^6 cells.

References

1. Beutler, E. (1986). *Red Cell Metabolism: A Manual of Biochemical Methods*, 1888 pp. Grune and Stratton, Orlando, USA.
2. Urata, Y., Yamamoto, H., Goto, S., Tsushima, H., Akazawa, S., Yamashita, S., Nagataki, S., and Kondo, T. (1996). Long exposure to high glucose concentrations impairs the responsive expression of γ-glutamylcysteine synthetase by interleukin-1β and tumor necrosis factor-α in mouse endothelial cells. *J. Biol. Chem.*, **271**, 15146–52.

25

Measurement of Mn-SOD and Cu,Zn-SOD

KEIICHIRO SUZUKI

Introduction

Superoxide dismutase (SOD) catalyses the dismutation of superoxide anion to produce O_2 and hydrogen peroxide:

$$2O_2\cdot^- + 2H^+ \rightarrow H_2O_2 + O_2$$

There are three kinds of SOD in mammalian systems—Cu,Zn-SOD, Mn-SOD and extracellular SOD (EC-SOD).

Activity assay of SOD

Because the enzyme substrate and the products are unstable, it is difficult to measure the disappearance or the formation of products as is usual in enzymatic assays. Routine assays for SOD usually employ an indirect assay in which one unit of enzyme activity is defined as the amount of enzyme that inhibits the reaction of $O_2\cdot^-$ with the indicator by 50%. The most frequently used method for measuring SOD activity employs the xanthine/xanthine oxidase reaction for $O_2\cdot^-$ generation and reduction of cytochrome-*c*, or nitroblue tetrazolium, for $O_2\cdot^-$ detection. In this section we describe the nitroblue tetrazolium (NBT) method (1).

Protocol

Reagents

Solution A
Sodium carbonate buffer (50 mM) containing xanthine (0.1 mM), NBT (0.025 mM) and EDTA (pH 10.2 at 25°C, 0.1 mM)

Solution B
Xanthine oxidase solution (diluted with 2 M ammonium sulfate and 1 mM EDTA).

Protocol *Continued*

Procedure

1. Place solution A (2.9 mL) in the cuvette.
2. Add solution B (50 μL) and sample (or blank) (50 μL); final volume 3.0 mL.
3. Mix, and monitor at 560 nm in the spectrophotometer for several minutes at 25°C. Solution B (xanthine oxidase solution) should be diluted if the absorbance of the blank increases at a rate of 0.015 min^{-1} or more.

Calculation

The amount of enzyme at which 50% inhibition is achieved is defined as one unit. Each enzyme sample is prepared at several different dilutions, and 50% inhibition is calculated by use of logit paper, Fig. 1 (2).

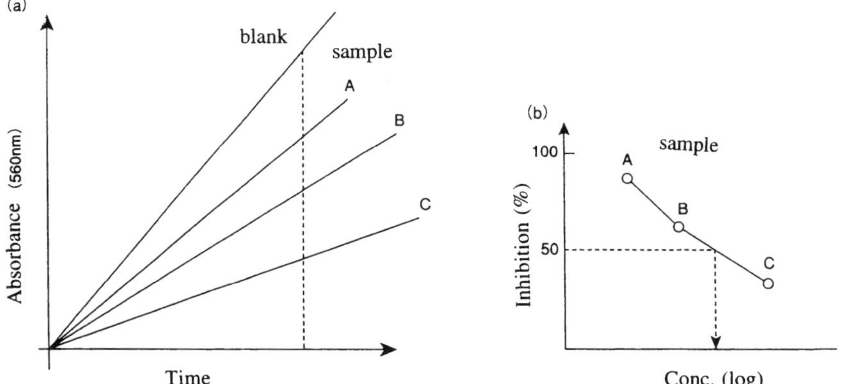

Figure 1. Assay of SOD activity.

50% inhibition can also be calculated by the method of Kuthan *et al.* (3):

$$\text{SOD activity} = ([b/a] - 1) \times (\text{dilution})$$

where a is the increase in the absorbance of a sample and b is the increase in the absorbance of the blank. This calculation should be used when b/a is approximately equal to 2.

Comments

Because Cu,Zn-SOD activity is inhibited by cyanide, it is possible to distinguish between Cu,Zn-SOD and Mn-SOD (for example, 50 μL of 0.25 M NaCN will be enough for a 3-mL cuvette). Inhibition is not, however, complete. The value (85–95%) depends on the time of pre-incubation of samples with cyanide).

Careful controls are essential when using crude samples. Tissue homogenate can, for example, contain various substances which have SOD-like activity or inhibit SOD activity.

Immunochemical assay of SOD

Compared with enzymatic methods, immunochemical assays of SODs are more reliable and reproducible, because the determinations are specific to the protein moiety, and immunogenicity will not change in the presence of inhibitors or activators of SOD in the tissues. Here we will describe an ELISA (enzyme-linked immunosorbent assay) system using polyclonal antibodies (4).

Protocol

Biotination of the antibody

1. Dialyse the antibody solution with a solution of $NaHCO_3$ (pH 8.0, 0.1 M) at 4°C.
2. Dissolve biotin-*N*-hydroxysuccinimide in DMSO (1 mg mL^{-1}).
3. Mix the antibody and biotin solutions (1:1).
4. Stand for 4 h at room temperature.
5. Dialyse with PBS (pH 7.4, 20 mM) at 4°C.

Reagents

1. Prepare phosphate-buffered saline (PBS) dilution buffer (pH 7.4, 20 mM) containing BSA (0.1%). This buffer is used for dilution of standard SOD, and samples containing it.
2. Prepare PBS washing buffer (pH 7.4, 20 mM) containing Tween 20 (0.05%).
3. Prepare PBS blocking buffer (pH 7.4, 20 mM) containing BSA (1%).
4. Prepare peroxidase-labelled avidin solution from horseradish peroxidase avidin D solution (commercially available) diluted 5000-fold with dilution buffer.
5. Prepare 1st Ab solution by diluting an antibody (Ab) with 50 mM $NaHCO_3$, pH 9.6 (the usual concentration is 1–10 μg mL^{-1}).
6. Prepare 2nd Ab solution by diluting a biotinated antibody with dilution buffer (the usual concentration is 1 to 5 μg mL^{-1}).
7. Prepare OPD solution by mixing citrate buffer (pH 5.0, 0.1 M, 15 mL), *o*-phenylenediamine (OPD; 10 mg), and H_2O_2 (30%, 6 μL).

Protocol *Continued*

Assay

1. Place 1st Ab solution (100 μL) in each well of a microtiter plate (96 wells).
2. Leave overnight at 4°C.
3. Discard 1st Ab solution and wash three times with washing buffer.
4. Add blocking buffer (100 μL).
5. Leave overnight at 4°C (or at 37°C for 1 h).
6. Discard blocking buffer and wash three times with washing buffer.
7. Add standard or sample (100 μL).
8. Leave at room temperature for 2 h (or at 37°C for 1 h).
9. Discard and wash three times with washing buffer.
10. Add 2nd Ab solution (100 μL).
11. Leave at room temperature for 2 h (or at 37°C for 1 h).
12. Discard 2nd Ab solution and wash three times with washing buffer.
13. Add peroxidase-labelled avidin solution (100 μL).
14. Leave at room temperature for 10 – 15 min.
15. Discard and wash 4 or 5 times.
16. Add OPD solution (100 μL).
17. After 10–15 min add H_2SO_4 (0.5 M, 50 μL).
18. Measure the absorbance of each well by means of a microplate reader (490 nm).

Calculation

A typical standard curve is shown in Fig. 2 (*X*-axis; log.). Each sample should be measured at least in duplicate and diluted as measured at mid portion of the standard curve.

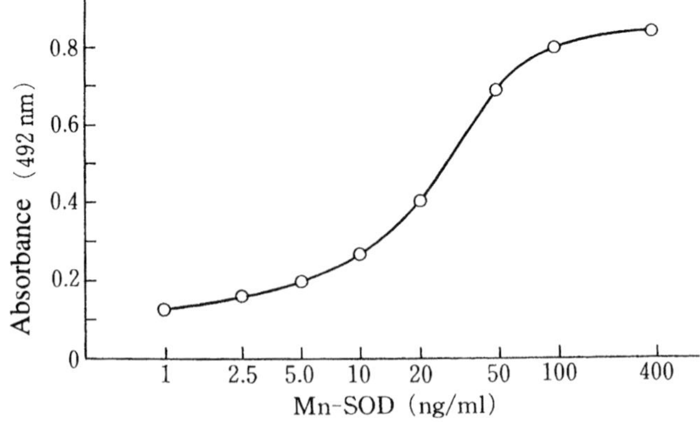

Figure 2. A typical standard curve for Mn-SOD ELISA.

Comments

After step 6 of the assay the microtiter plate can be kept for approx. 1 week at 4°C and for approx. 1 year at −20 to −30°C if it is completely sealed. Be careful not to allow the samples to dry between steps.

References

1. Beauchamp, C. and Fridovich, I. (1978). Superoxide dismutase: Improved assays and an assay applicable to acrylamide gels. *Anal. Biochem.*, **24**, 75–8.
2. Kobayashi, Y., Okahata, S., Tanabe, K., and Usui, T. (1978). Use of logit paper in determination of superoxide dismutase activity in human blood cells. *J. Immunol. Methods*, **24**, 75–8.
3. Kuthan, H., Haussmann, H., and Werringloer, J. (1986). A spectrophotometric assay for superoxide dismutase activities in crude tissue fractions. *Biochem. J.*, **237**, 175–80.
4. Suzuki, K., Miyazawa, N., Nakata, T., Seo, H. G., Sugiyama, T., and Taniguchi, N. (1993). High copper and iron levels and calculation of expression of Mn-superoxide dismutase in mutant rats displaying hereditary hepatitis and hepatoma (LEC rats) *Carcinogenesis*, **14**, 1881–4.

26

EC-SOD activity assay/immunochemical assay

TETSUO ADACHI

Introduction

Extracellular-superoxide dismutase is a secretory tetrameric copper- and zinc-containing glycoprotein. For assay of EC-SOD activity, it should be separated from other SOD isozymes such as copper and zinc-containing SOD (Cu,Zn-SOD) and manganese-SOD (Mn-SOD) by concanavalin A (Con A)–Sepharose column chromatography (EC-SOD binds to a Con A column unlike other SODs), because EC-SOD occurs in tissues at much lower concentrations than the other SODs. EC-SOD activity is assayed by a direct spectrophotometric assay with potassium superoxide (KO_2) which is approx. 40 times more sensitive than the conventional cytochrome-*c* method. The concentration of EC-SOD is determined directly with high sensitivity and without a separation step by enzyme linked-immunosorbent assay (ELISA).

Activity assay

Protocol

Separation of EC-SOD by Con A column chromatography

1. Equilibrate a Con A-Sepharose column (vol. 0.5 mL) with phosphate-buffered saline (PBS).
2. Apply sample (200–500 μL) and let it enter into the gel bed, collect the effluent[1].
3. Leave at room temperature for 5 min.
4. Add PBS (2 mL) and collect the effluent[1].
5. Wash the column with PBS (10 mL).
6. Add PBS containing α-methyl-D-mannoside (0.5 M, 1 mL × 5 times) and collect the eluate (1 mL × five fractions)[2].

Activity assay by KO_2 method (1)

1. Prepare DTPA stock solution (0.1 M) by suspending diethylenetriamine pentaacetic acid (DTPA) in distilled water. This will dissolve during titration to pH 9.5 with NaOH. Store under refrigeration.
2. Prepare AMP stock solution (2 M) by dissolving 2-amino-2-methyl-1-propanol (AMP) in distilled water. Store under refrigeration.
3. Prepare AMP.HCl (pH 9.5, 50 mM)–DTPA (0.2 mM) by mixing AMP stock solution (2 M, 25 mL), DTPA stock solution (0.1 M, 2 mL), and distilled water (900 mL). Titrate with HCl to pH 9.5 and dilute to 1 L. Filter (glass-fibre) and degas (e.g. by water suction) the buffer immediately before use.
4. Prepare NaOH (50 mM)–DTPA (0.5 mM) by mixing immediately before use and keep on ice.
5. Prepare catalase solution by washing crystals of Boehringer crystalline suspension of bovine liver catalase 5–10 times with distilled water, in which native catalase does not dissolve, and dissolving in AMP-HCl (pH 9.5, 50 mM) in which it dissolves very slowly. Prepare a stock solution (30 μM, $\varepsilon_{405} = 324\,000\ M^{-1}\ cm^{-1}$ and store aliquots in the freezer.
6. Place 3 mL of AMP.HCl (pH 9.5, 50 mM)–DTPA (0.2 mM) in a cuvette.
7. Add catalase (5 μL).
8. Add sample (*x* μL of above fractions from Con A column chromatography).
9. Mix cuvette contents with spatula. Place in a cell holder in a photometer, adjust the zero and record signal.
10. Dissolve KO_2 (5–10 mg; need not be weighed) in 3 mL ice-cold NaOH (50 mM)–DTPA (0.5 mM). When the intense bubbling begins to lessen, transfer 5–15 μL of this to a cuvette, mixing rapidly with a spatula.
11. Follow the reaction on a recorder (absorbance at 250 nm, range 0–0.05, paper speed 1–5 mm s^{-1}).
12. When the baseline is reached, add more KO_2 solution (10–20 μL). When a baseline has again been obtained, add a third aliquot of KO_2 solution (15–25 μL) (three subsequent assays).

[1] Cu,Zn-SOD and Mn-SOD are collected in these fractions because these SODs do not bind to a Con A column.
[2] EC-SOD is recovered in these fractions.

Results and calculations

As in Fig. 1A, the half-life for $O_2{\cdot}^-$ disproportionation (ΔTs) is measured from $A_{250} = 0.008$–0.004. The mean half-lives are then calculated from the results of the three subsequent assays.

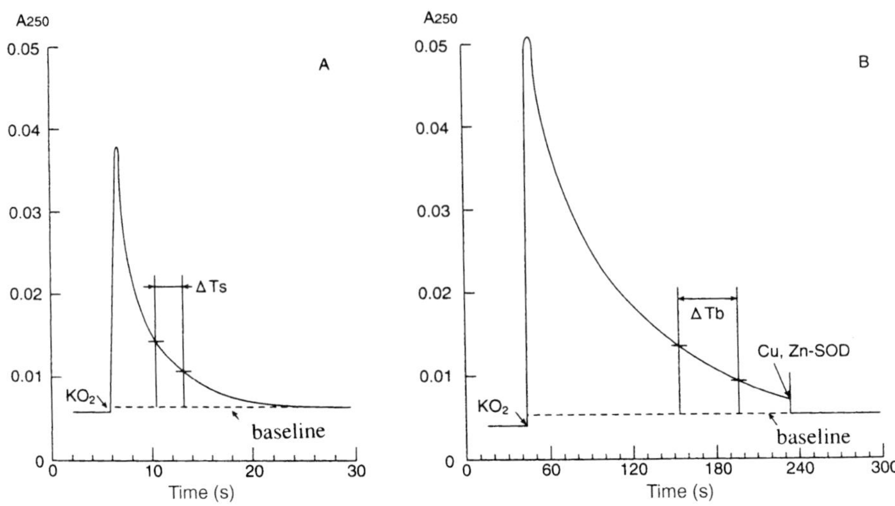

Figure 1. Determination of SOD activity by KO_2 method.

For the blank assay follow the same procedure as above except that no sample is added and a lower recorder speed is used. The KO_2 solution is added only once. The baseline is approached very slowly. The baseline is obtained by adding of Cu,Zn-SOD (0.3 μg mL^{-1}, 10 μL) after a suitable time, as shown in Fig. 1B. The half-life (ΔTb) is measured from A_{250} = 0.008–0.004.

One unit in this assay is arbitrarily defined as the activity that brings about disproportionation of O_2^- at a rate of 0.1 s^{-1}, in the volume and under the conditions described here. The SOD activity will be obtained as:

$$[(0.693/\Delta T\text{s}) - (0.75/\Delta T\text{b})] \times 10 \times [1000\ (\mu\text{L})/x\ (\mu\text{L})]\ \text{units mL}^{-1}$$

Immunochemical assay

Protocol

Enzyme-linked immunosorbent assay (ELISA) 1 (2)

1. Prepare PBS–Tween reagent containing Tween 20 (500 μg mL^{-1}) and sodium azide (200 μg mL^{-1}).
2. Prepare 1% BSA–PBS–Tween by dissolving bovine serum albumin (BSA) to 1% in PBS–Tween. Store under refrigeration.
3. Prepare a polyclonal antibody coating plate by placing rabbit anti-EC-SOD polyclonal antibody (20 μg mL^{-1}, 80 μL) dissolved in carbonate buffer (pH 9.5, 50 mM) containing sodium azide (200 μg mL^{-1}) in each

well of immunoplate and storing at 4°C overnight. Recover the antibody solution and wash each well three times with PBS–Tween and then add 1% BSA–PBS–Tween (300 μL). Store under refrigeration until use (leave for at least several days)[1].

4. Remove 1% BSA–PBS–Tween from polyclonal antibody coating plate.
5. Add standard (50 pg mL^{-1} to 50 ng mL^{-1}, 70 μL) or sample diluted with 1% BSA–PBS–Tween.
6. Store at 4°C overnight or for 2 h at room temperature.
7. Wash the plate five times with PBS–Tween.
8. Add anti-EC-SOD monoclonal antibody (approx. 5 μg mL^{-1}, 80 μL) dissolved in 1% BSA–PBS–Tween[2].
7. Store for 2 h at room temperature.
8. Wash the plate five times with PBS–Tween.
9. Add alkaline phosphatase (ALP)-conjugated anti-(mouse IgG) antibody (Zymed, San Francisco, CA, USA; 90 μL) at 1:2000 dilution in 1% BSA–PBS–Tween.
10. Store for 2 h at room temperature.
11. Wash the plate five times with PBS–Tween.
12. Add ALP substrate solution (150 μL) (dissolve *p*-nitrophenyl phosphate at 1 mg mL^{-1} in 0.1 M diethanolamine.HCl, pH 9.8, containing 0.5 mM $MgCl_2$ and 200 μg mL^{-1} sodium azide. Prepare immediately before use).
13. Incubate for 15 to 30 min at room temperature.
14. Add NaOH (5 M, 50 μL).

15) Assay the absorbance at 415 nm.

Enzyme-linked immunosorbent assay (ELISA) 2 (3)

1. Prepare the monoclonal antibody coating plate by placing mouse anti-EC-SOD monoclonal antibody (50 μg mL^{-1}, 80 μL) dissolved in carbonate buffer (pH 9.5, 50 mM) containing sodium azide (200 μg mL^{-1}) in each well of an immunoplate and store at 4°C overnight. Recover the antibody solution, wash each well three times with PBS–Tween, and then add 1% BSA–PBS–Tween (300 μL). Store under refrigeration until used (leave for at least several days)[1].
2. Remove 1% BSA–PBS–Tween from monoclonal antibody coating plate.
3. Add standard (50 pg mL^{-1} to 50 ng mL^{-1}, 70 μL) or sample diluted with 1% BSA–PBS–Tween.
4. Store at 4°C overnight or for 2 h at room temperature.
5. Wash the plate five times with PBS–Tween.

Protocol *Continued*

6. Add ALP-conjugated anti-EC-SOD monoclonal antibody (80 μL) suitably diluted with 1% BSA–PBS–Tween.
7. Store for 2 h at room temperature.
8. Wash the plate five times with PBS–Tween.
9. Add ALP substrate solution (150 μL).
10. Incubate for 30 to 60 min at room temperature.
11. Add NaOH (5 M, 50 μL).
12. Measure the absorbance at 415 nm.

[1] Recovered antibody solution can be used several times.
[2] Monoclonal antibodies were purified from ascitic fluids by ammonium sulfate (40% saturated) precipitation followed by DEAE-cellulose column chromatography.

Results and calculations

Standard curves for ELISA 1 and ELISA 2 are shown in Fig. 2.

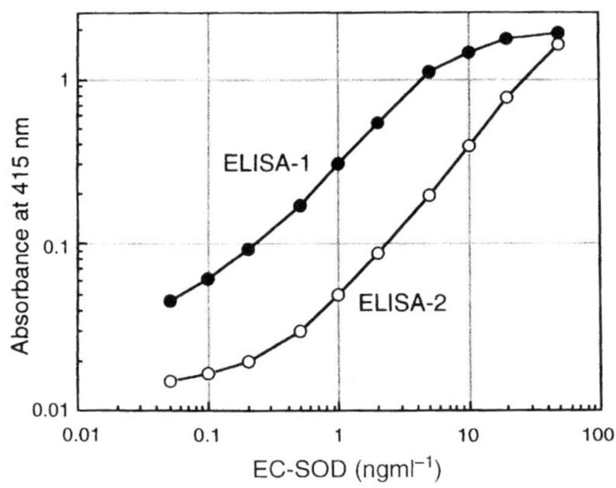

Figure 2. Standard curves for ELISA 1 and ELISA 2.

References

1. Marklund, S. L. (1985). In *CRC Handbook of Methods for Oxygen Radical Research* (ed. by R. A. Greenwald), pp. 249–55. CRC Press, Boca Raton, Florida.
2. Adachi, T., Ohta, H., Hirano, K., Hayashi, K., and Marklund, S. L. (1991). Non-enzymic glycation of human extracellular superoxide dismutase. *Biochem. J.*, **279**, 263–7.
3. Adachi, T., Ohta, H., Yamada, H., Futenma, A., Kato, K., and Hirano, K. (1992). Quantitative analysis of extracellular-superoxide dismutase in serum and urine by ELISA with monoclonal antibody. *Clin. Chim. Acta*, **212**, 89–102.

27

Nitric oxide synthase (NOS) activity assay: RI method to detect L-citrulline

JUNICHI FUJII

Introduction

There are several methods of assaying nitric oxide synthase (NOS) activity. They differ in sensitivity and the product to be measured. Because NOS converts L-arginine to L-citrulline, NOS activity can be measured by assaying L-[U-^{14}C] citrulline produced from L-[U-^{14}C] arginine (1). Among three NOS isozymes, inducible NOS (NOS II) has Ca^{2+}-independent activity and it is thus possible to distinguish NOS II activity from brain NOS (NOS I) and endothelial NOS (NOS III) by assaying the activity in the presence of a Ca^{2+} chelator. Although the method is simple and highly sensitive, special attention should be paid to arginase activity because this enzyme also produces L-citrulline from L-arginine. It is recommended that the result be verified by another method.

Protocol[2]

1. Homogenize cells or tissues in homogenization buffer prepared by mixing Tris-HCl (pH 7.4, 50 mM), sucrose (250 mM), dithiothreitol (1 mM), EDTA (100 μM), EGTA (100 μM), leupeptin (5 μM), pepstatin (2 μM), and *p*-amidino PMSF (20 μM).
2. Centrifuge at 15000 g for 30 min at 4°C.
3. Prepare assay mixture from diluted L-[U-^{14}C] arginine (final concentration 1–2 mM, 10 μL), reaction mixture (70 μL), and water (10 μL). To determine the effect of reagents on NOS activity, they should be substituted with water.
4. Add assay mixture (10 μL) to reaction mixture (90 μL) prepared by mixing Hepes–NaOH (pH 7.8, 50 mM), dithiothreitol (1 mM), $CaCl_2$ (1 mM), tetrahydrobiopterin (0.1 mM), NADPH (1 mM), FAD (10 μM), and calmodulin (10 μg mL^{-1}).

Protocol *Continued*

5. Incubate for 10 min at 37 °C.
6. Terminate the reaction by addition of stop buffer (200 μL) prepared by mixing Hepes–NaOH (pH 5.1, 100 mM) and EDTA (10 mM).
7. Apply to a 0.25 mL Dowex 50W-X8 column (Na^+ form) pre-equilibrated with a stop buffer at room temperature.
8. Wash the column with stop buffer (500 μL)
9. Add scintillator (3 mL)
10. Count (in disintegrations min^{-1}, dpm) the ^{14}C radioactivity by means of a liquid scintillation cocktail counter.

Calculations

NOS activity is calculated from:

$$\text{NOS activity (pmol min}^{-1}\text{)} = [\text{count (dpm)}]/a \times 60$$

where a is the specific activity of L-[U-^{14}C] arginine (GBq $mmol^{-1}$)

Comments

Because NOS is quickly inactivated, assays should be conducted as soon as possible after homogenization.

Crude homogenate or microsomal fractions should be used for NOS III assays because NOS III binds to the membrane fraction (3).

References

1. Bredt, D. S. and Snyder, S. H. (1990). Isolation of nitric oxide synthetase, a calmodulin-requiring enzyme. *Proc. Natl. Acad. Sci. USA*, **87**, 682–5.
2. Seo, H. G., Fujii, J., Asahi, M., Okado, A., Fujiwara, N., and Taniguchi, N. (1997). Roles of purine nucleotides and adenosine in enhancing NOS II gene expression in interleukin-1β-stimulated rat vascular smooth muscle cells. *Free Radic. Res.*, **26**, 409–18.
3. Pollock, J. S., Forstermann, U., Mitchell, J. A., Warner, T. D., Schmidt, H. H., Nakane, M., and Murad, F. (1991). Purification and characterization of particulate endothelium-derived relaxing factor synthase from cultured and native bovine aortic endothelial cells. *Proc. Natl. Acad. Sci. USA*, **88**, 10480–4.

28

Nitric oxide synthase (NOS) activity assay: determination of produced NADP$^+$

JUNICHI FUJII

Introduction

Nitric oxide synthases (NOS) consume L-arginine, NADPH, and O_2 and produce NO, L-citrulline, and NADP$^+$. This chapter describes a method of measuring NOS activity by quantitating the NADP$^+$ produced. The enzymatic reaction is terminated by addition of HCl, and the enzymes and NADPH are inactivated by heating. The NADP$^+$ produced is converted to a strongly fluorescent compound, and the fluorescence is measured. If the activity is low, the NADP$^+$ signal can be enzymatically amplified by NADP-cycling. The specific NOS activity can be determined as N^G-nitro-L-arginine (LNA)-inhibitable activity.

Protocol [(1)]

NOS assay by measuring NADP$^+$-derived fluorescence

1. Add homogenization solution[1] (5 to 10 vol.) to the tissue and homogenize.
2. Centrifuge at 100 000 g for 1 h at 4 °C. The supernatant can be lyophilized and used for NADP$^+$ cycling assay.
3. Prepare reaction mixture[2] (100 μL) with or without N^G-nitro-L-arginine (500 μM) in a disposable 3 mL Pyrex tube and pre-incubate at 25 °C for 5 min. Prepare NADP$^+$ (0.1–3 nmol) in the same reaction mixture as standards.
4. Add supernatant (2 μL).
5. Incubate for 30 min at 25 °C.
6. Add HCl (5 M, 5 μL) to stop the reaction.
7. Boil for 3 min to denature enzymes and destroy NADPH.
8. Add alkaline solution[3] (1 mL) and mix well.

Protocol *Continued*

9. Stand the tubes in the dark for 1 h at room temperature to quench non-specific fluorescence.
10. Measure the fluorescence by means of a fluorimeter (Ratio Fluorimeter, Farrand, Valhalla, USA).

[1] Homogenization solution is prepared by mixing Tris-HCl (pH 7.5, 50 mM), EDTA (1 mM), dithiothreitol (1 mM), antipain (10 μg mL^{-1}), pepstatin (10 μg mL^{-1}), leupeptin (10 μg mL^{-1}), and phenylmethyl sulfonyl fluoride (0.1 mM).
[2] Reaction mixture is prepared by mixing PIPES–NaOH (pH 7.0, 100 mM), L-arginine (100 μM), NADPH (15 μM), $CaCl_2$ (1.25 mM), calmodulin (62.5 nM, 1 μg mL^{-1}), dithiothreitol (1 mM), EDTA (1 mM), nicotinamide (3 mM), antipain (10 μg mL^{-1}), pepstatin (10 μg mL^{-1}), leupeptin (10 μg mL^{-1}), and phenylmethyl sulfonyl fluoride (0.1 mM).
[3] Alkaline solution is a mixture of NaOH (6.6 M) and imidazole (low fluorescence; 2 mM).

NADP Cycling Assay

1. Reaction mixture (1.25 μL), with or without LNA (500 μM) is placed at the bottom of Terasaki plate wells, and the wells are filled with paraffin oil to prevent evaporation. Prepare 1–5 pmol $NADP^+$ in the same reaction mixture as standards.
2. Add lyophilized sample to the reaction mixture to start the reaction.
3. Incubate for 30 min at 20 °C.
4. Add HCl (0.3 M, 1.25 μL) and heat the plate at 95 °C for 15 min.
5. Cool to 20 °C, and add the mixture (2 μL) to NADP cycling mixture[1] (50 μL) in a Pyrex tube on ice.
6. Incubate for 1 h at 38 °C, boil for 3 min, then cool under tap water.
7. Add indicator reaction mixture[2] (1 mL) and incubate for 30 min at 38 °C.
8. After cooling the tube to room temperature, the fluorescence is measured.

[1] NADP cycling mixture is prepared from Tris-HCl (pH 8.0, 100 mM), α-oxo-glutarate (5 mM), glucose-6-phosphate (1 mM), ammonium acetate (25 mM), ADP (0.1 mM), bovine serum albumin (0.02%), glucose 6-phosphate dehydrogenase (3 μg mL^{-1}), and glutamate dehydrogenase (32 μg mL^{-1}).
[2] Indicator reaction mixture is a mixture of Tris-HCl (pH 8.0, 100 mM), $NADP^+$ (100 μM), EDTA (0.1 mM), and 6-phosphogluconate dehydrogenase (0.2–0.5 μg mL^{-1}).

Results

The amount of $NADP^+$ produced, which corresponds to the amount of NO produced, can be calculated from calibration curves obtained from standards.

$$\text{NOS activity} = \text{Total activity} - \text{Activity}\ (+\ \text{LNA})$$

Comments

1. Concentrations of NADPH and $NADP^+$ contaminating samples are usually negligibly low. Because crude biological samples generally contain $NAD(P)^+$ nucleosidase, however, the $NADP^+$ produced is converted to nicotinamide and ADP. In this method, therefore, nicotinamide is added to the assay mixture to inhibit the nucleosidase.
2. In crude samples, the background is sometimes too high to measure NOS activity correctly.
3. If the alkaline solution is fluorescent, the fluorescence should be eliminated by standing the solution under daylight for 1 day.
4. The reaction solution can be kept overnight in the dark at room temperature.

Reference

1. Wang, W., Inoue, N., Nakayama, T., Ishii, M., and Kato, T. (1995). An assay method for nitric oxide synthase in crude samples by determining product NADP. *Anal. Biochem.* **227**, 274–80.

29

Myeloperoxidase

MASAO NAKAMURA

Introduction

Myeloperoxidase (MPO), a haemoprotein present at high concentrations in polymorphonuclear leukocytes (PMNs), plays a role in the microbicidal activity of PMNs (1). MPO reacts with H_2O_2 to form Compound I, which catalyses two-electron oxidations of halogens (Cl^-, Br^-, I^-) (Fig. 1, Path 1) or a consecutive one-electron oxidation of substrates (AH_2) through Compound II (Fig. 1, Path 2) (2, 3).

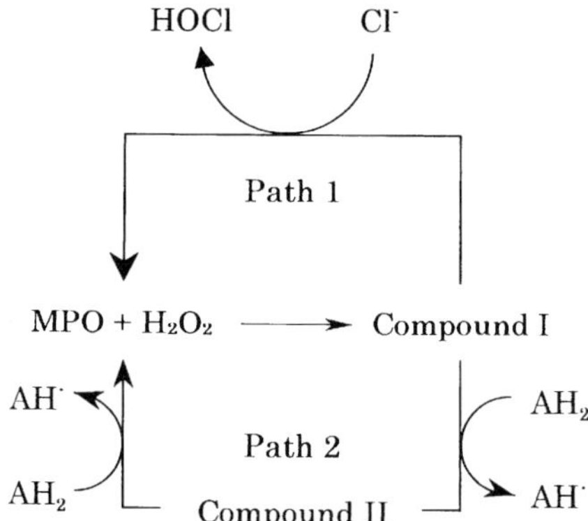

Figure 1. Mechanism of reaction of myeloperoxidase. Myeloperoxidase activity is measured optically by monitoring the chlorination of monochlorodimedone (Path 1) or the formation of final oxidation products of substrates (AH_2) (Path 2).

Protocol

Measurement of MPO activity[1,2]

Reactions are conducted at 25°C in potassium phosphate (pH 7.0, 0.1 M). A 1-cm light-path cuvette contains 2.5 mL of the assay solution.

MPO-dependent chlorination of monochlorodimedone (MCD) (Path 1)

1. Place assay solution[3] (2.5 mL) in 1-cm light-path cuvette.

2. Add cell extract (10 μL containing 0.1 mg mL^{-1} protein)[4] or MPO[5] (100 nM).
3. Mix rapidly.
4. Record decrease of absorbance at 290 nm for 4 to 5 min and determine the change in absorbance min^{-1} from the initial velocity.
5. Estimate rate of MPO-dependent chlorination from the decrease of the MCD concentration min^{-1} (ε_{290} 19 000 M^{-1} cm^{-1}).

ABTS [2, 2′-Azino-bis-(3-ethyl benzothiazoline-6-sulfonic acid)] peroxidase reaction of MPO (Path 2)[6]

1. Place assay solution[7] (2.5 mL) in 1-cm light-path cuvette.
2. Add cell extract (10 μL containing 0.1 mg mL^{-1} protein) or MPO (100 nM).
3. Mix rapidly.
4. Record increase of absorbance at 414 nm for 4 to 5 min. and determine the change in absorbance min^{-1} from the initial velocity.
5. Estimate the rate of the peroxidase reaction from the increase of oxidation product of ABTS (ε_{414} 36 000 M^{-1} cm^{-1}).

[1] One unit of enzyme is the amount that forms 1 μmol product min^{-1}.
[2] MPO or H_2O_2 is added last to initiate the reaction. When both MPO and H_2O_2 are present in the absence of substrate, inactivation of MPO occurs.
[3] Assay solution (2.5 mL) contains MCD (100 μM), NaCl (100 mM), and H_2O_2 (100 μM). Hydrogen peroxide solutions are prepared daily by diluting stock solution and concentrations are determined by using $\varepsilon_{240} = 43.6\ M^{-1}\ cm^{-1}$.
[4] The concentration of MPO in the cell extract is estimated from the increase in the absorbance at 472 nm and by using $\varepsilon_{472} = 75000\ M^{-1}\ cm^{-1}$ (reduced MPO − oxidized MPO). Reduced MPO is prepared with a few crystals of sodium dithionite under anaerobic conditions.
[5] The concentration of human MPO is estimated by using $\varepsilon_{430} = 91000\ M^{-1}\ cm^{-1}$ for haem.
[6] The guaiacol-peroxidase activity of MPO is measured at 470 nm and the rates are estimated from the increase of oxidation product ($\varepsilon_{470} = 26600\ M^{-1}\ cm^{-1}$). The assay mixture contains guaiacol (30 mM) and H_2O_2 (0.2 mM).
[7] Assay solution (2.5 mL) contains ABTS (100 μM) and H_2O_2 (50 μM).

References

1. Klebanoff, S. J. (1991). Myeloperoxidase: occurrence and biological function. In *Peroxidases in Chemistry and Biology*, Vol. 1 (ed. J. Everse, K. E. Everse, and M. B. Grisham), pp. 1–35. CRC Press, Boca Raton, Florida.
2. Bolscher, B. G. J. M. and Wever, R. (1984). A kinetic study of the reaction between human myeloperoxidase, hydroperoxides and cyanide: Inhibition by chloride and thiocyanate. *Biochim Biophys Acta*, **788**, 1–10.
3. Kettle, A. J. and Winterbourn, C. C. (1988). Superoxide modulates the activity of myeloperoxidase and optimizes the production of hypochlorous acid. *Biochem J.*, **252**, 529–36.

30

Haem oxygenase

TADASHI YOSHIDA

Introduction

Haem oxygenase (HO) is an amphipathic microsomal protein which catalyses the regiospecific oxidative degradation of haem to biliverdin, CO, and Fe in the presence of NADPH-cytochrome P-450 reductase as an electron donor (1–4). Haem degradation is physiologically important because it is the process not only of dissimilation of haem but also of biosynthesis of CO and bile pigments. CO acts as a physiological messenger and bile pigments can scavenge reactive oxygen species.

HO has two isoforms, referred to as HO-1 and HO-2. HO-1, an inducible form with a molecular mass of 33 kDa, is distributed mainly in reticuloendothelial cell-rich tissues such as the spleen and liver. HO-2, with a molecular mass of 36 kDa, is expressed constitutively and is distributed mainly in the brain and testis (5).

Protocol

Preparation of microsomes

1. Homogenize tissues or cells in homogenizing buffer[1] (4 vol.).
2. Centrifuge at 105 000 g for 60 min at 4 °C.
3. Wash the precipitate with KCl (1 M) containing potassium phosphate buffer (pH 7.4, 0.02 M) and EDTA (10 mM) and centrifuge at 105 000 g for 60 min at 4 °C.
4. Suspend the precipitate in potassium phosphate buffer (pH 7.4, 50 mM) to give a protein concentration of approximately 10 mg mL^{-1}

Preparation of NADPH-cytochrome P-450 reductase and biliverdin reductase

NADPH-cytochrome P-450 reductase is purified from rat liver microsomes by a method reported elsewhere (6). Biliverdin reductase (10 units) is a Step 3 enzyme (7).

Preparation of haem solution

Haem solution (1 mM) is prepared by dissolving crystalline chlorhaemin (6.5 mg) in NaOH (0.1 M, 2.5 mL) and adding Tris base solution (6 mg mL^{-1}, 1 mL) and HCl (0.03 M, 6.5 mL) sequentially.

Assay conditions

1. The final reaction mixture is prepared by mixing potassium phosphate buffer (pH 7.4, 1 M, 0.2 mL), haem (1 mM, 0.04 mL), biliverdin reductase, bovine serum albumin (10 mg mL^{-1}, 0.2 mL), microsomes, and water to a total volume of 2 mL. NADPH (10 mM, 0.1 mL) is added to the assay, but not to the control.
2. The reaction is performed for 5–10 min in a test tube placed in a shaking water bath at 37 °C.
3. The reaction is started by addition of NADPH after pre-incubation for 2 min.
4. Stop the reaction by immersing the tube in ice water.
5. Measure the difference in absorbance at 468 nm.

[l] Homogenizing buffer is prepared from KCl (0.134 M) and potassium phosphate buffer (pH 7.4, 0.02 M)

Calculations

For calculation of the amount of bilirubin formed, the value of 43.5 mM^{-1} cm^{-1} is adopted as the molar absorption extinction coefficient of bilirubin at 468 nm (2).

Comments

Biliverdin reductase and bovine serum albumin are not essential requirements for the HO reaction, but these are included in the assay system for the convenience of measuring the HO activity by means of the formation of bilirubin.

The activity of NADPH-cytochrome p-450 reductase is high in liver microsomes but the content is lower in other tissues. It is, therefore, desirable to add an appropriate amount of the reductase as a supplement to obtain the full activity of HO.

References

1. Tenhunen, R., Marver, H., and Schmid, R. (1969). Microsomal heme oxygenase. Characterization of the enzyme. *J. Biol. Chem.*, **244**, 6388–94.
2. Yoshida, T. and Kikuchi, G. (1978).Purification and properties of heme oxygenase from pig spleen microsomes. *J. Biol. Chem.*, **253**, 4224–9.

3. Yoshida, T. and Kikuchi, G. (1978). Features of the reaction of heme degradation catalysed by the reconstituted microsomal heme oxygenase system. *J. Biol. Chem.*, **253**, 4230–6.
4. Kikuchi, G. and Yoshida, T. (1998). Heme degradation by the microsomal heme oxygenase system. *Trends Biochem. Sci.*, **5**, 323–5.
5. Maines, M. D. (1988). Heme oxygenase: function, multiplicity, regulatory mechanisms, and clinical applications. *FASEB J.*, **2**, 2557–68.
6. Yasukochi, Y. and Masters, B. S. S. (1976). Some properties of a detergent-solubilized NADPH-cytochrome c (cytochrome P-450) reductase purified by biospecific affinity chromatography. *J. Biol. Chem.*, **251**, 5337–44.
7. Tenhunen, R., Ross, M. E., Marver, H. S., and Schmid, R. (1970). Reduced Nicotinamide-adenine dinucleotide phosphate dependent biliverdin reductase: partial purification and characterization. *Biochemistry*, **9**, 298–303.

31

Caeruloplasmin

TOSHIHIRO SUGIYAMA, YOSHIHIKO KAWARADA, and FUJIKO HIRASAWA

Introduction

The procedures most commonly used are based on the oxidase activity of the protein towards *p*-phenylenediamine (1, 2) or *o*-dianisidine dihydrochloride (3). These methods are more sensitive than measurement of the intrinsic blue colour (4) or measurement of the caeruloplasmin content by an immunological method, and are more precise than measuring ferroxidase activity (5).

The enzymatic activity is measured in international units ($U\ L^{-1}$) and protein content ($mg\ L^{-1}$) when *p*-phenylenediamine is used as a substrate, and international units when *o*-dianisidine dihydrochloride is used as a substrate. The normal ranges are, respectively, ~100 units L^{-1} and ~50 mg dL^{-1} for human serum, and ~150 units L^{-1} and 30 mg dL^{-1} for rat serum.

p-Phenylenediamine end-point method

Protocol

1. Place sodium acetate buffer (pH 5.4, 0.1 M, 2 mL) into two tubes, reaction (R) and blank (B).
2. Add serum (0.1 mL) to each tube.
3. Preincubate in a water bath at 37 °C.
4. Add warmed (37 °C) *p*-phenylenediamine solution[1] (27 mM, 1 mL) to both tubes.
5. After 5 min, add sodium azide solution (1.5 M, 50 μL) to tube B.
6. Exactly 30 min later, add sodium azide solution (1.5 M, 50 μL) to tube R.
7. Measure the absorbance (A_R and A_B) at 530 nm.

[1] Prepared by dissolving *p*-phenylenediamine (0.5 g) in acetate buffer (pH 5.4, 0.1 M, 100 mL).

Calculation

$$\text{Caeruloplasmin (g L}^{-1}) = 0.752 \times (A_R - A_B)$$
$$\text{Oxidase activity (units L}^{-1}) = 660 \times (A_R - A_B)$$

Comments

The *p*-phenylenediamine oxidase assay method requires purification of the substrate owing to substrate autooxidation by light and metals. The solution of the substrate is stable for 3 h. Oxidase activity is inhibited by ascorbic acid, citrate, chloride ion, and EDTA.

p-Phenylenediamine kinetic method

Protocol

1. Place *p*-phenylenediamine solution (0.99 mL) in a cuvette.
2. Add serum (10 mL).
3. Measure the change of absorbance at 530 nm for 10 min.
4. Read the change in absorbance min^{-1}.

[1] Prepared by dissolving *p*-phenylenediamine (0.5 g) in acetate buffer (pH 5.4, 0.1 M, 100 mL).

Calculation

$$\text{Oxidase activity (units L}^{-1}) = (A \times V \times d \times v)/\varepsilon = 52.4 \times 103 \times A$$

where A is the change in absorbance min^{-1} at 530 nm, V is the total volume of reaction mixture (1 mL), $\varepsilon = 1910$ M^{-1} cm^{-1}, d is the light path-length (1.0 cm) and v is the volume of serum (10 μL).

Comments

The activity of rat caeruloplasmin is markedly inhibited by phosphate (0.1 M), unlike that of human caeruloplasmin. To avoid interference by inhibitors such as ascorbic acid, citrate, and EDTA, the reaction should be started 5 min after addition of the sample for routine assays.

o-Dianisidine dihydrochloride end-point method

Protocol

1. Place sodium acetate buffer (pH 5.0, 0.1 M, 0.75 mL) and serum (50 μL) into two tubes, reaction (R) and blank (B).
2. Place the tubes in a water bath at 30 °C.

3. After 5 min pre-incubation, add *o*-dianisidine dihydrochloride solution (7.9 mM, 0.2 mL; pre-incubated at 30 °C) to each tube.
4. After 5 min, add sulfuric acid (9 M, 2.0 mL) to tube B and mix immediately.
5. After 15 min, add sulfuric acid (9 M, 2.0 mL) to tube R and mix immediately.
6. Measure the absorbance of reagent and blank (A_R and A_B) of the purplish–red solution at 540 nm, in a cuvette with a 1-cm light path and against de-ionized water as reference.

Calculation

$$\text{Activity (units L}^{-1}) = (\text{Absorbance} \times V)/\varepsilon \times t \times l = 625 \times (A_R - A_B)$$

where ε is the molar absorption coefficient of the substrate in the coloured solution (9.6 mM^{-1} cm^{-1}), t is the incubation time (10 min), l is the optical path length (1.0 cm), and V is the volume correction.

Comments

The *o*-dianisidine dihydrochloride oxidase activity assay method takes special precautions against inhibitors in the serum. The substrate solution is stable at 4 °C for 3 months. Inhibition by ascorbic acid results in a lag phase dependent on its concentration in the serum. The enzyme activity is calculated by measuring the activity in the interval between 5 and 15 min incubation.

References

1. Rice, E. W. (1962). Standardization of ceruloplasmin activity in terms of international enzyme units: Oxidative formation of 'Bandrowski's base' from *p*-phenylenediamine by ceruloplasmin. *Anal. Biochem.*, **3**, 452–6.
2. Sunderman, F. W. (1970). Measurement of human serum ceruloplasmin by its *p*-phenylenediamine oxidase activity. *Clin. Chem.*, **16**, 903–10.
3. Schosinsky, K. H., Lehmann, P., and Beeler, M. F. (1974). Measurement of ceruloplasmin from its oxidase activity in serum by use of *o*-dianisidine dihydrochloride. *Clin. Chem.*, **20**, 1556–63.
4. Scheinberg, I. H. and Gitlin, D. (1952). Deficiency of ceruloplasmin in patients with hepatolenticular degeneration (Wilson's disease). *Science*, **116**, 484.
5. Johnson, D. A., Osaki, S., and Frieden, E. A. (1967). A micromethod for the determination of ferroxidase (ceruloplasmin) in human serums. *Clin. Chem.*, **13**, 142.

32

Caeruloplasmin ferroxidase activity

J. M. C. GUTTERIDGE and G. J. QUINLAN

Introduction

The major copper-containing protein of human plasma, caeruloplasmin, can catalyse the oxidation of ferrous ions to the ferric state (ferroxidase activity). This activity is considered to be an important part of the antioxidant defence of the plasma (see Chapter 11). When ferric ions bind to apotransferrin they form a pink iron–transferrin complex absorbing at 460 nm.

Protocol

1. Into clean micro cuvettes place sodium acetate buffer (pH 5.5, 1 M, 0.17 mL), Chelex resin-treated distilled water (0.27 mL), apotransferrin (10 mg mL^{-1}, 0.25 mL), and plasma (0.01 mL)
2. Mix the contents of cuvette and place in spectrophotometer block at 30 °C for 5 min.
3. Add ammonium ferrous sulfate[1] (0.4 mM, 0.3 mL).
4. Measure the change in absorbance at 460 nm for 6 min.

[1] The ferrous salt is prepared in distilled water at pH < 7.0, and purged with oxygen-free nitrogen.

Calculation

The change in absorbance at 460 nm (Abs) is measured from time 20 to 220 s. The activity (international units mL^{-1}) is given by:

$$\frac{\Delta \text{Abs}}{\Delta \text{t}}$$

$$\frac{\dfrac{\Delta \text{Abs sample}}{\text{t}} - \dfrac{\Delta \text{Abs blank}}{\text{t}}}{\gamma \text{Et}} = \text{IU/ml}$$

where E is the absorption coefficient (0.00250), γ (μL) is the amount of sample used, and t is the time (min). The blank sample contains no plasma and a control is also included which contains plasma plus sodium azide (3.3 mg mL^{-1}, 0.02 mL).

Comments

Stored plasma can contain a ferroxidase II activity which is not caeruloplasmin. Unlike caeruloplasmin, ferroxidase II is not inhibited by azide. Azide controls are essential. A low-pH buffer is used to prevent the rapid *autoxidation* of ferrous ions.

Conalbumin (ovotransferrin) can be used as a cheaper alternative to apotransferrin.

References

1. Johnson, D. A., Osaki, S., and Frieden, E. (1967). A micro method for the determination of ferroxidase (caeruloplasmin) in human serum. *Clin. Chem.* **13**, 142–6.
2. Gutteridge, J. M. C., Winyard, P., Blake, D. R. *et al.* (1985). The behaviour of caeruloplasmin in stored human extracellular fluids in relation to ferroxidase II activity, lipid peroxidation and phenanthroline-detectable copper. *Biochem. J.*, **230**, 517–23.

33

Ferritin

JUNJI KATO, KOSHI FUJIKAWA, and YOSHIRO NIITSU

Introduction

Iron is essential for life and is involved in various physiological functions such as respiration, electron transport, and cell proliferation. Conversely, iron overload, especially free iron, causes cell damage by production of free radicals, oxidation of lipids, and DNA fragmentation. For these reasons, intracellular iron is strictly regulated. Part of that regulation is provided by ferritin, an iron-storage protein that consists of twenty-four L and H subunits. Each molecule of ferritin stores approx. 4500 atoms of iron and thus indirectly regulates free-radical production. Here we introduce a method for measurement of tissue ferritin.

Protocol

Preparation of samples

1. Place tissue sample (1 g) in a Dounce homogenizer and add ice-cold double-distilled water (DDW; 4 mL).
2. Homogenize with 20 strokes using pestle B.
3. Transfer the homogenized sample to a 15-mL conical tube and heat for 10 min at 80 °C.
4. Cool to 4 °C for 10 min.
5. Centrifuge at 15 000 g for 20 min at 4 °C.
6. Isolate supernatant (heat-treated sup.)[1].

Two-site immunoradiometric assay (IRMA)—the paper disc method

Preparation of standard ferritin[2]

1. Prepare heat-treated sup. from a piece of human liver as described in Section 1.
2. Filter heat-treated sup. through gauze.
3. Centrifuge at 78 000 g for 1 h at 4 °C.

4. Re-suspend precipitate in ice-cold DDW (1–2 mL).
5. Centrifuge at 7000 g for 1 h at 4 °C and isolate supernatant.
6. Centrifuge at 95 000 g for 1 h at 4 °C and re-suspend precipitate in DDW (1 mL).
7. Measure concentration of human liver ferritin by means of protein assay (Bio-Rad, Hercules, CA, USA).
8. Dilute the protein standard in PBS (e.g., 800, 400, 200, 100, 50, 25, 12.5, 6.25, 3.13, and 1.56 ng mL^{-1}).

^{125}I-labelling of anti-ferritin antibody (chloramine-T method)[3]

1. Mix sodium phosphate buffer (pH 7.5, 0.5 M, 50 mL), polyclonal anti-ferritin antibody (Dako, Carpinteria, CA, USA; 20 μg) , $Na^{125}I$ (ICN, Costa Mesa, CA, USA) 1 mCi chloramine-T (Sigma, St. Louis, MO, USA; 20 μg) and incubate for 1 min at room temperature.
2. Add sodium metabisulfite (20 mg mL^{-1}, 50 μL) and potassium iodide (1 M, 5 μL). Separate the iodinated antibody by gel filtration through a Sephadex G25 (Pharmacia Biotech, Piscataway, NJ, USA) column eluting with TEN buffer (20 mM Tris HCl, pH 7.6, 10 mM EDTA and 100 mM NaCl) containing 1 mg mL^{-1} BSA.
3. Collect 500-mL fractions.
4. Count the radioactivity in each fraction.
5. Pool the fractions containing the iodinated antibody (peak radioactivity), and store at 4 °C.

Preparation of anti-ferritin antibody-coated paper discs[4]

1. Dip 500 pieces of 10-mm diameter Whatman #1 filter disc in 100 mL DDW.
2. Add cyanogen bromide solution (33.3%, i.e. 10 g cyanogen bromide in 300 mL DDW; 300 mL).
3. Adjust pH to 10.5 with NaOH (1 M).
4. Wash with $NaHCO_3$ solution (5 mM, 2 × 250 mL).
5. Wash with DDW (2 × 250 mL).
6. Wash with 25% acetone (2 × 250 mL).
7. Wash with 50% acetone (2 × 250 mL).
8. Wash with 75% acetone (2 × 250 mL).
9. Wash with 100% acetone (2 × 250 mL).
10. Dry at room temperature.
11. Add anti-human ferritin antibody solution (1.5 mg mL^{-1}, 200 mL; unlabelled antibody in 0.1 M $NaHCO_3$ containing 0.5 M NaCl).
12. Incubate for 3 h at 4 °C.
13. Wash with $NaHCO_3$ (0.1 M) containing NaCl (0.5 M) (2 × 250 mL).

Protocol *Continued*

14. Incubate with ethanolamine (0.05 M) for 3 h at room temperature.

15. Wash with $NaHCO_3$ (0.5 M; 2 × 400 mL).

16. Wash with acetate buffer (pH 4.0, 0.1 M, 2 × 400 mL).

17. Wash with $NaHCO_3$ (0.1 M) containing NaCl (0.5 M) (250 mL).

18. Wash with 250 mL incubation buffer (sodium phosphate (pH 7.4, 0.1 M), NaCl (0.15 M), BSA (0.1%), normal rabbit serum (Dako; 1 mL L^{-1}), and sodium azide (0.02%)).

19. Store in incubation buffer at 4 °C.

Two-site IRMA

1. Dip an anti-ferritin antibody-immobilized paper filter in incubation buffer (100 μL).

2. Add standard ferritin solution or sample (heat-treated sup.; 50 μL).

3. Rock at 37 °C for 3 h.

4. Remove solution and wash the paper disc with incubation buffer (4 × 2.5 mL).

5. Add 20 000 counts min^{-1} ^{125}I labelled anti-ferritin antibodies[5].

6. Rock at 4 °C for 24 h.

7. Remove solution, wash with NaCl (0.85%, 4 × 2.5 mL) and dry.

8. Count dry filter in a γ counter.

9. Plot a standard curve and determine the ferritin concentration in samples.

[1] Ferritin resists heating at 80 °C for 10 min, and can be recovered in the supernatant after centrifugation.
[2] Human liver ferritin is commercially available (Dako, Carpinteria, CA, USA).
[3] Lactoperoxidase labelling is an option.
[4] Polystyrene tubes coated with unlabelled antibody can be also used instead of paper discs.
[5] A radioimmunoassay kit for ferritin can be purchased from Ramco Laboratories (Houston, TX, USA) or Diagnostics Products (Los Angeles, CA, USA).

References

1. Miles, L. E., Lipschitz, D. A., Bieber, C. P., and Cook, J. D. (1974). Measurement of serum ferritin by a 2-site immunoradiometric assay. *Anal. Biochem.*, **61**, 209–24.
2. Niitsu, Y., Kohgo, Y., Yokota, M., and Urushizaki, I. (1975). Radioimmunoassay of serum ferritin in patients with malignancy. *Ann. NY Acad. Sci.*, **259**, 450–2.
3. Rouault, T. and Klausner, R. (1997). Regulation of iron metabolism in eukaryotes. *Curr. Topics Cell Regul.*, **35**, 1–19.

34

Histochemical staining of ROS: NBT and $O_2\cdot^-/H_2O_2$ and cerium methods

KEI-ICHI HIRAI

Introduction

$O_2\cdot^-$ generated by phagocytes can be histochemically demonstrated by light microscopy using a nitro blue tetrazolium (NBT) method (1). The presence of H_2O_2 can be demonstrated cytochemically by electron microscopy by use of the cerium method of Ohno *et al.* (2) modified from Briggs *et al.* (3). Originally, these methods were applied to living cells; it was, therefore, difficult to demonstrate ROS-generating activity in secretory vesicles which contain cytochrome *b*558 as NADPH oxidase (4) because vesicular fusion to the phagosomes or plasma membranes had already advanced during the incubation process.

These methods have recently been used for glutaraldehyde-fixed cells (5, 6) as the NADPH–cerium method. The reaction proceeds as follows:

$$NADPH + 2O_2 \rightarrow NADP^+ + H^+ + 2O_2\cdot^-$$

$$O_2\cdot^- + NBT \rightarrow O_2 + \text{deformazan} \downarrow \text{(bluish–purple) ...... light microscopy}$$

$$2O_2\cdot^- + 2H^+ \rightarrow H_2O_2 + O_2$$

$$H_2O_2 + Ce^{3+} \rightarrow Ce(OH)_3OOH \downarrow \text{(cerium perhydroxide)} + H_2O \text{ ...... electron microscopy}$$

Protocol

Demonstration of $O_2\cdot^-$ in living cells: NBT method (1)

1. Harvest physiologically intact granulocytes or macrophages in PBS at room temperature.
2. Suspend 10^5 cells in NBT reaction medium[1] (1 mL) in a conical tube.
3. Add stimulant solution[2] (10 mL) and incubate for 10 min at 37 °C in suspension.

Protocol *Continued*

Stimulants for human cells

Opsonized zymosan (heat-killed yeast). Add zymosan A (Sigma, St. Louis, MO, USA; 10 mg) to autologous fresh serum (1 mL) and incubate for 30 min at 37 °C. Centrifuge at 2000 rpm for 10 min and wash in phosphate-buffered saline.

Calcium ionophore A23187 (20 mM)

Formyl-leucyl-phenylalanine (FMLP; 1 mM).

12-*O*-Tetradecanoylphorbol-13-acetate (TPA; 8–80 mM).

Scavengers

p-Benzoquinone (0.1 mM); cytochrome-*c* (type IV, horse heart; Sigma; 16 mM); superoxide dismutase (240 units mL^{-1}).

Inhibitor

Diphenyliodonium chloride (PI; ICN Biochem, Cleveland, OH, USA; 0.1 mM).

4. Add one drop of conventional glutaraldehyde (20 or 25% *v*/*v*) to fix for 10 min.
5. Spin at 100 rpm for 5 min at room temperature.
6. Wash the cells once or twice with distilled water.
7. Smear the cells on a glass slide and leave to dry. Mount with a cover slip and a mounting agent before observation.

Demonstration of H_2O_2: the cerium method of Ohno et al. (2)

1. Suspend 10^5 cells in PBS (1 mL) with $CaCl_2$ (0.09 mM), $MgCl_2$ (0.05 mM), glucose (0.1%), and stimulant solution (10 μL) in a conical tube.
2. Incubate cells at 37 °C for appropriate time periods.
3. Centrifuge at 1000 rpm for 5 min at 4 °C.
4. Wash the stimulated cells in 2 changes of 3 mL of Tris-maleate (0.1 M)–7% (*w*/*v*) sucrose buffer at pH 7.2 and 4 °C (as living-cell method). As fixed-cell method, wash the stimulated cells in 2 changes of Tris-maleate–sucrose buffer and then immediately fix the re-suspended cells for 10 min with pure glutaraldehyde (2% *v*/*v*; purity ratio at A_{235}/A_{280} nm should be below 0.22) in cacodylate buffer (0.1 M) at pH 7.2. Wash the fixed cells in 3 changes of Tris-maleate buffer.
5. Incubate living cells in the Ohno *et al.* cerium medium[2] (3 mL) (2). Incubate fixed cells in the NADPH–cerium medium[3] for fixed cells (3 mL) (3, 4) in a conical tube for 10–20 min at 37 °C.

Scavengers

o-Benzoquinone (0.1 mM); ferricytochrome-*c* (type IV, horse heart; Sigma; 16 mM); crystallized catalase (S-100, bovine liver, Sigma; 4000 units mL^{-1})

Inhibitor

Diphenyliodonium chloride (DPI; ICN Biochem., Cleveland, OH, USA; 0.1 mM)

6. Centrifuge at 100 rpm for 5 min at 4°C and discard the incubation solution.
7. Gently cover centrifuged living or fixed cell pellet at the bottom of conical tube with glutaraldehyde (2% *v/v*) in cacodylate buffer for 1 h at 4°C.
8. Remove cell clot, dice into 5 or 6 pieces ~1 mm cube and refix in OsO_4 (1% *w/v*, 1 mL) in phosphate buffer (0.1 M) at pH 7.2 in a screw-capped 5-mL vial for 1 h at 4°C.
9. Wash the diced samples in 2–3 changes of phosphate buffer.
10. Dehydrate in ethanol series (50% to anhydrous), and embed in resin according to general procedures for preparing samples for electron microscopy.

[1] PBS (pH 7.4, 10 mM) containing $CaCl_2$ (0.09 mM), $MgCl_2$ (0.05 mM), NaN_3 (10 mM), sucrose (0.1%) and NBT (0.05 mM).

[2] The Ohno *et al.* cerium medium contains $CeCl_3$ (1 mM), glucose (0.1%) and NaN_3 (10 mM) in Tris-maleate (0.1 M)–7% (*w/v*) sucrose buffer at pH 7.5.

[3] The NADPH–cerium medium for fixed cells contains NADPH (3 mM), $CeCl_3$ (1 mM), and NaN_3 (10 mM) in tris-maleate buffer (pH 7.5, 0.1 M).

Results

Demonstration of $O_2\cdot^-$ and H_2O_2 in living human neutrophils (1)

In neutrophils engaged in phagocytosis, $O_2\cdot^-$ (diformazan) and H_2O_2 (cerium perhydroxide) production is restricted to the cell–particle interface and on the inner surface of phagosomes (Figs 1a and 2a). The remaining free surface of the plasma membrane is devoid of reaction deposits. In neutrophils stimulated with A23187, the production of $O_2\cdot^-$ and H_2O_2 is apparent over the whole surface of the plasma membrane (Figs 1b and 2b). When stimulated with FMLP, the production of $O_2\cdot^-$ and H_2O_2 is restricted to the cell–cell interface of aggregated neutrophils (Figs 1c and 2c).

Demonstration of H_2O_2 in fixed human neutrophils (6)

Neutrophils are mixed with opsonized zymosan A for 0–20 min and fixed with pure glutaraldehyde for 10 min before reaction incubation in an NADPH–cerium medium. No deposition of reaction products is apparent in unstimulated cells (Fig. 3a). H_2O_2 production in the secretory vesicles is apparent in the particle-engulfing cells at several time periods (Fig. 3b). Some of the

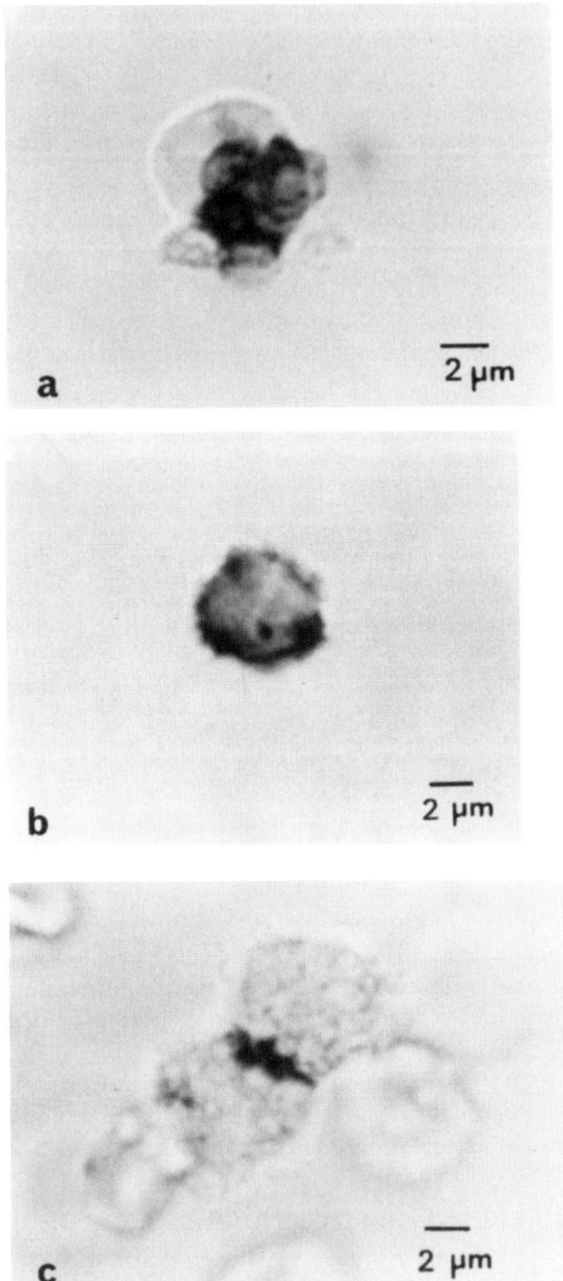

Figure 1. The production of $O_2^{\cdot -}$ in normal human living neutrophils advanced for 10 min in total (NBT method). (a) Cells in phagocytosis; (b) stimulated with 20 mM A23187; (c) stimulated with 1 mM FMLP.

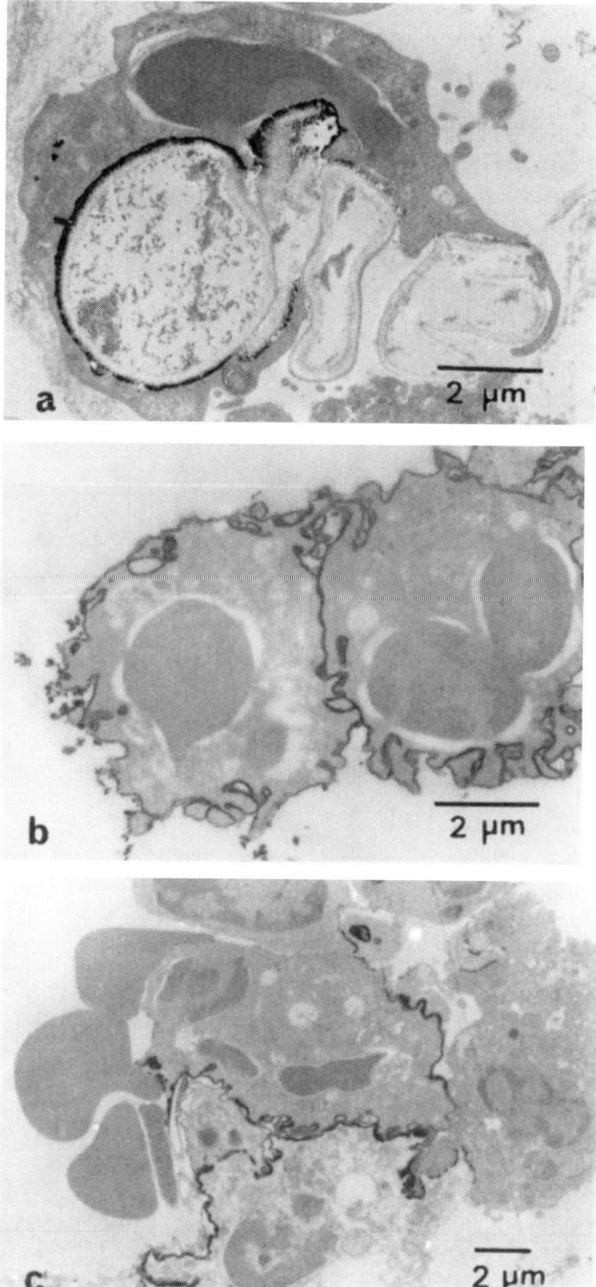

Figure 2. The production of H_2O_2 in normal human living neutrophils advanced for 20 min in total (Ohno *et al.* cerium method). (a) Cells in phagocytosis; (b) stimulated with 20 mM A23187; (c) stimulated with 1 mM FMLP.

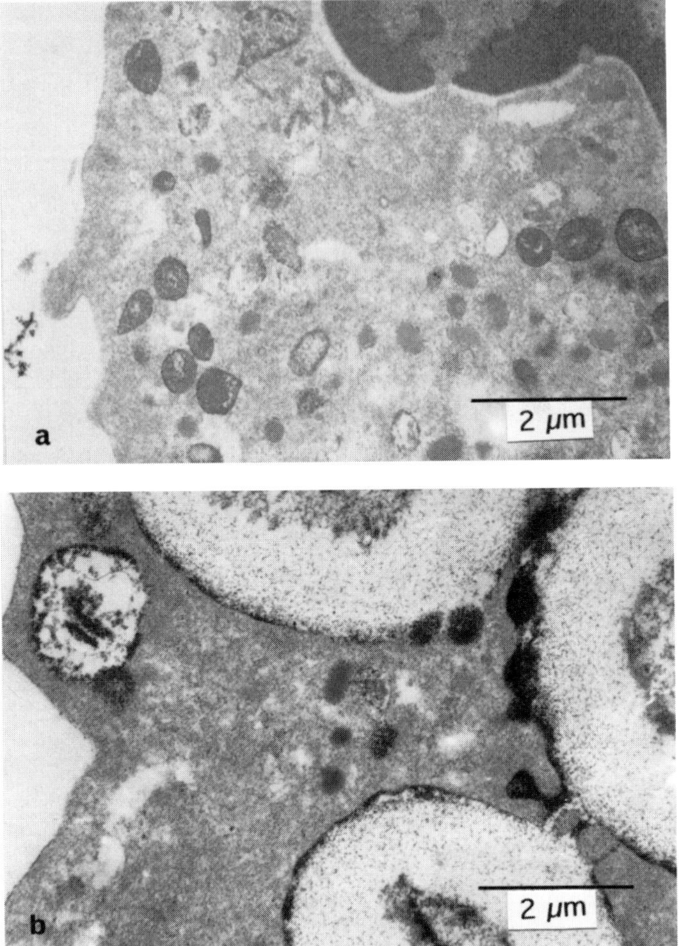

Figure 3. The production of H_2O_2 in normal human fixed neutrophils in phagocytosis (NADPH–cerium method). (a) Dormant cell without reaction; (b) phagocytosed for 10 min and then fixed with glutaraldehyde before reaction–incubation for 10 min.

vesicles are fused to the phagosomal membranes, depending on the time period, and release ROS into the lumens.

Comments

When living cells are used for reaction incubation, do not attach cells on to a glass slide as this causes artificial staining, as previously observed (1). Cells can be placed on a matrix base, polyvinylidine chloride films or uncoated plastic culture dishes without artefacts (Moriguchi *et al.*, unpublished data). In the experiments described above, exclude NADPH or NADH from the reaction medium, because these are unable to penetrate the plasma membranes but can be synthesized by cells from added glucose.

When cells are fixed before reaction–incubation, use photometrically pure glutaraldehyde, otherwise contaminating impurities inactivate all enzyme activity. Add NADPH as the substrate into this reaction medium, because fixed cells are unable to supply this endogenously.

References

1. Hirai, K. I., Moriguchi, K., and Wang, G. Y. (1991). Human neutrophils produce free radicals from the cell–zymosan interface during phagocytosis and from the whole plasma membrane when stimulated with calcium ionophore A23187. *Exp. Cell Res.*, **194**, 19–27.
2. Ohno, Y. I., Hirai, K. I., Kanoh, T., Uchino, H., and Ogawa, K. (1982). Subcellular localization of H_2O_2 production in human neutrophils stimulated with particles and an effect of cytochalasin-b on the cells. *Blood*, **60**, 253–60; Ohno, Y. I., Hirai, K. I., Kanoh, T., Uchino, H., and Ogawa, K. (1982). Subcellular localization of hydrogen peroxide production in human polymorphonuclear leukocytes stimulated with lectsins, phorbol myristate acetate, and digitonin: An electron microscopic study using $CeCl_3$. *Blood*, **60**, 1195–202
3. Briggs, R. T., Drath, D. B., Karnovsky, M. L., and Karnovsky, M. J. (1975). Localization of NADH oxidase on the surface of human polymorphonuclear leukocytes by a new cytochemical method. *J. Cell Biol.*, **67**, 566–86.
4. Borregaard, N., Kjeldsen, L., Lollike, K., and Sengeløv, H. (1993). Granules and vesicles of human neutrophils. The role of endomembranes as source of plasma membrane proteins. *Eur. J. Haematol.* **51**, 318–22.
5. Kobayashi, T., Garcia del Saz, E., Hendry, J., and Seguchi, H. (1999). Detection of oxidant producing-sites in glutaraldehyde-fixed human neutrophils and eosinophils stimulated with phorbol myristate acetate. *Histochem. J.*, **31**, 181–94.
6. Hirai, K. I., Moriguchi, K., and Ohno, N. (1996). Vesicular origin of superoxide-producing NADPH oxidase in the human neutrophil phagosomes. *Acta Histochem. Cytochem.*, **29** (Suppl), 280–1

35

Immunostaining of Mn–SOD, Cu,Zn–SOD and NOS

MASAHIRO NAKAMURA, HARUYUKI TATSUMI, and HIROKI NOGAWA

Introduction

To elucidate the mechanisms of carcinogenesis and senescence, it is important to investigate the distribution of key enzymes, such as superoxide dismutase (SOD) (1, 2), nitric oxide synthase (NOS) and others. Among these two are several isoenzymes, which differ from each other in their function and location. For location, immunohistochemistry is a useful approach, although a stereotyped approach is sometimes insufficient for detection of individual enzymes because of the unusual properties of the proteins. To overcome these problems, we shall overview the methods, explain the key points, and demonstrate the possibility of semi-quantitative measurement by use of confocal laser-scanning microscopy and image processing.

Protocol

It is necessary to have a knowledge of general immunohistochemistry to perform these methods. In this section, we describe our methods with Mn–SOD, Cu,Zn–SOD and NOS antibodies. These methods are also applicable to other histochemistry techniques.

For light microscopy

Fixation

Paraformaldehyde (PFA) fixative[1] or Zamboni's fixative[2] can be used. The latter can be stored for twelve months in 4 °C but the former is not so durable. We use freshly prepared PFA solution. To preserve the structure of tissues and cells for immunohistochemistry, perfusion fixation is desirable. If fixation is too long immunoreactivity is reduced.

1. Perfuse fixative from left ventricle.
2. Cut tissues into small pieces (5–10 mm).

3. Immerse in the same fixative for 4–6 h at 4°C (ideally with stirring or shaking)

Rinsing and Infiltration

Wash off the fixative and avoid forming ice crystals during the process

1. Wash with 10% (*w/v*) sucrose in PBS for 4 h at 4°C.
2. Wash with 15% (*w/v*) sucrose in PBS for 4 h at 4°C.
3. Wash with 20% (*w/v*) sucrose in PBS for 4 h or overnight at 4°C.

Embedding and Freezing

Choose cryo-embedding medium according to the method of freezing. For rapid freezing with liquid nitrogen, Microm cryo-embedding compound Cat. No. 350100 is recommended. For slow freezing, Microm cryo-embedding compound Cat. No. 358100 or O.C.T. compound (Miles) are suitable.

Cryo-sectioning

Cut sections 4 μm in thickness, mount a section on a slide and air-dry for 30 min. Use clean glass slides coated with poly L lysine to avoid detachment of the sections from the slides.

Permeabilization

Triton X-100 is added to 0.01 M PBS (3:1000 *v/v*) for rinsing and immersing sections before false-positive suppression.

1. Rinse three times with PBS, each time for 5 min, at room temperature to remove the cryo-embedding medium.
2. Immerse for 1 h in PBS containing Triton X-100 for permeabilization.

Endogenous peroxidase blocking

If endogenous peroxidase activity does not hinder observation, this step can be omitted. Blocking can reduce immunoreaction.

1. Incubate in H_2O_2 (3% *v/v*) for 10 min.
2. Rinse three times with PBS, each time for 5 min.

EABA (endogenous avidin binding activity) blocking (2)

1. Immerse in 0.01% (*w/v*) (strept)avidin (Sigma; St Louis, MO, USA) in PBS for 15–20 min
2. Rinse three times with PBS, each time for 5 min.
3. Immerse in 0.001% (*w/v*) biotin (Sigma; St Louis, MO, USA) in PBS for 15–20 min.
4. Rinse three times with PBS, each time for 5 min.
5. Immerse in 4% (*w/v*) buffered PFA for 1 h.
6. Rinse three times with PBS, each time for 5 min.

Protocol *Continued*

Blocking of nonspecific binding
Immerse in normal goat serum (1:10 or 1:100 diluted) for 30 min

Immunostaining
The dilution ratio and incubation time vary depending on the specimens, the antibodies, and the conditions. For example, we use rabbit anti-rat Mn-SOD antiserum diluted 1:500 and incubated for 2 h at room temperature (approx. 20 °C) or 1:3000 dilution overnight at 4 °C. Use of room temperature facilitates immunoreaction whereas 4 °C is recommended for tissue preservation, even though a longer reaction time is required. Control experiments are performed by substituting normal rabbit serum for specific antibodies or by omitting the first antiserum.

First layer

1. Immerse in anti-SOD antibody × 500–1000, anti-NOS antibody × 1000–5000, and control serum (normal rabbit serum) for 2 h at room temperature or 12–18 h at 4 °C in a moist chamber.
2. Rinse three times with PBS, each time for 5 min.

Peroxidase-conjugated streptavidin–biotin immunoglobulin detection system (Histofine SAB-PO kit; Nichirei Company, Japan) is used to locate rabbit IgG.

Second layer at room temperature

1. Immerse in biotinylated anti-rabbit IgG goat antibody for 15–30 min.
2. Rinse three times with PBS, each time for 5 min.

Third layer at room temperature

1. Immerse in peroxidase-conjugated streptavidin for 10–30 min
2. Rinse three times with PBS, each time for 5 min.

Visualization with DAB reaction
Incubate sections in freshly prepared reaction medium[3] until adequate under microscopic observation. The duration should not exceed 3 min.

For confocal laser-scanning microscopy

1. 1st layer for 2 h at room temperature.
2. Rinse three times with PBS, each time for 5 min.
3. 2nd layer at room temperature FITC-conjugated anti-rabbit IgG antiserum for 2 h.
4. Rinse three times with PBS, each time for 5 min.

5. Put a cover glass on the specimen in glycerine–PBS (9:1) and paint the periphery of the cover glass with lacquer.
6. Examine with a laser-scanning confocal microscope (BioRad MCR 600).

The digital image was processed on Nextstep (3.3J) workstations to demonstrate differences between the location of Mn-SOD and Cu,Zn-SOD in the rat stomach. The application software (ConfocalViewer.app) is available at ftp://ftp.sapmed.ac.jp/pub/NeXT/ConfocalViewer. The application is a four-way object (Motorola, Intel, Sparc, HP version). (Development was partly supported by Special Coordination Funds of the Science and Technology Agency of the Japanese Government)

For immunoelectron microscopy

Specimen preparation and fixation
Rat stomach and A549 (human lung adenocarcinoma cell line)[4] were used in this study (see notes). Briefly fix with PFA (3% *w/v*) and glutaraldehyde (0.1% *v/v*) buffered with Millonig's phosphate (0.1 M) at pH 7.4.

Rapid freezing and freeze-substitution (1)

1. Freeze specimens by pressing them rapidly against a copper block cooled to 4 °C with liquid helium.
2. Freeze-substitute the blocks in a vial containing glutaraldehyde (0.1% *v/v*) in acetone. Keep at −80 °C for 3 days then successively at −20 °C, 4 °C, and room temperature, each for 2 h.
3. Rinse with pure acetone.

Infiltration

1. Rinse five times with pure ethanol.
2. Immerse in a 1:1 mixture of Lowicryl K4M and ethanol (1:1) for 2 h at −20 °C.
3. Immerse in pure Lowicryl K4M overnight at −20 °C.

Polymerization
Polymerize the blocks at −20 °C with ultraviolet light.

Ultra-thin sectioning
Mount sections on nickel grids.

Immunostaining

1. Block non-specific binding by immersing in bovine serum albumin (0.1% *w/v*) in PBS for 1 h at room temperature.
2. Immerse in anti Mn-SOD antibody × 100–300 for 1 h at room temperature.

Protocol *Continued*

3. Rinse three times with PBS, each time for 5 min.
4. Immerse in colloidal gold conjugated anti-rabbit immunoglobulins × 100 for 1 h at room temperature (AuroProbe EM GAR G10, Janssen Life Sciences).
5. Rinse with distilled water and dry.

Contrast staining

1. Immerse in uranyl acetate (3% *w/v*) solution for 3 min.
2. Rinse with distilled water.
3. Immerse in Reynolds' lead citrate[5] for 1 min.
4. Rinse with distilled water and dry

[1] To prepare PFA fixative make up solutions (a) and (b) and mix in equal quantities. (a) PFA solution (8% *w/v*). Dissolve PFA powder (20 g) in distilled water (200 mL) at 60 °C with a few drops of NaOH (1 M). It is recommended that the minimum quantity of NaOH be used to dissolve the powder. Heating the solution above 60 °C weakens the fixation activity. After cooling and filtering the solution, dilute to 250 mL with distilled water. (b) Millonig's phosphate buffer (0.2 M). Dissolve $NaH_2PO_4.H_2O$ (11.04 g) in NaOH (10 M, 6.7 mL) dilute to 400 mL with distilled water.
[2] To prepare Zamboni's fixative make up solutions (a) and (b) and dilute to 1000 mL with phosphate buffer (c). (a) Picric acid, saturated solution, 50 mL. (b) Paraformaldehyde (20% *w/v*) solution, 100 mL. (c) Phosphate buffer (320 mOsm). Dissolve $NaH_2PO_4.2H_2O$ (3.31 g) and $Na_2HPO_4.12H_2O$ (33.7 g) in distilled water (1000 mL).
[3] To prepare reaction medium mix Tris HCl buffer (pH 7.6, 0.05 M, 100 mL), 3,3'-diaminobenzidine.4HCl (DAB; 40 mg), and hydrogen peroxide solution (1% *v/v*, 1 mL).
[4] A549 cells were grown in Isacove's modification of Dulbecco's medium (IMDM) containing foetal calf serum (Gibco; 10% *w/v*) and glutamine (3% *w/v*), penicillin (100 units mL^{-1}), and streptomycin (100 μg mL^{-1}). IMDM contained 2-mercaptoethanol (0.05 mM). The cell line was plated in culture flasks (Falcon) and routinely maintained in a CO_2 incubator at 37 °C. A549 is one of the cell lines resistant to TNF cell-killing activity and TNF stimulation results in an increase in Mn-SOD levels. To increase the Mn-SOD content, the cells were incubated with TNF for 0, 3, 6, 12, 24, and 48 h. The human recombinant TNF-α (rh-TNF) was from Ube Industries. The specific activity of the purified rh-TNF was 2.0×10^6 units/mg protein. After stimulation, a solution of trypsin (5%) and EDTA (1 mM) in PBS was added and the plates were incubated for 5 to 10 min. The enzyme reaction was stopped by adding culture medium. Cells were removed by centrifugation at 3000 rpm for 5 min.
[5] To prepare Reynolds' lead citrate dissolve lead nitrate (1.33g) and tri-sodium citrate dihydrate (1.76 g) in distilled water (30 mL), add NaOH (1 M, 8 mL) to dissolve completely, and dilute to 50 mL with distilled water.

Results and comments

The antibody against Mn-SOD is specific, but not reactive to the antigen in paraffin sections. Being preserved in frozen sections, Mn-SOD immunoreactivity is very weak. Consequently we used Triton X-100 and a labelled-streptavidin-biotin immunoglobulin detection system to enhance the visualization of Mn-SOD. When using the combination of Triton X-100, frozen

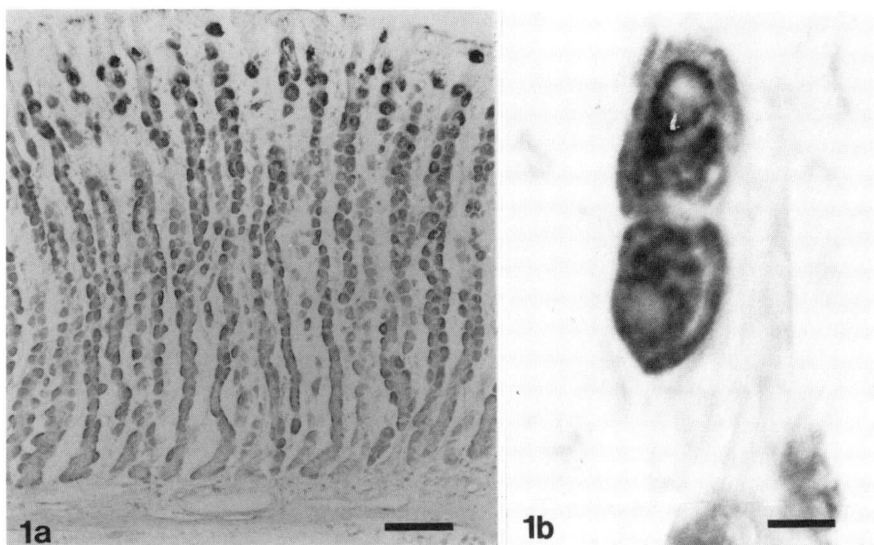

Figure 1. Rat stomach treated with false-positive suppression and anti-rat Mn-SOD antibody. (a) In the frozen section treated with Triton X-100, immunopositive cells are distributed in gastric glands. (b) Immunonegative area surrounds the nucleus of the immunopositive cells in a longitudinal section of a gastric gland. Original magnification: (a) × 25; (b) × 250; bar = 100 μm (a); 10 μm (b).

sections, and the labelled-streptavidin–biotin system, we encountered strong false-positive staining. Although the false-positive staining was abolished by ethanol dehydration of the frozen section, Mn-SOD immunoreactivity was also extinguished as a result of the dehydration. Consequently, in addition to the combination described above, a false-positive suppression technique was employed and the location of Mn-SOD was investigated in the rat stomach (Fig. 1). The immunopositive cells, which are located in the gastric gland, especially in the upper portion of the gland, are large and round in shape, with characteristic intracellular canaliculi around the nucleus. The other types of cell have no strong immunoreaction and no immunopositive reactions were observed in control sections.

Because antigens located in membranous organelles are difficult to detect it is recommended that Triton X-100 is added to PBS (3:1000, *v*/*v*) for rinsing sections and diluting antibodies. The non-ionic detergent dissolves phospholipid in membranes and increases the permeability of antibodies. The detergent enhances not only the immunoreactivity but also the endogenous avidin binding activity (EABA) on frozen sections. Either formalin fixation or processing for paraffin embedding usually abolishes EABA, because the binding activity of endogenous biotin to (strept)avidin is destroyed by formalin fixation or alcohol dehydration. Within the membranous cell components, probably mitochondria, the EABA on the cryosection seems to survive

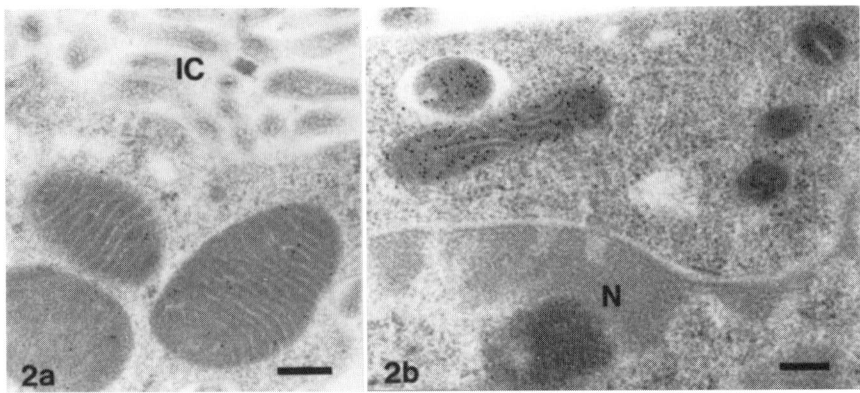

Figure 2. Post-embedding immunoelectron micrographs for Mn-SOD using rapid-freeze freeze-substitution technique. (a) Immunogold particles are seen on the mitochondria of the rat parietal cells. IC: intracellular canaliculi. (b) A549 (human lung adenocarcinoma cell line) was stimulated with TNF-α to increase Mn-SOD. High concentrations of immunogold particles are seen on the mitochondria of the cell. N: nucleus. Original magnification: (a) × 30 000; (b) × 20 000; bars = 0.25 μm.

fixation and is revived by the permeabilization with Triton X-100, which causes the false positive staining. Therefore EABA must be blocked sufficiently, when the (strept)avidin–biotin immunoglobulin detection system is used for frozen sections treated with Triton X-100.

For immunoelectron microscopy, dehydration of specimens with ethanol or acetone and Epon epoxy resin for embedding also make it impossible to detect Mn-SOD. To circumvent this obstacle and demonstrate the location by post-embedding immunoelectron microscopy, a rapid freezing and freeze-substitution technique was employed using Lowicryl K4M embedding medium. It was so effective that immunogold particles for Mn-SOD were observed specifically on the mitochondria of parietal cells in the rat stomach (Fig. 2a) and A549 cells (Fig. 2b). In the latter cells stimulated with TNF-α, immunogold particles were highly concentrated on the mitochondria.

Because the antigen–antibody reaction occurs under physiological conditions, which is an aqueous environment, the dehydration process usually affects antigen–antibody reactivity. Although dehydration impaired the reactivity of Mn-SOD in the conventional method, our rapid freezing and freeze-substitution method preserved the antigen–antibody binding activity of Mn-SOD even after dehydration. This post-embedding immunoelectron microscopy with rapid freezing and freeze-substitution technique is extremely useful and might facilitate further studies on Mn-SOD. It could, moreover, be employed for other antigens which have been difficult to detect by conventional methods.

Although the functional diversity between the Mn-SOD and Cu,Zn-SOD in the stomach has not yet been clarified, the pseudocolour presentation

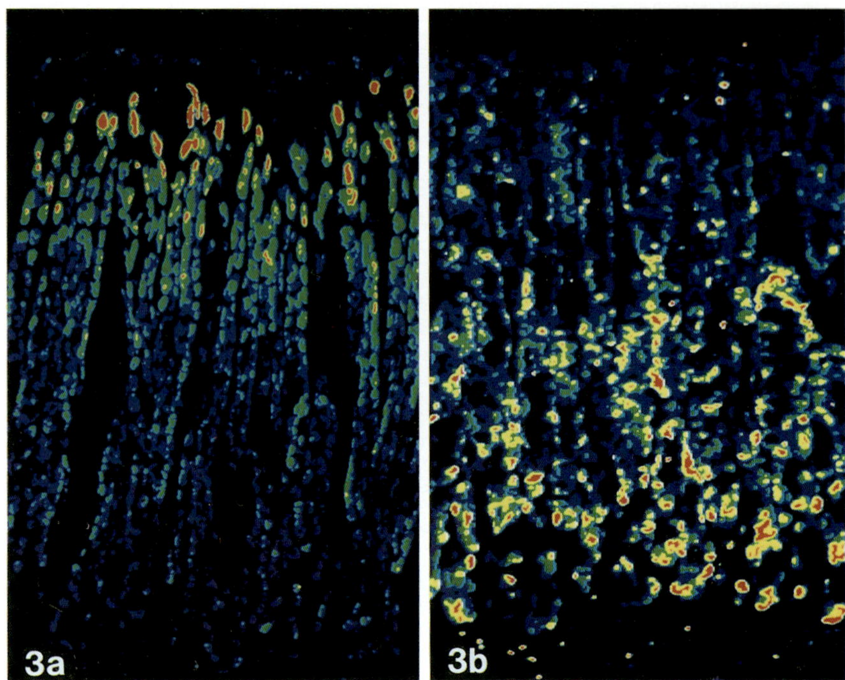

FIGURE 3. Confocal laser-scanning micrographs of Mn-SOD (a) and Cu,Zn-SOD (b) immunostaining of the rat stomach. Pseudocolour presentation by image processing clearly reveals the different locations of the enzymes in the gastric glands. A high concentration of Mn-SOD is apparent at the top of the glands, whereas Cu,Zn-SOD exists mainly in the lower part of the glands. The relative concentration gradient is: red > yellow > green > blue.

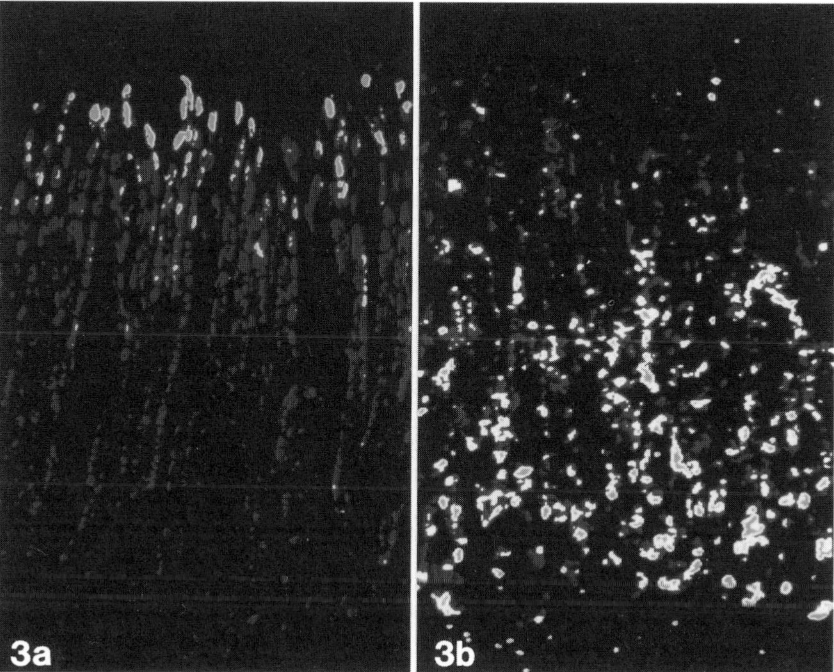

Figure 3. Confocal laser-scanning micrographs of Mn-SOD (a) and Cu,Zn-SOD (b) immunostaining of the rat stomach. Pseudocolour presentation by image processing clearly reveals the different locations of the enzymes in the gastric glands. A high concentration of Mn-SOD is apparent at the top of the glands, whereas Cu,Zn-SOD exists mainly in the lower part of the glands. The relative concentration gradient is: red > yellow > green > blue.

based on the FITC-immunostaining intensity elucidates the distinctive difference in the location of the enzymes in the gastric gland (Fig. 3). This is an example of a semi-quantitative investigation, which is difficult by conventional immunohistochemistry techniques. With regard to neuronal nitric oxide synthase (NOS), surface mucous cells are specifically stained with the antibody (Fig. 4), in contrast with the parietal cells which are devoid of the positive immunoreaction.

These characteristic enzyme distributions could throw light on cell differentiation mechanisms, and three-dimensional reconstructions of gastric glands with computer graphics and spatial mapping of the enzyme distribution by immunohistochemistry (3, 4) will enable us to obtain more precise information. Aiming at elucidation of the mechanisms of cell differentiation from another point of view, we have been developing an integrated three-dimensional reconstruction system for morphological studies (5, 6) in combination with histochemical techniques mentioned in this section.

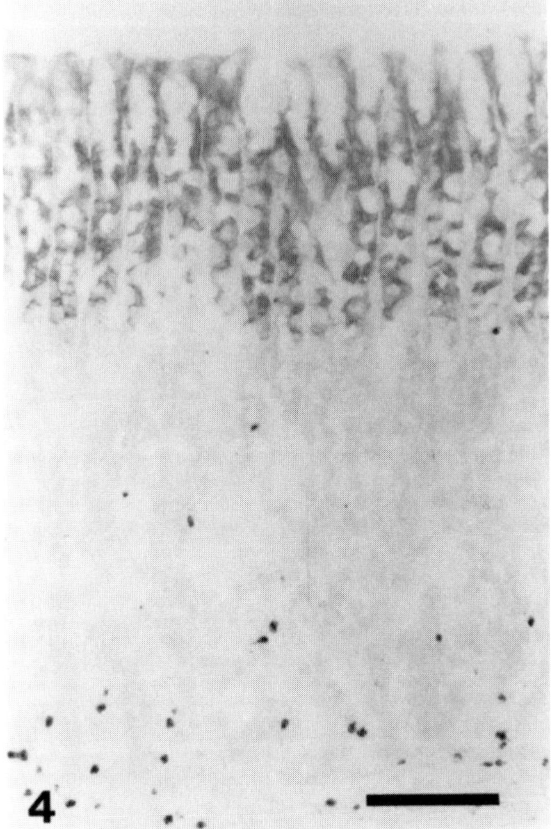

Figure 4. Immunostaining of the rat stomach treated with anti-rat NOS antibody. The surface mucous cells are specifically stained, in contrast with the adjacent immunonegative parietal cells. Original magnification: × 50; bars = 100 μm.

References

1. Tatsumi, H., Suzuki, K., Senda, T., Satoh, S., Fujita, H., and Taniguchi, N. (1991). Post-embedding immuno-electron microscopy with rapid freezing and freeze-substitution technique for localization of manganese superoxide dismutase. *Arch Histol. Cytol.*, **54**, 465–9.
2. Satoh, S., Tatsumi, H., Suzuki, K., Taniguchi, N. (1992). Distribution of manganese superoxide dismutase in rat stomach: Application of triton X-100 and suppression of endogenous streptavidin-binding activity. *J. Histochem. Cytochem.*, **40**, 1157–63.
3. Tatsumi, H., Takaoki, E., Omura, K., Fujita, H. (1990). A new method for three-dimensional reconstruction from serial sections by computer graphics using 'metaballs': Reconstruction of 'hepatoskeletal system' formed by Ito cells in the cod liver. *Comp. Biomed. Res.*, **23**, 37–45.

4. Tatsumi, H., Satoh, S., and Takaoki, E. (1990). Application of a reconstruction method using computer graphics to study cell differentiation. *Proceedings of the 4th Sapporo International Computer Graphics Symposium*, pp. 132–6.
5. Tatsumi, H. (1994). Computer utilization in anatomy: a renaissance in new anatomy. *J. Clin. Exp. Med. (IGAKU NO AYUMI)*, **170**, 281–5, (in Japanese).
6. Tatsumi, H., Ohkawa, Y., Nakamura, M., and Hayashi, K. (1996). Computerized anatomy—aiming at a renaissance of anatomy. *Med. Imag. Tech.*, **14**, 115–23 (in Japanese).

36

Xanthine oxidase

TAKAAKI AKAIKE, YOICHI MIYAMOTO, and
TOMOHIRO SAWA

Introduction

Xanthine oxidoreductase is widely distributed among mammalian species and is expressed mainly in vascular endothelial cells, hepatocytes, epithelial cells of the intestine, and the secretory cells of the mammary gland. Expression of the enzymatic activity of XO/XD depends on the animal species. XO/XD activity is, for example, high in the blood of mice and rats, whereas very low levels of XO/XD activity are detected in human blood. This enzyme, however, is also distributed in human tissues, particularly in endothelial cells. We must, therefore, employ a highly sensitive and selective method for detection of the activity of this enzyme in human specimens.

It is well known that this enzyme produces the active oxygen species ($O_2\cdot^-$ and H_2O_2) which play important roles in the pathogenesis of ischaemia–reperfusion injury, and inflammatory and virus-induced tissue injury. It has, moreover, been suggested that these active oxygen species interact with nitric oxide produced by the endothelial type of nitric oxide synthase. One of the most intriguing features of the enzyme xanthine oxidoreductase is that under physiological conditions it exists in the form xanthine dehydrogenase (XD), which utilizes NAD^+ as a coenzyme, and is converted to xanthine oxidase (XO) which utilizes O_2 as an electron acceptor. Xanthine oxidoreductase also catalyses the metabolism of hypoxanthine and xanthine to uric acid[1] and metabolizes some other naturally occurring heterocyclic compounds.

Widely employed assays for xanthine oxidoreductase activity are spectrophotometric quantitation of uric acid, the final product of the reaction (1), quantitation of hypoxanthine, xanthine or uric acid by high performance liquid chromatography (HPLC) (2), fluorimetric quantitation of isoxanthopterin, the metabolite of pterin by XO/XD (2, 3), and quantitation of radiolabelled uric acid after thin-layer chromatography (TLC) of the reaction mixture of radioactive hypoxanthine/xanthine and the enzyme (2, 4). These methods are introduced briefly in this section.

Note

[1]XO/XD catalyses two steps of the reactions as shown in the equation:

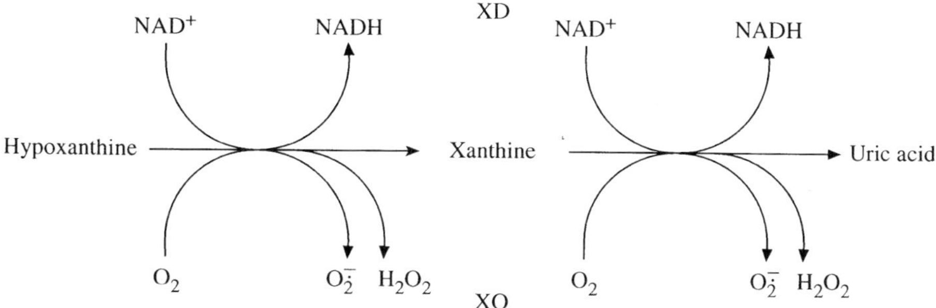

Xanthine is not only the reaction product derived from hypoxanthine but also the substrate for uric acid formation. The XO/XD activity can, therefore, be determined by measuring the amount of final product—uric acid or both xanthine and uric acid—from hypoxanthine. XD is converted to XO by limited proteolysis. Oxidation of the thiol moieties in the XD molecule also causes conversion to XO. Conversion of XD to XO induces a dramatic change in the dependence of redox coenzymes for its enzymatic reaction.

Spectrophotometric quantification of uric acid formation (1)

Protocol

XO/XD activity is easily evaluated by quantifying the formation of uric acid by measuring the absorbance at 290 nm (A_{290}), an absorption maximum of uric acid. This method is, however, not sufficiently sensitive to be applicable to the measurement of biological samples which contain only low levels of XO/XD activity (see above). Because XO/XD occurs in the cytosol fractions of cells, we usually quantify the enzymatic activity of some tissues by using the cytosol fractions obtained by ultracentrifugation (100 000 g) of the tissue (organ) homogenates[1].

1. Mix sample or XO/XD enzyme solution[2], the substrate (hypoxanthine, 50 μM final concentration), and sodium phosphate buffer (pH 7.8, 50 mM), total volume 1.0–3.0 mL, in a cuvette.
2. Record the time-dependent increase in the absorbance at 290 nm during incubation of the reaction mixture at 25 or 37 °C.

Protocol *Continued*

3. Determine XD activity in a sample by adding NAD^+ to the reaction mixture[3].

[1] Because, as mentioned above, XD is readily converted into XO, the use of protease inhibitors and dithiothreitol is highly recommended during preparation of the tissue homogenate to avoid artificial XO conversion (2).
[2] Because commercially available XO prepared from bovine milk has been treated with proteases such as pancreatin to digest casein during the process of purification, these enzymes have been irreversibly converted to XO. The purity of XO from commercial sources is, moreover, not always sufficient for biochemical studies. One should pay attention to the impurity mainly derived from the pancreatin preparations.
[3] See footnote to the introductory section.

Calculation

Because 1 unit of XO/XD activity is defined as the velocity of the formation of 1 μmol uric acid min^{-1}, the enzymatic activity in each sample can be calculated by use of the equation:

$$\text{units/mg protein} = (\Delta A/\text{min} \times 1000)/(1.22 \times 10^4 \times \text{mg mL}^{-1} \text{ reaction mixture})$$

in which the molar absorption coefficient of uric acid is 1.22×10^4 M^{-1} cm^{-1}.

Measurement of XO/XD activity by HPLC (2)

The sensitivity and specificity of spectrophotometric detection of uric acid formation for XO/XD assay (see above) can both improved by use of HPLC (column, e.g., Asahipak GS-320; mobile phase, sodium phosphate buffer (pH 7.4, 0.01 M) containing NaCl (0.15 M) (PBS); the elution time of uric acid can be adjusted by modifying the concentration of sodium chloride in the mobile phase). The detection limit of uric acid in this method is 1 μM or more.

Fluorimetric assay (2, 3)

The sensitivity of the XO/XD assay can be improved further without the use of radiolabelled compounds (substrates). As mentioned above, XO/XD efficiently metabolizes pterin (2-amino-4-hydroxypteridine) into a fluorescent product, isoxanthopterin (2-amino-4,7-pterinediol), which enables the highly sensitive quantitation of the activity of the enzyme.

It is necessary to remove endogenous substrates from biological samples, because these often interfere with the oxidation of pterin by XO/XD, by dialysis or molecular sieve column chromatography on, e.g., Sephadex G25.

Protocol

1. Mix sample (without endogenous substrates) or enzyme solution, the substrate (pterin; Sigma; 10 μM final concentration), and sodium phosphate buffer (pH 7.8, 50 mM), total volume 1.0 mL.
2. Incubate the reaction mixture for 10–60 min at 37 °C.
3. Quantify the amount of isoxanthopterin produced by spectrofluorimetry.

The fluorescence intensity of isoxanthopterin is measured with excitation at 345 nm and emission at 390 nm and quantitation achieved by use of a calibration curve constructed by use of authentic isoxanthopterin (Aldrich, Milwaukee, WI, USA).
Because XD cannot use oxygen in the solution efficiently as a coenzyme, little pterin is oxidized by XD. XD activity is, therefore, evaluated by measuring the increase in the total activity after addition of methylene blue (final concentration 9 μM) as a coenzyme (an electron acceptor) of XD (2). It is also recommended that inhibition by allopurinol, a specific inhibitor of XO/XD, is to examined to confirm the specificity of the reaction.

Radiochemical method (2, 4)

Although the procedure is somewhat complicated, the method using radiolabelled hypoxanthine ([^{14}C]hypoxanthine) is the most sensitive way of determining XO/XD. The principle of this method is essentially the same as those of the spectrophotometric quantitation of uric acid formation and the measurement of XO/XD activity by HPLC. In this method the reaction products [^{14}C]xanthine and [^{14}C]uric acid are separated from the substrate by TLC (Cellulose F; E. Merck, Darmstadt, Germany). Fig. 1 shows the chromatogram

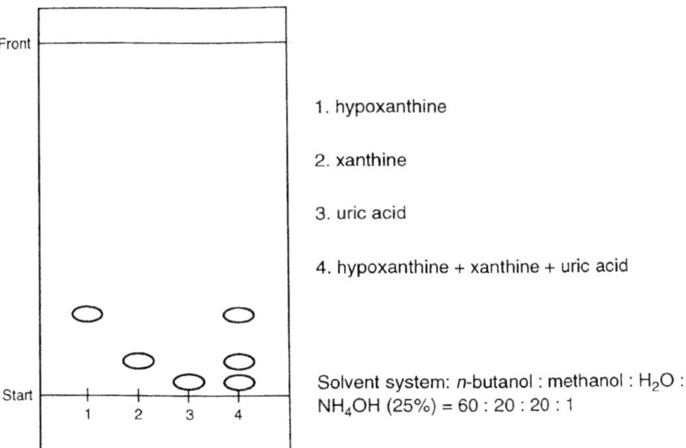

Figure 1. Typical TLC chromatogram of hypoxanthine, xanthine, and uric acid.

obtained from these compounds. To enable identification of the spot of each compound on the plate under UV illumination, non-radiolabelled hypoxanthine, xanthine and uric acid should also be applied to the plate.

Protocol

1. Mix [^{14}C]hypoxanthine (57 mCi mmol^{-1}, 200 μM final concentration), sample, and Tris-HCl buffer (pH 7.8, 50 mM), total volume 100 μL.
2. Stop the reaction by adding perchloric acid (0.15 M)[1].
3. Centrifuge to remove proteins.
4. Apply a mixture of non-radiolabelled hypoxanthine, xanthine and uric acid at each sample application site of the TLC plate (at levels sufficient to be detected under UV illumination). After drying the plate, apply the reaction mixture to each spot.
5. Develop the plate with 60:20:20:1 (*v*/*v*) *n*-butanol–methanol–water–25% (*v*/*v*) aqueous ammonia.
6. Determine the positions of hypoxanthine, xanthine and uric acid separated on the TLC plate under UV illumination then measure the radioactivity of [^{14}C]xanthine and [^{14}C]uric acid on the plate by use of a scintillation counter.

[1] It is recommended that the perchloric acid-treated sample be neutralized with NaOH (1 M) before application to the TLC plate for a better chromatographic performance.

The enzymatic activity is calculated according to the method described in the section on spectrophotometric quantitation of uric acid formation.

Other methods

Expression of XO/XD in the organs and tissues can be detected by Western blotting or by immunohistochemical techniques—by use of polyclonal and/or monoclonal antibodies against XO/XD. Because the cDNAs for XO/XD have been cloned by Amaya, *et al.* (5) and Terao, *et al.* (6), expression of mRNA of this enzyme can be detected by RT-PCR or Northern blotting.

References

1. Xanthine oxidase (milk). In *Worthington Enzyme Manual*, pp. 30–31. Worthington Biochemical, Freehold, NJ. 2. 1971
2. Akaike, T., Ando, M., Oda, T., Doi, T., Ijiri, S., Araki, S., and Maeda, H. (1990). Dependence on O_2^- generation by xanthine oxidase of pathogenesis of influenza virus infection in mice. *J. Clin. Invest.*, **85**, 739–45.
3. Beckman, J. S., Park, D. A., Pearson, J. D., Marshall, P. A., and Freeman, B. A.

(1989). A sensitive fluorimetric assay for measuring xanthine dehydrogenase and oxidase in tissues. *Free Radic. Biol. Med.*, **6**, 607–15.

4. Begleiter, A., Glazer, R. I., Israels, L. G., Pugh, L., and Johnston, J. B. (1987). Induction of DNA strand breaks in chronic lymphocytic leukaemia following treatment with 2′-deoxycoformycin *in vivo* and *in vitro*. *Cancer Res.*, **47**, 2498–503.
5. Amaya, Y., Yamazaki, K., Sato, M., Noda, K., Nishino, T., and Nishino T. (1990). Proteolytic conversion of xanthine dehydrogenase from the NAD-dependent type to the O_2-dependent type. *J. Biol. Chem.*, **265**, 14170–5.
6. Terao, M., Cazzaniga, G., Ghezzi, P., Bianchi, M., Falciani, F., Perani, P., and Garattini, E. (1992). Molecular cloning of a cDNA coding for mouse liver xanthine dehydrogenase. *Biochem. J.*, **283**, 863–70.

37

Plasma free fatty acids

YOSHIHIRO YAMAMOTO, AKIO FUJISAWA, YUICHIRO NAGATA, and SATOSHI YAMASHITA

Introduction

Polyunsaturated fatty acids (PUFA) are highly susceptible to oxidation and their levels can rapidly decrease under oxidative stress (1). A concomitant increase in monounsaturated fatty acids such as oleic and palmitoleic acids is often observed probably as a result of the activation of Δ^9-desaturase (1). Analysis of fatty acid profiles is therefore of importance. Here we describe the analysis of plasma free fatty acids (FFA) only, because the measurement of fatty acids in neutral lipids, phospholipids, and total lipids has already been well documented. Plasma FFA can, furthermore, be produced by oxidatively damaged tissues because hydrolysis of phospholipids is stimulated by oxidative stress (2).

Protocol (2)

1. Mix plasma (50 μL) with methanol (200 μL) containing margaric acid (12.5 μM; internal standard) and centrifuge at 12 000 rpm for 3 min.
2. Dry the supernatant (50 μL) under a stream of N_2 and mix the residue with a solution of monodansylcadaverine in *N,N*-dimethylformamide (2 mg mL^{-1}, 50 μL) and with diethyl phosphorocyanidate (1 μL).
3. Stand for 20 min at room temperature in the dark.
4. Analyse aliquots (5 μL) by HPLC with fluorescence detection. Column: 3.3 cm × 4.6 mm i.d., 3 μm, octadecylsilyl (Supelco) and 25 cm × 4.6 mm i.d., 5 μm, pKb-100 (Supelco) in series. Mobile phase: 17.5:65.0:17.5 (*v/v*) acetonitrile–methanol–water. Flow rate: 1.5 mL min^{-1}. Column oven: 40°C. Excitation: 320 nm. Emission: 520 nm.

Calculation

Concentrations of plasma FFA can be calculated by use of the equation:

$$[\text{Plasma FFA}] \text{ in } \mu\text{M} = \frac{(50 \times \text{peak area for FFA})}{(\text{peak area for margaric acid})}$$

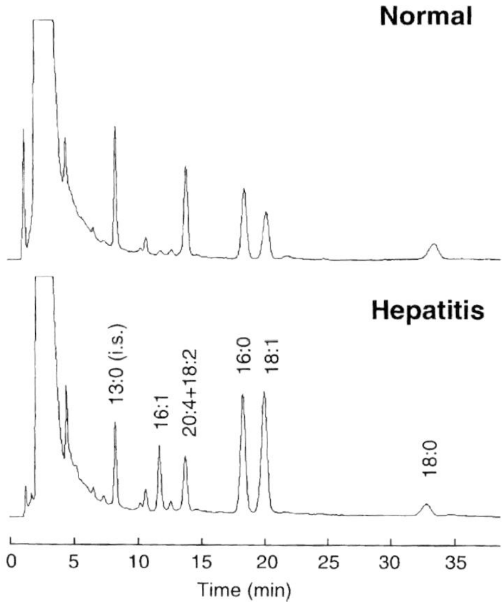

Figure 1. Typical HPLC chromatograms of derivatized human plasma FFA from a normal subject and from a patient with hepatitis. There is a large difference between the content of PUFA and monounsaturated fatty acids.

Comments

Only FFA are measured because fatty acids in phospholipids and neutral lipids are not derivatized.

References

1. Gutteridge, J. M., Quinlan, G. J., and Yamamoto, Y. (1998). Hypothesis: Are fatty acid patterns characteristic of essential fatty acid deficiency indicative of oxidative stress? *Free Radical Res.*, **28**, 109–14.
2. Yamamoto, Y., Nagata, Y., Katsurada, M., Sato, S., and Ohori, Y. (1996). Changes in rat plasma-free fatty acid composition under oxidative stress induced by carbon tetrachloride: Decrease of polyunsaturated fatty acids and increase of palmitoleic acid. *Redox Report*, **2**, 121–5.

38

Phospholipids and their hydroperoxides

YOSHIHIRO YAMAMOTO, YASUHIRO KAMBAYASHI, and SATOSHI YAMASHITA

Introduction

Biomembranes contain phosphatidylcholine, phosphatidylethanolamine, phosphatidylserine, and phosphatidylinositol as major phospholipids. Here we describe the separation of these phospholipids by HPLC, and the simultaneous detection of their primary oxidation products, hydroperoxides, by means of hydroperoxide-specific chemiluminescence detection (1, 2).

Protocol

1. Prepare tissue homogenate. For example, homogenize minced rat liver (0.5 g) in Chelex 100-treated Tris-HCl (5 mM, pH 7.4, 5 mL) containing sucrose (0.25 M) by means of a Teflon homogenizer.
2. Discard solids after centrifugation at 3000 rpm for 10 min.
3. Extract oxidized and unoxidized phospholipids with 2 vol. chloroform–methanol (2:1, *v*/*v*) containing 2,6-di-*tert*-butyl-4-methylphenol (100 μM; to prevent oxidation).
4. Centrifuge at 12000 rpm for 3 min.
5. Analyse aliquots (20 μL) of the organic phase by HPLC equipped with UV and chemiluminescence detection. Column: 250 mm × 4.6 mm i.d., 5 μm, aminopropylsilyl (Supelco) with a 20 mm × 4.6 mm i.d., 5 μm, silica gel guard column (Supelco). Mobile phase: 6:3:1 (*v*/*v*) methanol–*tert*-butyl alcohol–aqueous monobasic sodium phosphate (40 mM) at a flow rate of 1.0 mL min^{-1}. Chemiluminescence reagent: aqueous borate buffer (100 mM; 38.14 g sodium tetraborate decahydrate L^{-1}) is prepared, and the pH is adjusted to 10 with sodium hydroxide. Isoluminol (177.2 mg, final concentration 1 mM) is dissolved in methanol (500 mL) and borate buffer (500 mL), and microperoxidase (5 mg) is added. The

flow rate for the chemiluminescence reagent is 1.5 mL min^{-1}. The UV detector is operated at 210 nm.

6. Analyse aliquots (10 μL) of methanolic phosphatidylcholine hydroperoxide (PC-OOH: standard, 10 μM).

Results and calculations

Fig. 1 depicts the separation and detection of 10 pmol each of PC-OOH, phosphatidylethanolamine hydroperoxide (PE-OOH), phosphatidylserine hydroperoxide (PS-OOH), and phosphatidylinositol hydroperoxide (PI-OOH) (3).

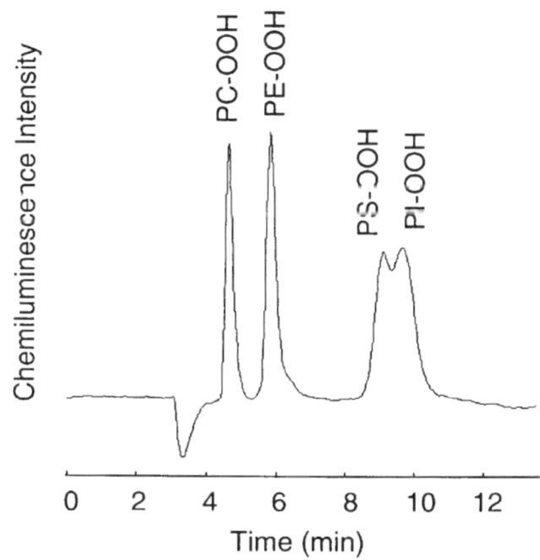

Figure 1. Typical chromatogram of the hydroperoxides.

The concentration (μM) of hydroperoxides in the homogenate can be calculated by use of the equation:

$$[\text{hydroperoxide}] = \frac{[(100/20 \times 3 \times 0.46) \times \text{peak area for hydroperoxide}]}{(\text{peak area for standard})}$$

because the organic and aqueous phases are present in the ratio 46:54 (*v*/*v*), respectively.

Comments

1. Levels of phospholipids can be estimated from the absorption at 210 nm.
2. PC-OOH and PE-OOH were the major oxidation products of rat liver homogenate (3).

References

1. Yamamoto, Y., Brodsky, M. H., Baker, J. C., and Ames, B. N. (1987). Detection and characterization of lipid hydroperoxides at picomole levels by high-performance liquid chromatography. *Anal. Biochem.*, **160**, 7–13.
2. Yamamoto, Y. (1994). Chemiluminescence-based high-performance liquid chromatography assay of lipid hydroperoxides. *Methods Enzymol.*, **233**, 319–24.
3. Kambayashi, Y., Yamashita, S., Niki, E., and Yamamoto, Y. (1997). Oxidation of rat liver phospholipids: Comparison of pathways in homogeneous solution, in liposomal suspension and in whole tissue homogenates. *J. Biochem.*, **121**, 425–31.

39

Cholesterol and its oxides

TERUO MIYAZAWA

Introduction

Several cholesterol oxidation products might play important roles in the development of atherosclerosis (1). This concept is fostered by mounting interest in the role of oxidatively modified lipoproteins and cholesterol in atherogenesis. Cholesterol oxides are widely encountered in foods and biological tissues. Here the analysis of plasma cholesterol oxidation products by high-performance liquid chromatography–mass spectrometry (HPLC–MS) (2, 3) is described.

Protocol

1. Blood is drawn into 10-mL Vacutainer tubes containing EDTA (15% *w/v*, 0.10 mL).
2. Lipids are extracted from plasma (1.0-mL) in 12-mL borosilicate screw cap tubes by addition of chloroform–methanol (2:1, *v/v* 6.0 mL) containing BHT (0.01% *w/v*).
3. Internal standard stock solution (5α-cholestane in toluene, 1.0 mg mL^{-1}, 100 μL) is added to each sample and the tubes are sealed under argon and mixed in a rotary drum for 30 min at room temperature.
4. After centrifugation at 3000 rpm for 15 min, the organic phase is isolated and the aqueous phase is re-extracted with chloroform–methanol (3.0 mL).
5. The pooled organic phases are evaporated to dryness under nitrogen and the residue, dissolved in ethanol or hexane–isopropanol (1:1, *v/v*), is analysed by HPLC–MS. (HPLC–MS analysis was performed by means of a Hewlett–Packard HPLC 1090A chromatograph, a particle-beam interface, and a Hewlett–Packard 5988A mass spectrometer. Compounds were separated on an Altex-Ultrasphere-Si 250 column with 96:4 (*v/v*) hexane–isopropanol as mobile phase at a flow rate of 0.4 mL min^{-1}.)

Results

The principal ions monitored for analysis of cholesterol oxides are listed in Table 1.

TABLE 1. The principal ions monitored for analysis of cholesterol oxides by HPLC–MS with the particle-beam interface.

Compound name	Major ions	Minor ions	Detection limit (ng)
Cholesterol	386, 275	301, 368, 353	0.70
Cholestanetriol	402, 384	369, 303, 331	1.50
Cholesterol-5β,6β-epoxide	402	384, 369, 275	1.20
Cholesterol-7-hydroperoxide	400	367, 382, 287	0.75
7-Ketocholesterol	400	367, 382, 287	0.75
5α-Cholestane	372	–	–
7α-Hydroxycholesterol	384	402, 369	1.0
Cholesta-3,5-diene-7-one	174	382, 367, 398	1.0
Cholesterol-5α,6α-epoxide	402	384, 369, 275	1.20
25-Hydroxycholesterol	402	400	1.0

Comments

Together with HPLC–MS, capillary gas chromatography–mass spectrometry (GC–MS) can also be used for the analysis of cholesterol oxides after derivatization, usually to the trimethylsilyl ethers.

References

1. Peng, S. K., Hu, B., and Morin, R. J. (1992). Effects of cholesterol oxides on atherogenesis. In *Biological Effects of Cholesterol Oxides*, (ed. S. K. Peng and R. J. Morin), pp. 167–89. CRC Press, London.
2. Sevanian, A. and McLeod, L. L. (1987). Cholesterol autoxidation in phospholipid membrane bilayers. *Lipids*, **22**, 627–36.
3. Sevanian, A., Seraglia, R., Traldi, P., Rossato, P., Ursini, F., and Hodis, H. (1994). Analysis of plasma cholesterol oxidation products using gas- and high-performance liquid chromatography / mass spectrometry, *Free Radic. Biol. Med.*, **17**, 397–409.

40

LDL isolation and copper-catalysed oxidation

MASAYUKI YOKODE and TORU KITA

Introduction

A characteristic feature of atherosclerotic lesions is macrophage-derived foam cells, filled with cholesteryl esters, in the intima. Recent studies have shown that macrophages take-up oxidatively-modified low density lipoprotein (LDL) (oxidized LDL) by means of the scavenger receptor and are converted into foam cells. Thus, it is important to understand how oxidized LDL is produced in the body.

Although there is increasing evidence that oxidized LDL occurs in the body, little is known about how the lipoproteins are oxidatively modified. In this protocol we describe a method of preparation of oxidatively modified LDL using a copper salt *ex vivo*. Probucol, a lipid-soluble antioxidant which protects LDL from copper-induced modification, has been shown to inhibit atherosclerosis in hypercholesterolaemic animals. The sensitivity of the LDL towards copper-induced modification is, therefore, likely to correlate with its atherogenicity.

Protocol

Isolation of LDL[1] (1)

1. Add EDTA.Na_2 (pH 7.4, 0.25 M, 7.5 mL) to blood (500 mL).
2. Obtain 80 ml plasma fraction by low-speed centrifugation.
3. Add potassium bromide (density 1.019 g mL^{-1}, 3.81 g)[2].
4. Ultracentrifuge in Beckman Polyallomer centrifuge tubes[3] at 100 000 g for 20 h at 4 °C.
5. Slice tube and discard top fraction. Save bottom fraction (80 mL) and add KBr (5.27 g).
6. Ultracentrifuge in a Beckman Polyallomer centrifuge tube at 100 000 g for 20 h at 4 °C.

Protocol *Continued*

7. Slice tube and save top fraction (10 mL; density 1.040 g mL^{-1}). Add KBr (density 1.063, 0.34 g).
8. Ultracentrifuge in a Beckman Polyallomer centrifuge tube at 100 000 g for 20 h at 4 °C.
9. Slice tube and save top fraction (10 mL).
10. Dialyse against 12 L PBS for 48 h at 4 °C (two changes of 6 L each).
11. Sterilize through a 0.45 μm filter.
12. Determine protein content by the Lowry method.

Preparation of oxidized LDL (2)

1. Dissolve LDL (5 mg) in PBS (1 mL).
2. Add PBS (1 mL) containing $CuSO_4$ (10 μM).
3. Transfer to 3.5 cm sterile plastic culture dish and incubate for up to 48 h at 37 °C.
4. Dialyse against 12 L PBS for 48 h at 4 °C (two changes of 6 L each).
5. Determine protein content by the Lowry method.

Measurement of LDL thiobarbituric acid-reactive substances (TBARS) in oxidized LDL (2)

1. Add LDL (50 μg) to NaCl (150 mM, 1.5 mL) in a 13 × 10 cm culture tube.
2. Add TCA (20% *w/v*, 0.5 mL) and TBA reagent (0.5 mL)[4].
3. Boil at 95 °C for 1 h.
4. Cool with tap water.
5. Add butan-1-ol (2.0 mL) and vortex mix vigorously.
6. Centrifuge at 4000 g for 15 min.
7. Remove top layer and measure fluorescence (excitation at 515 nm, emission at 550 nm).

[1] All procedures should be conducted at 4 °C.
[2] The plasma (initial density; d_i) is adjusted to the final density (d_f) by addition of potassium bromide as indicated by the equation: $X = V \times (d_f - d_0)/(1 - 0.312 \times d_f)$ where X is the mass of potassium bromide (g) and V is the volume of plasma (mL).
[3] Do not use centrifuge brake during deceleration of the ultracentrifuge.
[4] The preparation of this reagent is described on page 291.

References

1. Goldstein, J. L., Basu, S. K., and Brown, M. S. (1983). Receptor-mediated endocytosis of low-density lipoprotein in cultured cells. *Methods Enzymol.*, **98**, 241.

2. Yokode, M., Kita, T., Kikawa, Y., Ogorochi, T., Narumiya, S., and Kawai, C. (1988). Stimulated arachidonate metabolism during foam cell transformation of mouse peritoneal macrophages with oxidized low density lipoprotein. *J. Clin. Invest.*, **81**, 720.

41

Diene conjugation

HIROYUKI SHIMASAKI

Introduction

Lipid peroxidation begins with the abstraction of a hydrogen atom from an unsaturated lipid, and involves the formation and propagation of lipid radicals, uptake of oxygen, and rearrangement of double bonds in the polyunsaturated fatty acid (PUFA). Rearrangement of the double bonds results in the formation of conjugated dienes (lipid hydroperoxides), which are characterized by intense absorption near 233 nm (1).

Protocol

Spectrophotometric detection of diene conjugation is covered in detail elsewhere for plasma (2) and cellular (3) preparations. The method described here is a simple procedure that eliminates the lipid extraction step for direct assay of conjugated diene formation in *in vitro* (or *in vivo*).

1. Into screw-capped tubes (1.3 × 10 cm, Pyrex) place membrane sample (such as oxidized-tissue homogenate, plasma, mitochondria, microsomal suspension, lipoproteins, or PC-liposomes; 0.5 mL) and ethanol–ether (3:1, *v*/*v*, 1.5 mL).
2. Mix on a vortex mixer for 1 min.
3. Centrifuge for 10 min at 3000 rpm (4 °C).
4. Pipette the solution (1.5 mL) into a quartz cuvette.
5. Measure absorbance at 233 nm against an equivalent amount of non-peroxidized sample (control solution) by UV spectrophotometry.

Comments

Lipid hydroperoxides (conjugated dienes) are relatively unstable in biological membranes or plasma, particularly in the presence of transition metals and haemoproteins. The diene conjugation peak might be undetectable in tissue homogenates or plasma freshly obtained by the procedure described here.

Diene conjugation

The absorbance of peroxidized lipids at 233 nm can be obtained as a difference spectrum between a peroxidized sample and an equivalent amount of non-peroxidized control.

References

1. Halliwell, B. and Gutteridge, J. M. C. (1985). *Free Radicals in Biology and Medicine*, p. 162. Oxford University Press.
2. Recknagel, R. O. and Glende, E. A. Jr. (1984). Spectrophotometric detection of lipid conjugated dienes. *Methods Enzymol.* **105**, 331–7.
3. Corongiu, F. P. and Banni, S. (1994). Detection of conjugated dienes by second derivative ultraviolet spectrophotometry. *Methods Enzymol.* **233**, 303–10.

42

Fluorescent products

HIROYUKI SHIMASAKI

Introduction

The reaction of lipid peroxidation products, such as malonaldehyde, with amino groups of proteins, free amino acids, aminophospholipids, or nucleic acid bases, produces fluorescent lipid-peroxidation products known as conjugated Schiff bases with the general structure RN=CHCH=CHNHR′ (*N,N′*-disubstituted 1-amino-3-iminopropenes) (1). When these products are excited at 360 nm the fluorescence maximum is in the region 430–440 nm.

Protocol

Spectrofluorimetric analysis of fluorescent lipid peroxidation products is covered in detail elsewhere (2, 3). The method described here is a simple procedure for detection of lipid-soluble and water-soluble fluorescent products formed during lipid peroxidation of biological membranes.

Lipid-soluble fluorescent products

1. Place membrane sample (e.g. oxidized tissue homogenate, plasma, mitochondria, microsomal suspension, or lipoproteins; 0.5 mL) and ethanol–ether (3:1, *v/v*, 1.5 mL) in screw-capped tubes (1.3 × 10 cm, Pyrex).
2. Mix on a vortex mixer for 1 min.
3. Centrifuge for 10 min at 3000 rpm (4 °C).
4. Wash the sediment at the bottom of tube twice with ethanol–ether (3:1, *v/v*, 2 mL) repeating assay steps 2 and 3.
5. Pipette the solution (1.5 mL) into a 1-mL quartz cuvette.
6. Measure fluorescence (emission) and excitation spectra by spectrofluorimetry. (The fluorescent lipid-peroxidation products have usually an excitation maximum in the region of 355–365 nm and an emission maximum at 430–440 nm.)

Water-soluble fluorescent products

1. Dissolve sediments from assay of lipid-soluble fluorescent products (above) in 2.0 mL 15% SDS–PBS (phosphate-buffered saline, pH 7.4) on a vortex mixer for 5 min.
2. Centrifuge for 10 min at 3000 rpm (room temperature).
3. Pipette the solution (1.5 mL) into a 1-mL quartz cuvette.
4. Measure fluorescence (emission) and excitation spectra by spectro-fluorimetry.

Calculation

The fluorescence intensity of quinine sulfate (1 mg mL^{-1}) in H_2SO_4 solution (0.05 M) is the standard for the relative fluorescence intensity of the sample (2). Quinine sulfate in the solution has a fluorescence maximum at 457 nm when excited at 360 nm.

Comments

Fluorescent products which bind to proteins are insoluble in common organic solvents and their structures (fluorophores) can be disrupted, in part, during extraction of lipids with 2:1 (*v*/*v*) chloroform–methanol. They are, however, stable in 3:1 (*v*/*v*) ethanol–ether, as described elsewhere (4). It should be noted that the lipid extracts from biological membranes and tissue homogenates contain retinol as one of interfering fluorescent compounds. Retinol, a lipid-soluble compound, has a fluorescence maximum at 478 nm when excited at 335 nm (5) and its fluorescence can be removed from the lipid extracts by exposing it to high-intensity ultraviolet light for 30 s (2).

References

1. Tappel, A. L. (1980). *Free radicals in Biology. IV*, (ed. W. A. Pryor), pp. 1–47. Academic Press, NY.
2. Dillard, C. J. and Tappel, A. L. (1984). β Fluorescent damage products of lipid peroxidation. *Methods Enzymol.*, **105**, 337–41.
3. Shimasaki, H. (1994). Assay of fluorescent lipid peroxidation products. *Methods Enzymol.*, **233**, 338–46.
4. Shimasaki, H., Ueta, N., and Privett, O. S. (1982). Covalent binding of peroxidized linoleic acid to protein and amino acids as models for lipofuscin formation. *Lipids*, **17**, 878–83.
5. Shimasaki, H., Ueta, N., and Privett, O. S. (1980). Isolation and analysis of age-related fluorescent substances in rat testes. *Lipids*, **15**, 236–41.

43

Lipid peroxides

ETSUO NIKI

Introduction

Lipid peroxides are major primary products of lipid peroxidation. Measurement of lipid peroxides in *in vitro* experiments or biological samples is often required to assess the extent both of oxidative damage and of changes in antioxidant activity. Various methods have been proposed and used, but each method has both merits and demerits. There still is no single method which is specific, quantitative and sufficiently accurate that can be applied widely. Numerous kinds of lipid peroxide can be formed during lipid peroxidation and these subsequently undergo secondary reactions Hence, measurement of lipid peroxides might not always reflect the extent of lipid peroxidation. There are, furthermore, several types of lipid peroxide, e.g. hydroperoxides, dialkyl peroxides and cyclic peroxides, and it is important to appreciate what is measured by the method employed.

Iodometric method

Introduction

This method is based on the capacity of a wide range of hydroperoxides to react with iodide to liberate iodine. Its advantage is that the reaction is quantitative, exactly stoichiometric, rapid, and reproducible; a disadvantage is that the sensitivity is often not high enough for biological samples. In the presence of excess iodide, iodine gives triiodide I_3^-, which can be measured by titration with thiosulfate, or spectrophotometrically or fluorometrically (1).

Protocol

1. Prepare methanol–acetic acid (2:1 *v/v*) containing KI (6% *w/v*)and EDTA (1 mM) and deoxygenate with a slow stream of inert gas (nitrogen or argon).
2. Place KI solution (2 mL) in a cuvette and purge for 5 min with either nitrogen or argon.

3. Introduce a sample (up to 100 μL) containing hydroperoxide (between 1 and 100 nmol) into the cuvette and leave to react with the KI.
4. Record the absorption of the solution at 358 nm until any further change is small.
5. Calculate the hydroperoxide concentration using the molar extinction coefficient $\varepsilon = 2.97 \times 10^4\ \text{M}^{-1}\ \text{cm}^{-1}$ at 358 nm.

Comments

The measurement is performed in an anaerobic cuvette with continuous monitoring at 358 nm in a UV spectrophotometer. Any cuvette which can be made anaerobic can be used.

TBA assay

Introduction

This TBA method is based on the acid-catalysed decomposition of lipid hydroperoxides to malondialdehyde (MDA), which reacts with thiobarbituric acid (TBA), to form a red chromogen. This is quite a sensitive method, but its major disadvantage is that different lipid hydroperoxides produce MDA in different yields and some potentially significant lipid hydroperoxides, e.g. cholesterol hydroperoxide, are not detected by this method.

Protocol

1. Place TBA reagent[1] (2 mL) and sample (1 mL) in a tube (10 mL) with a cap.
2. Add a solution of BHT[2] in ethanol (1% *v/v*, 50 mM; final BHT concentration 500 μM).
3. Heat at 100 °C in a water bath for 15 min.
4. Cool in ice-cold water.
5. Centrifuge at 3000 rpm for 10 min.
6. Measure the absorbance of supernatant at 535 nm.
7. Calculate the hydroperoxide concentration using the molar absorption coefficient $\varepsilon = 1.56 \times 10^5\ \text{M}^{-1}\text{cm}^{-1}$ at 535 nm.

[1] TBA reagent is prepared by dissolving trichloroacetic acid (15 g) in water and adding TBA (0.375g), HCl (1 M, 25 mL), and a solution of 2,6-di-*tert*-butyl-4-methylphenol (40 mg) in ethanol (22 mL). The solution is finally diluted to 100 mL with distilled water.
[2] Because the sample is treated with TBA reagent at high temperature for 15 min, it is advisable to add an antioxidant to minimize the formation of hydroperoxides during analysis.

Comment

It should be appreciated that different hydroperoxides and aldehydes give different yields of TBA-reactive substances (TBARS), making this method less quantitative.

High-performance liquid chromatography (HPLC)

Introduction

The major advantage of this HPLC method is that, when applied appropriately, different classes of lipid hydroperoxide (for example, hydroperoxides of phosphatidylcholine, phosphatidylethanolamine, phosphatidylserine, phosphatidylinositol, sphingomyelin, cardiolipin, and free fatty acids) can be analysed and quantified separately with high sensitivity. Cholesterol ester hydroperoxide, which has different fatty acid moieties and oxidation products from free cholesterol, can be separately measured. Furthermore, regio, stereo, and even optical isomers can be analysed after reduction and *trans* esterification. The separated hydroperoxides can be measured by chemiluminescence (2, 3), fluorescence (4), and by electrochemical detection. HPLC with post-column detection with diphenyl-1-pyrenylphosphine as a fluorescence reagent (4) is probably the most sensitive method at present.

The post-column HPLC–chemiluminescence method is described briefly in this section. After separation by HPLC, lipid hydroperoxides are decomposed by metal ion-containing enzymes such as microperoxidase and cytochrome-*c* and the resulting lipid alkoxyl radical reacts with isoluminol or luminol, resulting in emission of light. A schematic diagram of the HPLC–chemiluminescence system is shown in Fig. 1.

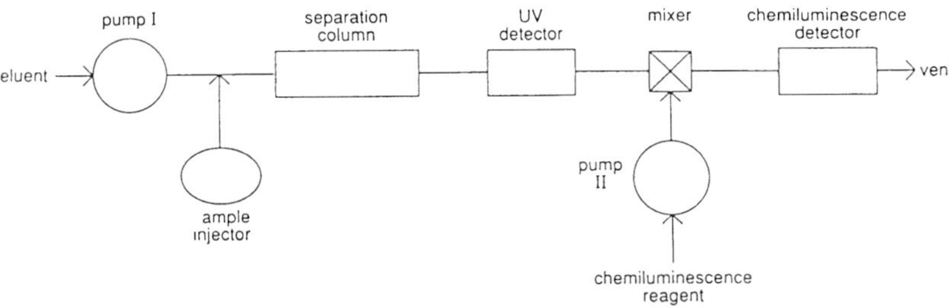

Figure 1. The HPLC–chemiluminescence system.

Protocol

1. Dissolve microperoxidase (250 mg) in water (30 mL) and add methanol (70 mL). This can be stored as a stock solution in an ice box.
2. Dissolve isoluminol (177.2 mg) in a mixture of borate buffer (pH 10.0,

100 mM, 500 mL) and methanol (500 mL) and add microperoxidase solution (2 mL).

3. Place plasma (0.5 mL) in a test tube with a cap.
4. Add methanol (2 mL) and mix with a vortex mixer.
5. Add hexane (10 mL) and shake the test tube vigorously.
6. Centrifuge at 1000 g for 10 min.
7. Take the upper hexane layer (9 mL) and evaporate to dryness by rotary evaporation or under a gas flow.
8. Add methanol–*tert*-butyl alcohol (1:1 *v*/*v*, 0.45 mL) and mix with a vortex mixer.
9. Analyse 50 μL by HPLC–chemiluminescence. Column: 4.6 mm i.d. × 250 mm octylsilyl-LC-8; mobile phase: 95:5 (*v*/*v*) methanol–*tert*-butyl alcohol; flow rate 1.0 mL min^{-1}; flow rate of chemiluminescence reagent: 1.5 mL min^{-1}.

Comments

This HPLC–chemiluminescence method is not specific to hydroperoxides. Ubiquinol gives a positive peak and tocopherols give negative peaks. These antioxidants should be separated from the lipid hydroperoxides on the HPLC column.

An inherent drawback of the HPLC–chemiluminescence method is that only a portion of the light emitted is actually measured; this depends on the type of hydroperoxide, the reaction conditions, and the apparatus used (5). A standard calibration curve should, therefore, be prepared for each hydroperoxide under the analytical conditions employed.

F2-Isoprostanes

Introduction

A series of prostaglandin F2-like compounds termed F2-isoprostanes is produced *in vivo* in man by non-cyclooxygenase free-radical-mediated peroxidation of arachidonic acid. It has been suggested that quantitation of F2-isoprostanes in plasma and urine is a reliable and useful non-invasive approach for assessing lipid peroxidation and oxidative stress *in vivo* (6, 7). Both free F2-isoprostanes and isoprostanes esterified to phospholipids are present *in vivo*. The esterified isoprostanes should first be extracted and then subjected to alkaline hydrolysis to release the free form. The protocol for analysis of urine is given below.

Protocol

1. Acidify urine (1 mL) to pH 3 with HCl (1 M).
2. Add [2H4]PGF2α (0.2–1 μg) as internal standard.
3. Perform solid-phase extraction on C_{18} and silica Sep-Pak cartridges.
4. Perform TLC of F2-isoprostanes as free acids.
5. Convert F2-isoprostanes to pentafluorobenzyl esters by treatment with a mixture of pentafluorobenzyl bromide and *N,N*-diisopropylethylamine.
6. Convert to trimethylsilyl (TMS) ether derivatives
7. Perform quantitative analysis by gas chromatography–negative ion chemical ionization mass spectrometry (GC–NICIMS).

Comment

F2-isoprostanes can be generated when samples containing arachidonate are left standing at room temperature or even during storage at −20 °C. Samples should, therefore, either be processed immediately or stored at −70 °C.

References

1. Jessup, W., Dean, R. T., and Gebicki, J. M. (1994). Iodometric determination of hydroperoxides in lipids and proteins. *Methods Enzymol.*, **233**, 289–303.
2. Yamamoto, Y. (1994). Chemiluminescence-based high-performance liquid chromatography assay of lipid hydroperoxides. *Methods Enzymol.*, **233**, 319–24.
3. Miyazawa, T., Fujimoto, K., Suzuki, T., and Yasuda, K. (1994). Determination of phospholipid hydroperoxides using luminol chemiluminescence–high-performance liquid chromatography. *Methods Enzymol.*, **233**, 324–32.
4. Meguro, H., Akasaka, K., and Ohrui, H. (1990). Determination of hydroperoxides with fluorimetric reagent diphenyl-1-pyrenylphosphine. *Methods Enzymol.*, **186**, 157–61.
5. Noguchi, N. and Niki, E. (1995) Dynamics of free radical formation from the reaction of peroxides with haemoproteins as studied by stopped-flow chemiluminescence. *Free Radical Res.*, **23**, 329–38.
6. Morrow, J. D., Hill, K. E., Burk, R. F., Nammour, T. M., Badr, K. F., and Roberts, L. J. (1990). A series of prostaglandin F2-like compounds are produced *in vivo* in humans by a non-cyclooxygenase free radical-catalysed mechanism. *Proc. Natl. Acad. Sci. USA*, **87**, 9383–7.
7. Morrow, J. D. and Roberts, L. J. (1994). Mass spectrometry of prostanoids: F2-isoprostanes produced by non-cyclooxygenase free radical-catalysed mechanism. *Methods Enzymol.*, **233**, 163–74.

44

Flow-cytometric analysis of lipid peroxides using DCFH-DA

YOSHIRO KAYANOKI

Introduction

Flow cytometry is a method which measures the fluorescence of the cells after they have been labelled with a fluorescent antibody or a fluorescent dye. We here describe how to measure intracellular peroxides of cultured cells after labelling them with a peroxide-sensitive dye, 2′,7′-dichlorofluorescein diacetate (DCFH-DA), using a FACScan (Becton Dickinson, San Jose, CA, USA). DCFH-DA, which is not a fluorescent compound, penetrates cell membranes and enters the cell. There it is deacetylated by esterases and oxidized by hydrogen peroxides or lipid peroxides to produce a fluorescent compound, 2′,7′-dichlorofluorescein (DCF); the amount of DCF produced depends on the amount of intracellular peroxides present. When DCF is exited by 488-nm light green fluorescence is detected at 530 nm (1). Because the relative intensity is measured in this assay, we can only show the fluorescence intensity of stimulated cells relative to that of unstimulated cells. See chapter 56.

Protocol

Labelling of intracellular peroxides by DCFH-DA

1. Culture cells in a culture dish (10 cm).
2. Add freshly prepared DCFH-DA[1,2] (Molecular Probes; 5 mM, 10 μL).
3. Culture for 30 min.
4. Remove medium by aspiration at room temperature.
5. Wash once with PBS(-): Mg^{2+}, Ca^{2+} free phosphate buffered saline.
6. Harvest cells with 0.25% trypsin, 0.02% EDTA PBS(-).
7. Centrifuge at 1000 rpm for 5 min at 4 °C.

Protocol *Continued*

8. Wash with PBS(-) (2 × 1 mL).
9. Suspend cells in PBS(-) (500 mL) and place in a Falcon 2052 tube.

Manipulation of FACScan instrument

1. Switch FACScan vent valve up.

[1] DCFH-DA (1 mg) in dimethyl sulfoxide (413 μL).
[2] DCFH-DA is toxic and must be handled carefully with gloves and mask.

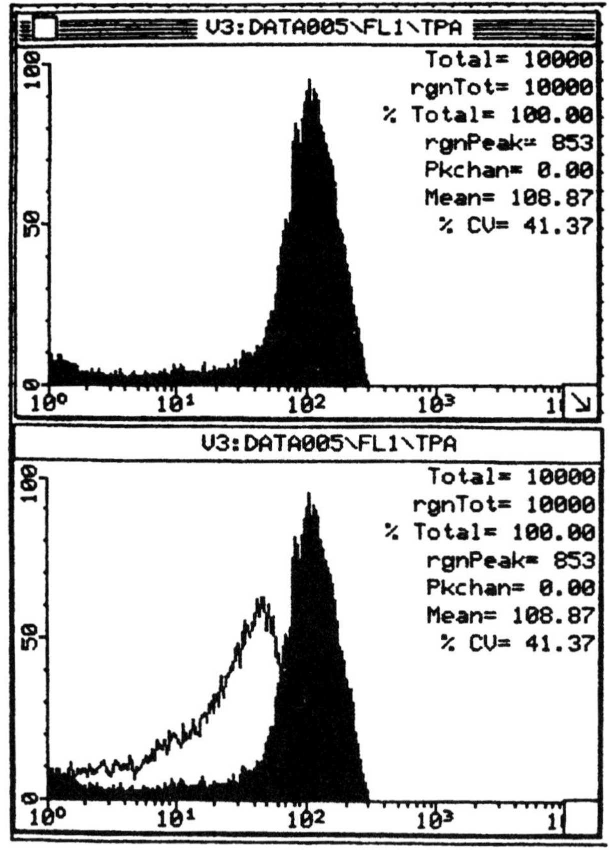

Figure 1. A typical pattern of increase of intracellular peroxides by phorbol 12-myristate 13-acetate.

Reference

1. Bass, D. A., Parce, J. W., Dechatelet, Lr, Szejda, P., Seeds, M. C., and Thomas, M. (1983). Flow cytometric studies of oxidative product formation by neutrophils: A graded response to membrane stimulation. *J. Immunol.*, **130**, 1910–7.

45

Lipid peroxidation models

NORIKO NOGUCHI and ETSUO NIKI

Introduction

Lipids are important targets of active oxygen species and free radicals and the oxidation products of lipids have a variety of biological effects. Accordingly, lipid peroxidation has often been studied *in vitro* to elucidate its mechanism and products and to assess antioxidant activity against it. Subjects which must be considered are how lipid peroxidation is induced in a particular type of reaction medium, and how the extent of lipid peroxidation is measured, because antioxidant efficacy is highly dependent on these factors.

Preparation of liposomes

Liposomes comprising phospholipids with and without free cholesterol are often used as a model of biological membranes (1).

Protocol

1. Prepare chloroform or methanol solution of phosphatidylcholine[1] (3 mM), free cholesterol (1 mM) and lipid-soluble additive, if any in a 50-mL pear-shaped flask.
2. Remove solvent under reduced pressure by rotary evaporation to obtain a thin film on the flask wall.
3. Add an equal volume of phosphate-buffered saline (pH 7.4) and slowly peel off the lipid film by shaking to obtain multilamellar vesicles.
4. If required, small unilamellar vesicles may be obtained by sonicating the above suspension in ice water[2].
5. Oxidize liposomes with a suitable oxidant.

[1] Phospholipids can be obtained from animal tissues, egg yolk, and soybeans. Synthetic phospholipids are also available commercially, but those with polyunsaturated lipids which are used for study of lipid peroxidation usually contain substantial amounts of peroxides.
[2] Some compounds can be oxidized during sonication. If iron or titanium probes are used for sonication, appropriate controls must be considered.

Preparation of liposomes for uptake by macrophages

Protocol

1. Prepare chloroform solution of dioleoyl phosphatidylcholine (DOPC; 25 mM), phosphatidylserine (PS, bovine brain; 2.5 mM), dicetyl phosphate (DCP; 1 mM) and lipid-soluble additive, if any (ex. fluorescent compound).
2. Place DOPC (40 μL), PS (40 μL), and lipid-soluble additive in a 50-mL pear-shaped flask.
3. Remove solvent under reduced pressure by rotary evaporation at 37 °C to obtain a thin film on the flask wall.
4. Add sterilized glucose solution (0.3 M, 1 mL; 37 °C) and peel off the lipid film by shaking to obtain multilamellar vesicles[1].
5. Sonicate the suspensions (20 W, 5 min) in an ice-cold water-bath.
6. Filter with a 0.22 μm Mille-GV filter (Millipore).

[1] Lipid film prepared from phospholipids containing unsaturated fatty acids is easy to peel at room temperature. Polyunsaturated lipid used for lipid peroxidation studies usually contains substantial amounts of peroxides, however.

Preparation of oxidized LDL (see also Chapter 40)

Protocol

1. Isolate LDL ($1.019 < d < 1.063$) from human or animal plasma by centrifugation.
2. Dialyse with PBS containing EDTA (100 μM) at 4 °C for more than 12 h.
3. Determine concentration of LDL by measuring protein concentration of LDL with a BCA kit (Pierce).
4. Place LDL (0.25–4.0 mg protein mL^{-1})[1], PBS (10 mM, pH 7.4), and EDTA (100 μM) in a brown bottle with a cap.
5. Add oxidizing species such as copper salt[2] ($CuCl_2$, $CuSO_4$ 1–50 μM), AAPH (0.5–5 mM), SIN-1 (20 μM–1 mM) or lipoxygenase (~5 nktal mL^{-1}) as PBS solution or MeO-AMVN (50–500 μM) as acetonitrile solution (less than 1% *v*/*v*).
6. Incubate at 37 °C with gentle mixing.
7. Remove an aliquot at appropriate time.

Protocol *Continued*

8. Stop oxidation by cooling on ice.
9. Remove oxidizing species by use of a PD-10 column (Pharmacia).
10. Sterilize by filtration with through a 0.22-μm Mille-GV filter (Millipore).

[1] When higher concentrations of LDL are needed for incubation with macrophages, higher concentrations of oxidized LDL are prepared. Otherwise, the LDL solution can be concentrated by ultrafiltration (Centricon-30, Amicon).
[2] Dialyse with PBS without EDTA at 4°C for more than 12 h before oxidation using a copper salt. Change PBS twice during dialysis to remove EDTA completely.

Preparation of macrophages

Protocol

1. Administer thioglycollate (40.5g L^{-1}, NIH Thioglycollate Broth, DIFCO; 1 mL) intraperitoneally (26-gauge needle) to mouse (ICR mouse, 6–8 weeks)[1].
2. After 3 days kill the mouse by cervical dislocation.
3. Sterilize abdomen with ethanol (70% *v/v*) and incise the abdominal skin.
4. Inject ice-cold RPMI 1640 (5 mL) into the peritoneum (22-gauge needle).
5. Massage abdomen to suspend peritoneal cells in the medium.
6. Collect the medium, as much as possible, by means of a syringe (22-gauge needle) and transfer the contents into a 50-mL Falcon tube (pretreated with FCS and cooled on ice to prevent adhesion of macrophages)[2].
7. Centrifuge at 250 g for 5 min at 4°C.
8. Wash cells 2 or 3 times with ice-cold PBS.
9. Incubate cells on non-fluorescent cover glass (1×10^5 cells cm^{-2}, 35 mm ϕ dish) with RPMI 1640 (5% FCS, 50 units mL^{-1} penicillin, 50 μg mL^{-1} streptomycin; 1 mL) for 1 h (37°C, 5% *v/v* CO_2 incubator).
10. Wash twice with PBS (37°C) to remove non-adherent cells.
11. Incubate (37°C, 5% *v/v* CO_2) with the same medium[3].

[1] Care should be taken not to injure the abdominal organs.
[2] By this method, 10^7 macrophages can be obtained from each mouse.
[3] Macrophages survive for at least 2 weeks.

Addition of liposomes

Protocol

1. Add one-twentieth of the volume of the liposome solution to the macrophages.
2. Incubate (37 °C, 5% *v*/*v* CO_2 incubator) for 6 h.
3. Wash twice with cold PBS.
4. Observe cells by fluorescence microscopy.
5. Dry the cells at room temperature for 10 min.
6. Add Triton X-100 (2%, 50 μL cm^{-2}) and dissolve the cells at room temperature for 20 min.
7. Measure the fluorescent intensity by spectrofluorimetry at an appropriate wavelength.

Addition of fluorescent probe (DiI)-labelled LDL

Protocol

1. Add LDL (2 mg protein mL^{-1}, 3 mL) to lipoprotein-deficient serum (Sigma; 5 mL).
2. Add DiI (3,3′-dioctadecylindocarbocyanine, Molecular Probes; 3 mg mL^{-1}, 100 μL)
3. Incubate at 37 °C overnight.
4. Isolate DiI-labelled LDL by ultracentrifugation.
5. Dialyse against PBS (100 μM EDTA pH 7.4) for 48 h.
6. Adjust LDL concentration to approx. 2 mg protein mL^{-1} PBS[1].
7. Oxidize LDL according to the Section *Preparation of oxidized LDL*, above.
8. Sterilize LDL by filtration through a Mille-GV 0.45 μm filter (Millipore).
9. Add LDL solution (1/200–1/500, *v*/*v*) to macrophage culture medium.
10. Incubate for 24 h (37 °C, 5% *v*/*v* CO_2).
11. Wash five times with PBS containing BSA (2 mg mL^{-1}) and once with PBS (pH 7.4).
12. Fix cells with 3% formalin in PBS (pH 7.4) at room temperature for 30 min.

Protocol *Continued*

13. Observe cells by microscopy with epifluorescent illumination and a rhodamine filter package.

[1] The DiI content of isolated LDL was determined from the fluorescence of an aliquot of LDL extracted with methanol, (excitation, 520 nm; emission, 570 nm) (2).

References

1. Noguchi, N., Gotoh, N., and Niki, E. (1993). Dynamics of the oxidation of low density lipoprotein induced by free radicals. *Biochim. Biophys. Acta*, **1168**, 348–57.
2. Pitas, R. E., Innerarity, T. L., Weinstein, J. N., and Mahley, R. W. (1981). Acetoacetylated lipoproteins used to distinguish fibroblasts from macrophages *in vitro* by fluorescence microscopy. *Arteriosclerosis*, **1**, 177–85.

46

4-Hydroxy-2-nonenal in plasma

G. J. QUINLAN and J. M. C. GUTTERIDGE

Introduction

4-Hydroxy-2-nonenal (HNE) is a peroxidation product of n-6 polyunsaturated fatty acids, which probably requires traces of catalytic iron for its formation.

Spectrophotometric, thin-layer chromatographic, and high-performance liquid chromatographic techniques for determination of HNE in fluids and cellular preparations are covered in detail elsewhere (1, 2). Here we describe a GC–MS method for use with plasma. HNE reacts with proteins to form stable complexes. This technique measures only 'free' HNE in plasma.

Protocol

1. Place freshly taken plasma (1.0 mL), BHT (2.0 mM, 0.010 mL), and nonanoic acid (internal standard; 0.006 mL) in new clean glass tubes.
2. Mix and divide into equal parts 'A' and 'B'.
3. Add chloroform–methanol (2:1 *v*/*v*, 4.0 mL) and NaCl (0.15 M, 0.4 mL) to each tube.
4. Purge with oxygen-free nitrogen for 2 min, cap, and vortex mix for 30 s.
5. Centrifuge at 5000 rpm for 6 min at 4°C.
6. Remove lower (chloroform) layer, and store at 4°C under N_2.
7. Extract the remaining (upper) residue with chloroform–methanol (2:1 *v*/*v*, 2 mL) by vortex mixing for 30 s.
8. Centrifuge at 5000 rpm for 6 min at 4°C.
9. Collect the lower phase and combine with previous extract.
10. Evaporate combined extracts to dryness under N_2 at 25°C.
11. Store at −20°C for less than 1 week.
12. Derivatize the samples:
    (a) Add dry acetone (0.2 mL) and bis(trimethylsilyl)trifluoroacetamide (BSTFA) containing 1% (*w*/*v*) trimethylchlorosilane (0.2 mL) to the dried extracts.

Protocol *Continued*

(b) Cap tubes and leave for 60 min at 25°C

(c) Centrifuge at 6000 rpm for 5 min to deposit residues.

13. Analyse by GC–MS:

The HNE-TMS derivatives are injected (2 μL) by splitless injection (purge time 5.5 min) on to a 50 m × 0.25 mm i.d. wall-coated open tubular 100% cyanopropylpolysiloxane column with helium as carrier gas at a column head pressure of 10 psig. The column temperature is maintained at 65°C for 3 min after injection then ramped in stages to 230°C over 25 min. Detection is by positive-ion electron-impact mass spectrometry with selective ion monitoring. Retention times of characteristic ions are used for identification.

Calculation

Selected ions characteristic of HNE-TMS are m/z 129, m/z 157, and m/z 199; for nonanoic acid-TMS the ions are m/z 117, m/z 129, and m/z 215. Quantitation is achieved by preparing a standard curve relating to m/z 117 for nonanoic acid-TMS and m/z 157 for HNE-TMS.

Results

A typical mass spectrum is shown in Fig. 1.

References

1. Rice-Evans, C. A., Diplock, A. T., and Symons, M. C. R. (1991). *Techniques in Free Radical Research*, pp. 155–74. Elsevier: Amsterdam.
2. Punchard, N. A. and Kelly, F. J. (ed.) (1996). *Free Radicals: A Practical Approach*, pp. 133–145. IRL Press, Oxford.
3. Esterbauer, H., Schaur, R. J., and Zollner, H. (1991). Chemistry and biochemistry of 4-hydroxynonenal, malonaldehyde and related aldehydes. *Free Rad. Biol. Med.*, **11**, 81–128.

(a)

(HNE) $\xrightarrow[\text{DERIVATISATION}]{\text{BSTFA}}$ (HNE-TMS)

C_5H_{11}, OH, O, H → C_5H_{11}, O-TMS, O-TMS, H

F W = 156.22 (156)

Molecular ion (M^+) 301 amu

$156 - 1H^+ + (2 \times TMS)$

$156 - 1 + (2 \times 73) = 301$

TMS (Trimethylsilyl) = $-Si(CH_3)_3$ (m/z 73)

Selective ions.

m/z 199 = (M^+) - CHOTMS (m/z 102)

m/z 157 = (M^+) - TMS (m/z 73) - $CH_3(CH_2)_5$ (m/z 71)

m/z 129 = (M^+) - C_5H_{11}CHO-TMS (m/z 173)

= $128 + H^+ = 129$.

(b)

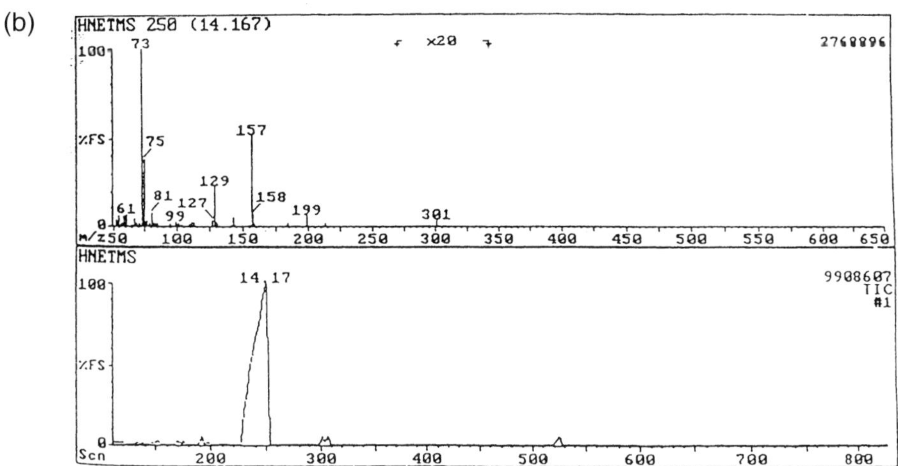

Figure 1. (a) The structures of HNE and its TMS derivative, and probable mechanisms for formation both of the ions used for selected-ion monitoring and of the molecular ion. (b) The total-ion chromatogram and mass spectrum of the HNE-TMS standard displayed clearly show the selected ions and the molecular ion. The mass-to-charge ratio (*m/z*) is given in atomic mass units (amu).

47

Total plasma iron

J. M. C. GUTTERIDGE and G. J. QUINLAN

Introduction

Because iron is an important catalyst of free-radical reactions, the body takes great care to sequester it in non- or poorly-reactive forms. Three major forms of iron are found in plasma—haem iron such as haemoglobin released from red blood cells, non-haem iron such as transferrin and ferritin, and low molecular mass iron (LMrFe) complexed in forms such as iron–citrate. The last occur only when transferrin is fully iron-saturated (see Chapter 49).

Total plasma iron, which includes haem iron, non-haem iron, and LMrFe (if present), can be measured by atomic absorption spectroscopy (AAS). In the absence of haemolysis total plasma iron will be close to total non-haem iron.

Protocol

Adequate methodology for use of AAS is given elsewhere (1). Additional information relating to the use of a graphite furnace, which requires considerably less sample, is given here.

1. Set the iron wavelength to 248.3 nm.
2. Prepare samples and standards in nitric acid (0.2% *v*/*v*) with magnesium nitrate (0.5% *w*/*v*) added as matrix modifier.
3. Use pyrocoated graphite tubes with the L'vov platform.
4. Increase stage temperature to 1400 °C (pre-treatment temperature).
5. Use an atomization temperature of 2400 °C.

Calculation

Prepare a ferric salt standard curve by analysis of standards dissolved in nitric acid (0.2% *v*/*v*) and magnesium nitrate (0.5% *w*/*v*) covering the concentration range required.

Reference

1. Rice-Evans, C. A., Diplock, A. T., and Symons, M. C. R. (1991). *Techniques in Free Radical Research*, pp. 109–110. Elsevier, Amsterdam.

48

Total plasma non-haem iron

J. M. C. GUTTERIDGE and S. MUMBY

Introduction

Assays for non-haem iron are performed at acid pH which enables release of iron from proteins such as transferrin, lactoferrin, and ferritin. The released ferric ions are reduced to the ferrous state and complexed with a ferroin-type chromogenic chelator such as ferrozine.

Protocol

1. Place plasma or ferric chloride standard (0.2 mL) and HCl (0.15 M, 0.3 mL) in new clean plastic Eppendorf tubes.
2. Mix and leave for 60 min at room temperature.
3. Add trichloroacetic acid (15% *w/v*, 0.3 mL), mix, centrifuge at 6000 rpm for 5 min and transfer supernatant (0.4 mL) to a new clean plastic tube.
4. Add buffered ascorbate (0.58 mL)[1].
5. Mix well and add ferrozine (3 mg mL^{-1}, 0.03 mL). Mix and leave for 15 min at 37 °C.
6. Read absorbance at 562 and 700 nm against blank (no plasma).

[1] Prepare Solution A by mixing saturated ammonium acetate (0.01 mL) with Chelex resin-treated water (0.57 mL). Prepare Solution B by dissolving sodium ascorbate (0.0345 g) in Solution A. Mix Solution B (0.5 mL) and Solution A (4.5 mL).

Calculation

The sample concentration is given by the formula:

$$(A_{562}\text{ test} - A_{562}\text{ blank})/(A_{562}\text{ iron standard} - A_{562}\text{ blank}) \times \text{concentration of iron standard}$$

If turbidity (slight) is present read the absorbance at 700 nm and subtract this value from the A_{562} reading.

Comments

The normal plasma non-haem iron value is approx. 15.7 ± 6.05 mM. Set up a standard curve to cover appropriate range.

References

1. Carter, P. (1971). Spectrophotometric determination of serum iron at the submicrogram level with a new reagent (Femozine). *Anal. Biochem.*, **40**, 450–8.
2. Rice-Evans, C. A., Diplock, A. T., and Symons, M. C. R. (1991). *Techniques in Free Radical Research*, pp. 109–110. Elsevier, Amsterdam.

49

Plasma unsaturated iron-binding capacity and percent saturation of transferrin

J. M. C. GUTTERIDGE and S. MUMBY

Introduction

The unsaturated iron-binding capacity (UIBC) of plasma is a measure of the available capacity of plasma to bind iron. Total iron-binding capacity (TIBC) is a measure of all the iron bound to transferrin when it has been fully saturated by adding iron.

Protocol

1. Combine plasma (0.025 mL), ferric chloride (89 μM, 0.050 mL), and Tris buffer (0.5 M, pH 8.1, 0.550 mL).
2. Mix well.
3. Add water (0.035 mL) to prepare the blank solution and water (0.025 mL) plus ferrozine (3 mg mL^{-1} 0.01 mL) to prepare the test solution.
4. Incubate at 37 °C for 10 min.
5. Read absorbance at 562 nm.

Calculation of unsaturated iron-binding capacity

A reagent blank (no iron, no sample) and iron standards are included (see Chapter 48).

$$\text{Concentration of iron standard } (\mu M) - \frac{\text{Abs test} - \text{Abs test blank} - \text{Abs blank} \times \text{Concentration iron standard } (\mu M)}{\text{Abs iron standard} - \text{Abs blank}} = \text{UIBC } \mu M$$

Calculation of percentage saturation of transferrin

% Saturation of transferrin = [Total plasma non-haem iron (TPNHI) − UIBC] × 100/TPNHI

Comments

At neutral pH iron tightly bound to transferrin will not react with ferrozine. Normal plasma has the capacity to bind 40–70 mmol iron. Normal human plasma transferrin is 25–30% saturated with iron.

50

Plasma transferrin

J. M. C. GUTTERIDGE and S. MUMBY

Introduction

Transferrin binds 2 moles ferric iron per mole protein. It can be measured by use of radial immunodiffusion plates. The amount in normal human plasma is 3–4 g L^{-1}.

Protocol

1. Open plates and leave to dry for 5 min.
2. Add plasma (0.05 mL) to each well.
3. Include human transferrin standards.
4. Close plates and leave at room temperature for 48 h.
5. Carefully measure the diameter of each precipitation ring around the well (opaque–white zone).
6. Read the concentration of transferrin from a calibration standard curve.

Calculation

The *Mr* of transferrin is approx. 76 500, so to convert μg L^{-1} to μM divide by 76 500. When fully saturated, transferrin binds 2 moles iron per mole protein; the saturated iron value is, therefore, 2 × μmol protein. This information, with the plasma non-haem iron value, can be used to calculate the percentage saturation of transferrin.

51

Low molecular mass chelatable iron: Reactive iron species (RIS)

J. M. C. GUTTERIDGE and S. MUMBY

Introduction

Normal plasma does not contain chelatable low molecular mass (LMr) forms of iron, because the transferrin has considerable iron-binding capacity (Chapter 49). When, however, transferrin becomes fully iron-saturated during situations of iron overload, LMrFe can be found in plasma and other extracellular fluids. Such forms of iron (often redox active) can be detected and measured in the following ways.

The bleomycin assay

Bleomycin chelates iron and binds to DNA. Redox cycling of iron on DNA releases malondialdehyde fragments from the deoxyribose sugar, which can be measured and related to the amount of iron present.

Protocol

1. Into new clean plastic tubes (heat resistant) place: calf thymus DNA (1 mg mL^{-1}, 0.40 mL) in NaCl (0.15 M); bleomycin sulfate (1.5 units mL^{-1}, 0.02 mL); plasma or ferric chloride standards (0.02 mL); magnesium chloride (50 mM, 0.10 mL); Tris buffer (pH 7.4, 1.0 M, 0.10 mL); and freshly prepared sodium ascorbate (7.5 mM, 0.05 mL).
2. Incubate for 30 min at 37 °C.
3. Add HCl (25% *v*/*v*, 0.5 mL) and thiobarbituric acid (1% *w*/*v*, 0.5 mL).
4. Heat for 10 min at 100 °C.
5. Cool, add butan-1-ol (1.5 mL), and vortex mix for 2 min.
6. Centrifuge at 6000 rpm for 5 min.
7. Measure absorbance (Abs) of upper organic phase at 532 nm against a reagent blank (no plasma).

Calculation

Prepare a standard curve between 0.5 and 10 μM with ferric chloride and calculate bleomycin iron values from this.

$$(\text{Abs test} - \text{Abs blank})/(\text{Abs standard} - \text{Abs blank}) \times \text{Concentration of iron standard} = \mu\text{M}$$

Comments

Bleomycin is a cytotoxic agent.

All reagents and chemicals contain adventitious iron which must be removed by treatment with Chelex resin and/or by dialysis (2). DNA is prone to microbial degradation, prepare in pyrogen-free water. Plasma containing no bleomycin-detectable iron will have a lower A_{532} value than the blank, because of the iron-binding properties of transferrin it contains.

The aconitase assay

The iron–sulfur cluster of aconitase is activated by ferrous ions to give an enzyme that catalyses the conversion of citrate to isocitrate. In the presence of isocitrate dehydrogenase and $NADP^+$, isocitrate is converted to 2-oxoglutarate, CO_2 and NADPH. The rate of change of A_{340} is related to the available iron that activates aconitase.

Protocol

1. Into new clean plastic microcuvettes place PIPES buffer (pH 7.4, 0.6 M, 0.37 mL); sodium citrate (50 mM, 0.10 mL); plasma or ferric chloride standards (0.05 mL); sodium azide (50 mM, 0.01 mL); tricarballyic acid (100 mM, 0.02 mL); magnesium sulfate (0.3 M, 0.02 mL); NADP (20 mM, 0.02 mL); isocitrate dehydrogenase (50 units mL^{-1}, 0.01 mL); aconitase (iron-requiring form; 126 units mL^{-1}, 0.005 mL); and sodium ascorbate (7.5 mM, 0.05 mL).
2. Place in spectrophotometer cell at 37 °C and record for 45 min.

Calculation

The rate of change of absorbance at 340 nm calculated, and related to iron standards similarly treated.

Comments

The iron-requiring form of aconitase (5) should be prepared for the following reaction to take place:

$$[3Fe\text{–}4S]^{+} + Fe^{2+} + e^{-} \rightarrow [4Fe\text{–}4S]^{2+}$$

Remove adventitious iron from reagents as described for the bleomycin assay.

References

1. Gutteridge, J. M. C., Rowley, D. A., and Halliwell, B. (1981). Superoxide-dependent formation of hydroxyl radicals in the presence of iron salts. Detection of 'free' iron in biological systems by using bleomycin-dependent degradation of DNA. *Biochem. J.*, **199**, 263–5.
2. Gutteridge, J. M. C. and Hou, Y. (1986). Iron complexes and their reactivity in the bleomycin assay for radical-promoting loosely-bound iron. *Free Rad. Res. Commun.*, **2**, 143–51.
3. Gutteridge, J. M. C., Mumby, S., Kiozumi, M., and Taniguchi, N. (1996). 'Free' iron in neonatal plasma activates aconitase: evidence for biologically reactive iron. *Biochem. Biophys. Res. Commun.*, **229**, 806–9.
4. Mumby, S., Koizumi, M., Taniguchi, N., and Gutteridge, J. M. C. (1998). Reactive iron species in biological fluids activate the iron-sulfur cluster of aconitase. *Biochim Biophys Acta*, **1380** 1., 102–8.
5. Fujita, T., Hamasaki, H., Fwukata, C., and Nonobe, M. (1994). New enzymatic assay of iron in serum. *Clin. Chem.*, **40**, 763–7.

52

Speciation of ferrous ions in biological fluids

J. M. C. GUTTERIDGE

Introduction

When plasma transferrin is fully iron-loaded and LMrFe is present (see Chapter 51) such iron would be kept in the ferric state by the ferroxidase activity of caeruloplasmin (see Chapter 31). If, however, ascorbate concentrations are high and caeruloplasmin levels low, plasma ferroxidase activity will be inhibited and LMrFe will be converted to the ferrous form.

Protocols

Use the bleomycin assay without adding ascorbate. The variability of ascorbate in the sample means that ferrous ions can be speciated but not quantitated.
Use the aconitase assay without adding ascorbate. This assay enables ferrous ions to be speciated and quantitated.

References

1. Berger, H. M., Mumby, S., and Gutteridge, J. M. C. (1995). Ferrous ions detected in iron-overloaded cord blood plasma from preterm and term babies: implications for oxidative stress. *Free Radical Res.*, **22**, 555–9.
2. Mumby, S., Koizumi, M., Taniguchi, N., and Gutteridge, J. M. C. (1998). Reactive iron species in biological fluids activate the iron–sulfur cluster of aconitase. *Biochim. Biophys. Acta*, **1380**, 102–8.

53

Tissue low molecular mass iron

J. M. C. GUTTERIDGE

Introduction

Unlike extracellular fluids, cells usually contain a low molecular mass (LMr) pool of iron, which is required for the synthesis of iron-containing proteins and for the synthesis of DNA. This pool of iron is a target for iron chelators, and can be measured with the fluorescent probe calcein.

Protocol

Calcein (CA) binds ferrous ions rapidly and stoichiometrically forming a fluorescence-quenched complex.

Cells are loaded with acetomethoxycalcein to an intracellular concentration of 10 μM, and the LMr iron pool is measured by assessing (a) the total intracellular concentration of calcein in loaded cells $[CA]_t$ estimated from fluorimetric measurement of CA in a cell suspension (number of cells and cell volume), and (b) the intracellular calcein-iron concentration [CA-Fe], which is the concentration of CA bound to metals (>95% iron), assessed from the relative rise in fluorescence at λ_{ex} 486 nm (range 480–490 nm), λ_{em} 517 nm (range 510–517 nm) obtained by adding high affinity chelators.

Calculations

The LMr cellular iron pool is derived from determination of $[CA]_t$ and [CA Fe], and from calculation of [Fe] by application of the mass law equation using the K_d value of [CA-Fe].

Comments

A value in the range 0.2–1.5 μM is obtained by use of this technique to measure the LMrFe cellular pool in resting erythroid and myeloid cells.

References

1. Breuer, W., Epsztejn, S., and Cabantchik, Z. I. (1995). Iron acquired from transferrin by K562 cells is delivered into a cytoplasmic pool of chelatable iron(II). *J. Biol. Chem.*, **270**, 24209–15.
2. Epsztejn, S., Kakhlon, O., Glickstein H. *et al.* (1997). Fluorescence analysis of the labile iron pool of mammalian cells. *Anal. Biochem.*, **248**, 31–40.
3. Cabantchik, Z. I., Glickstein, H., Milgram, P. *et al.* (1996). A fluorescence assay for assessing chelation of intracellular iron in a membrane model system and mammalian cells. *Anal. Biochem.*, **233**, 221–7.

54

Total plasma copper

J. M. C. GUTTERIDGE and G. J. QUINLAN

Introduction

An average adult human contains approx. 80 mg copper. The plasma concentration is approx. 17 mM, most of which (>98%) is associated with caeruloplasmin (see Chapter 31), which contains 6–7 copper ions per molecule of protein (*Mr* 132 000). This chapter describes the determination of total plasma copper by AAS with a graphite furnace.

Protocol

1. Set copper wavelength 324.8 nm.
2. Prepare samples and standards in nitric acid (0.2% *v/v*) containing magnesium nitrate (0.5% *w/v*) as matrix modifier.
3. Use pyrocoated graphite tubes and the L'vov platform.
4. Increase stage temperature to 1200 °C (pre-treatment temperature).
5. Use an atomization temperature of 2300 °C.

Calculations

Construct a suitable standard curve by use of cupric salt standards prepared in dilute nitric acid (0.2% *v/v*) containing magnesium nitrate (0.5% *w/v*).

55

Low molecular mass chelatable copper: Reactive copper species (RCS)

J. M. C. GUTTERIDGE

Introduction

Normal freshly taken human plasma does not contain chelatable forms of copper. Storage or mishandling of plasma does, however, lead to the formation of chelatable forms of copper by the proteolytic (rapid) degradation of caeruloplasmin. Redox active chelatable forms of copper can be detected and measured in biological fluids by use of the phenanthroline assay. Phenanthroline chelates copper and binds to DNA. Redox cycling of copper on DNA releases malondialdehyde from the deoxyribose sugar of DNA; this can be measured and related to the amount of copper present.

Protocol

1. Into new clean plastic tubes place: a solution of calf thymus DNA (1 mg mL^{-1}, 0.40 mL) in NaCl (0.15 M); 1,10-phenanthroline (0.364 mg mL^{-1}, 0.10 mL); sodium azide (100 mM, 0.05 mL); plasma or copper standard (0.05 mL); sodium phosphate buffer (pH 6.4, 0.25 M, 0.20 mL); and mercaptoethanol (0.4% *v/v*, 0.10 mL).
2. Incubate for 60 min at 37 °C.
3. Add EDTA (0.1 M, 0.1 mL); trichloroacetic acid (28% *w/v*, 0.5 mL); and thiobarbituric acid (1%, *w/v*, 0.5 mL).
4. Heat for 10 min at 100 °C.
5. Cool, add butan-1-ol (3.0 mL), and vortex mix for 2 min.
6. Centrifuge at 6000 rpm for 5 min.
7. Measure the fluorescence of the organic phase in a fluorimeter (λ_{ex} 532 nm, λ_{em} 553 nm).

Calculation

Include blanks (no sample, no copper standard), and standards (cupric chloride 0.5–10 μM).

$$(\text{Fluor. test} - \text{Fluor. blank})/(\text{Fluor. std} - \text{Fluor. blank}) \times \text{Concn copper std} = \text{Concn sample, } \mu\text{M}$$

Comments

Treat all reagents with Chelex resin (resin sealed in dialysis tubing and placed in reagent). Ferrous ions react with 1,10-phenanthroline to give a pink colour similar to that of the malondialdehyde thiobarbituric acid complex. This complex is not fluorescent, however, and will not be measured.

Loosely-bound (chelatable) copper has been detected in plasma from patients with Wilson's disease and patients with fulminant hepatic failure.

References

1. Gutteridge, J. M. C. (1984). Copper-phenanthroline-induced site-specific oxygen radical damage to DNA. Detection of loosely-bound trace copper in biological fluids. *Biochem. J.*, **218**, 983–5.
2. Gutteridge, J. M. C., Winyard, P., Blake, D. R. *et al.* (1985). The behavior of caeruloplasmin in stored human extracellular fluids in relation to ferroxidase II activity, lipid peroxidation and phenanthroline-detectable copper. *Biochem. J.*, **230**, 517–23.

Part 2

Gene Analyses

56

Confocal analysis for oxidized states in cells

KIYOSHI NOSE

Introduction

The redox state in cells changes significantly after stimulation with growth factors, and changes in environmental conditions. Such changes cause modulation of the activities of signalling molecules and gene expression, leading to cellular responses to such changes.

Oxidized states in cells can be measured in situ by use of a fluorescent dye, dichlorofluorescein diacetate, and a confocal microscope. Gene expression of early response genes, such as *c-fos* and *egr-1*, is one of the means for estimating nuclear events after oxidative stress (Fig. 1).

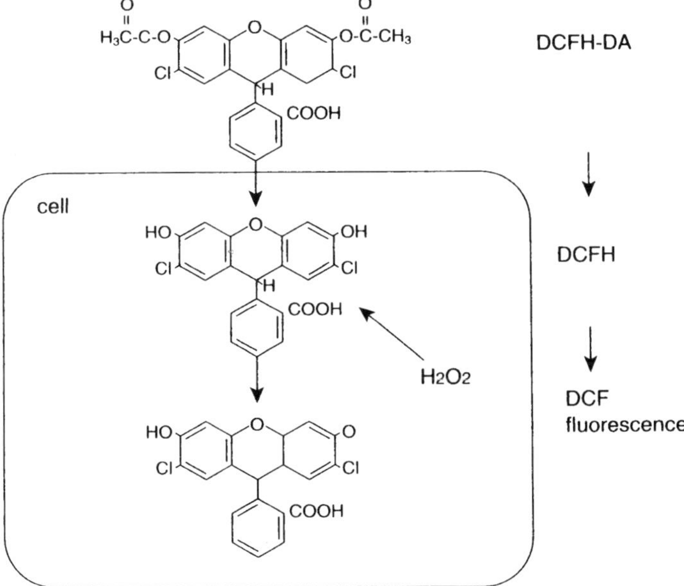

Figure 1. Oxidation of DCFH by oxidants.

Measurement of oxidized states with a confocal microscope

Protocol

1. Place culture cells in 60 mm plastic dishes (cells in pre-confluent state).
2. Stimulate (e.g. TNFα 10 ng mL^{-1} for 1 h) in a CO_2 incubator.
3. Rinse once with Hank's balanced salt solution (HBSS).
4. Add DCFH-DA solution[1] (5 mM; 2 mL/dish).
5. Incubate for 5 min[2] in a CO_2 incubator, and measure fluorescence under a confocal laser-scanning microscope[3,4] (e.g. BioRad, MRC500), with excitation at 513 nm and emission at 535 nm

[1] Stock solution: 1 mM in dimethylsulfoxide. Dilute before use in HBSS. DCFH-DA is a product of Eastman–Kodak.
[2] Incubation time of cells with DCFH-DA should be kept constant to ensure reproducible results.
[3] The confocal microscope should be adjusted to ensure clear cellular image (laser axis and focus).
[4] Fluorescent of DCFH is quenched by light. The samples should be kept in the dark, and the fluorescence measured promptly.

Comments

For floating cells such as leukaemia cells, fluorescent intensity can be measured by FACS.
The relationship between relative fluorescent intensity and the concentration of oxidants can be determined by use of cells treated with known concentrations of oxidants.

Expression of early response genes

Protocol

1. Treat cells with active oxygen species[1] at different concentrations for 10 to 120 min.[2]
2. Extract total RNA by the conventional guanidium/hot phenol method, and determine mRNA levels of genes by Northern blot hybridization.

[1] For production of superoxide anion use a stock solution containing 20 mM xanthine and 1–15 munits mL^{-1} xanthine oxidase (Sigma Chemicals). Add 1 part to 1000 parts (*v*/*v*) of cells. For production of H_2O_2 use a stock solution containing 20 mM glucose and 1–15 munits mL^{-1} glucose oxidase. Add 1 part to 1000 parts (*v*/*v*) of cells.
[2] Effects of oxidants vary significantly for different concentrations of cells. The number of cells and culture conditions should be kept constant in the experiments.

Comments

It is better to determine cell viability under treatment with oxidants by colony formation or by the dye-exclusion test.

References

1. Zamzami, N., Marchetti, P., Castedo, M., Zanin, C., Vayssiere, J. L., Petit, P. X., Kroemer, G. (1995). Reduction in mitochondrial potential constitutes an early irreversible step of programmed lymphocyte death in vivo. *J. Exp. Med.*, **181**, 1661–72.
2. Ohba, M., Shibanuma, M., Kuroki, T., Nose, K. (1994). Production of hydrogen peroxide by transforming growth factor beta 1 and its involvement in induction of *egr-1* gene in mouse osteoblastic cells. *J. Cell Biol.*, **126**, 1079–88.
3. Szejda, P., Parce, J. W., Seeds, M. S., Bass, D. A. (1984). Flow cytometric quantitation of oxidative product formation by polymorphonuclear leukocytes during phagocytosis. *J. Immunol.*, **133**, 3303–7.

57

Detection of nitric oxide synthase (NOS) expression: preparation of RNA

JUNICHI FUJII

Introduction

Nitric oxide (NO) is produced by nitric oxide synthase (NOS), L-arginine, and molecular oxygen as substrates. Three NOS isozymes, NOS I, also called brain NOS or bNOS, is expressed mainly in brain and muscle cells and has Ca^{2+}/calmodulin-dependent activity. NOS II, also called iNOS or macNOS, is induced in a variety of cells by stimuli such as interleukin-1, interferon-γ, and lipopolysaccharide. Constitutive expression in human tissues is reported. NOS III, also called ecNOS, is expressed in endothelial cells and has Ca^{2+}/calmodulin-dependent activity. They share about 50–60% homology and commonly have the same cofactor binding sites. A method is described here for preparing RNA suitable for Northern blotting and RT-PCR analysis.

Protocol

1. Place cells in 10-cm dish.
2. Wash with PBS(–) (Ca^{2+}-free PBS) (5 mL).
3. Add extraction solution (800 μL).[1]
4. Transfer the lysate to a 1.5-mL microcentrifuge tube.
5. Add chloroform–isoamyl alcohol (49:1, 80 μL).
6. Vortex-mix vigorously for 30 s.
7. Place on ice for 15 min.
8. Centrifuge at 15 000 rpm for 15 min at 4 °C.
9. Transfer the upper phase to a new 1.5-mL microcentrifuge tube.
10. Add an equal volume of 2-propanol.
11. Mix by inversion.
12. Cool for >1 h at –30 °C.

13. Centrifuge at 15000 rpm for 15 min at 4°C.
14. Wash the precipitate twice with EtOH (75% *v*/*v*).
15. Dry quickly.
16. Dissolve the precipitate in sterile, DEPC-treated water (40–100 μL).
17. Dilute the RNA 50-fold and quantitate the RNA by measuring absorbance at 260 nm (1 OD_{260} = 40 mg mL^{-1}).
18. Analyse mRNA by Northern hybridization or RT-PCR[2].

[1] The extraction solution (1), which should be kept at 4°C, is prepared by mixing solution D (10 mL), NaOAc (pH 4.0, 2 M, 1 mL, water-saturated phenol (10 mL), and 2-mercaptoethanol (2 mL). Solution D, which can be kept at room temperature for 6 months, is prepared by mixing guanidine thiocyanate (4 M), sodium citrate (pH 7.0, 25 mM), and sarcoxyl (0.5%).
[2] Northern blot analysis (2) and RT-PCR (3) can be performed by conventional methods. PCR primers for detection of human NOS isozymes by PCR are:
NOS I: TTCCGAAGCTTCTGGCAACA (base 4208–4676, sense) and ACTCAGATCTAAGGCGGTTG (base 4676–4689, antisense) amplifying a 469 bp product
NOS II: GCAAGCCCAAGGTCTATGTT (base 3148–3167 sense) and TGTCCTTCTTCGCCTCGTAA (base 3398–3167 antisense) amplifying a 270 bp product
NOS III: GAAGAGGAAGGAGTCCAGTA (base 1930–1949 sense) and GACTTGCTGCTTTGCAGGTT (base 2348–2367 antisense), amplifying a 438 bp product

Results

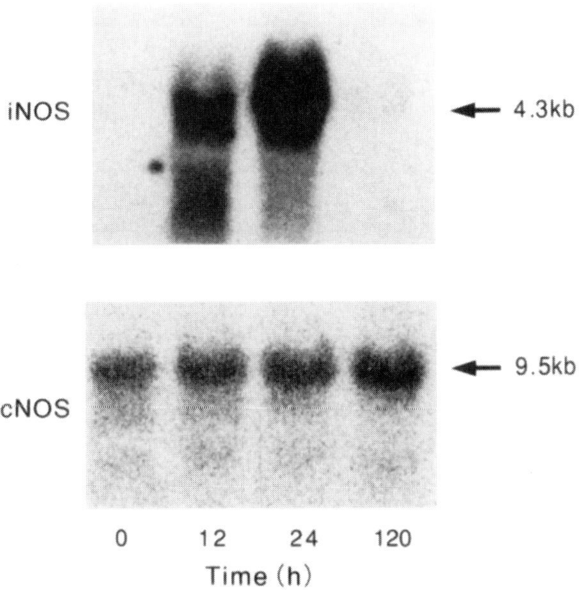

Figure 1. An example of Northern blot analysis for NOS II (upper panel) and NOS I (lower panel) in rat colitis tissue.

Comments

When total RNA is prepared from tissue samples, the tissues should be homogenized in 10 vol. extraction solution by means of a Polytron homogenizer. To prepare RNA from soft tissues such as brain and liver, the 2-propanol precipitate should be extracted once more.

NOS expression can also be detected by immunoblot using specific antibodies against each isozyme.

References

1. Chomczynski, P. and Sacchi, N. (1987). Single-step method of RNA isolation by acid guanidinium thiocyanate–phenol–chloroform extraction. *Anal. Biochem.*, **162**, 156–9.
2. Seo, H. G., Fujii, J., Asahi, M., Okado, A., Fujiwara, N., and Taniguchi, N. (1997). Roles of purine nucleotides and adenosine in enhancing NOS II gene expression in interleukin-1 β-stimulated rat vascular smooth muscle cells. *Free Radic Res.*, **26**, 409–18.
3. Jenkins, D. C., Charles, I. G., Baylis, S. A., Lelchuk, R., Radomski, M. W., Moncada, S. (1994). Human colon cancer cell lines show a diverse pattern of nitric oxide synthase gene expression and nitric oxide generation. *Br. J. Cancer*, **70**, 847–9.

58

Mn-SOD cell stimulation by cytokines: preparation of RNA

KEIICHIRO SUZUKI

Introduction

Mn-SOD expression is stimulated by various cytokines (TNF, IL-1), phorbol esters, lipopolysaccharides, etc. In this section we demonstrate Northern blot analysis and RNA preparation from cultured cells (1). Fig. 1 shows typical patterns of Northern blotting of human Mn SODs, and 1 kb in rat Mn-SOD. Many faint minor bands are apparent because of alternative splicing and different poly A binding sites.

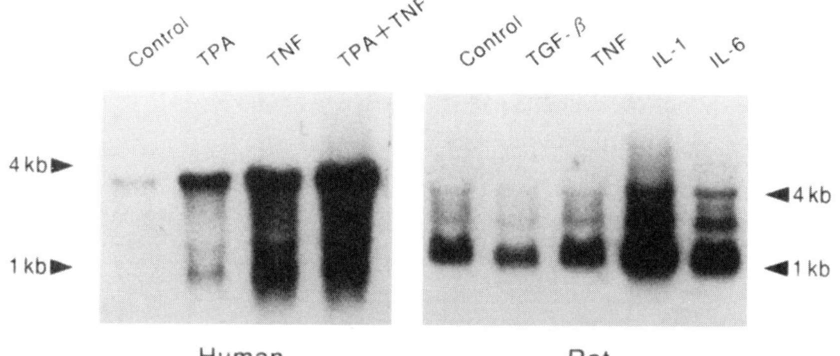

Figure 1. Examples of Northern blotting analysis of human and rat Mn-SOD.

Stimulation of cells

Protocol

1. Seed cells in 10-cm diameter dish.
2. Change medium containing reagents.
3. Prepare RNA after 4–6 h.

RNA preparation

Protocol

1. Wash cells with ice-cold PBS or TBS (2 × 4 mL).
2. Transfer cells to a 2 mL tube
3. Centrifuge at 8000 rpm for 5 min.
4. Remove supernatant.
5. Add solution D–phenol mixture[1] (1 mL) to the cells.
6. Vortex-mix mix the tube to dissolve cells.
7. Add chloroform (0.1 mL).
8. Vortex-mix mix the tube for 20 s.
9. Leave on ice for 20 min.
10. Centrifuge at 15 000 rpm for 15 min.
11. Prepare tube containing isopropanol (0.5 mL).
12. Transfer upper phase (0.5 mL) of centrifuged solution to the tube.
13. Leave for >30 min at –20 °C.
14. Centrifuge at 15 000 rpm for 20 min.
15. Wash twice with EtOH 75% (*v*/*v*).
16. Dry quickly.
17. Dissolve in water.
18. Quantitate RNA by measuring absorbance at 260 nm.
19. Perform Northern blot or RT-PCR analysis.

[1] Solution D is prepared by mixing guanidine thiocyanate (4 M), sodium citrate (pH 7.0, 25 mM), sarcocyl (0.5% *w*/*v*), and β-mercaptoethanol (0.1 M). The solution D–phenol mixture is prepared by mixing solution D, sodium acetate solution (pH 4.0, 2 M) and phenol, respectively, in the ratio 1:0.1:1

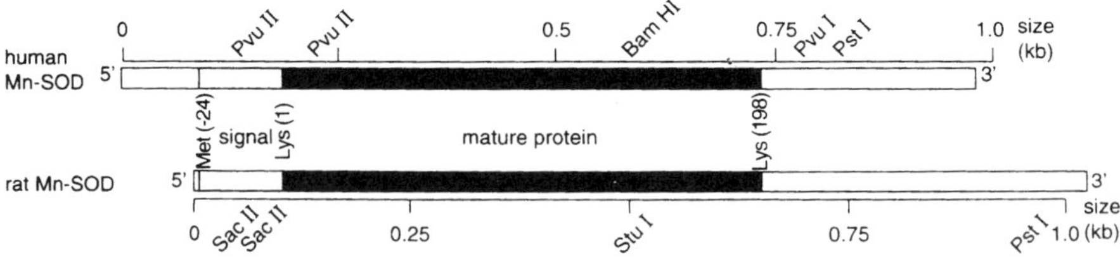

Figure 2. Structures of human and rat Mn-SOD cDNA.

Comments

Cu, Zn-SOD also exists in almost all cells. Because Cu, Zn-SOD expression is almost constant, even if stimulated by various cytokines, its expression is used as a control instead of β-actin.

Fig. 2 shows structures of human and rat Mn-SOD c DNA.

The Mn-SOD probes (human, rat) used in Figs. 1 and 2 were kind gifts from Dr Y.-S. Ho (Wayne State University).

Reference

1. Fujii, J., and Taniguchi, N. (1991). Phorbol ester induces manganese-superoxide dismutase in tumor necrosis factor-resistant cells. *J. Biol. Chem.*, **266**, 23142–6.

59

The expression of thioredoxin

YOSHIHISA TANIGUCHI

Introduction

Adult T cell-derived factor (ADF) is a 13-kDa protein and is a human homologue of thioredoxin present in *E. coli.* ADF/Trx has a variety of biological activities, such as a co-cytokine effect on the promotion of the cellular proliferation and reducing activity by which the disulfide bond of proteins is reduced and cleaved. Because, moreover, the expression of ADF/Trx is induced by oxidative stresses, e.g., irradiation of UV, oxidative agents, and reperfusion injury, ADF/Trx is a factor involved in protective system against oxidative stress. We present here a Western blotting method for ADF/Trx protein.

Protocol

Preparation of total cell lysates

1. Harvest the cells (ca 10^6), wash twice with PBS (pH 7.5) and measure the volume of the samples to be lysed.
2. Add the same volume of lysing buffer (Tris-HCl (pH 7.4, 1 mM), NaCl (150 mM), Nonidet P-40 (0.5% *w/v*), phenylmethylsulfonyl fluoride (1 mM)) and vortex-mix vigorously.
3. Stand on ice for 20 min.
4. Centrifuge the samples at 10 000 g for 10 min, and recover the supernatant.
5. Determine the protein concentration, for example, by use of the Bradford method (1).

Gel electrophoresis

1. Set the 10% *v/v* Tris/Glycine SDS PAGE separating gel (acrylamide (10% *w/v*), *N,N'*-methylene-bis-acrylamide (0.33%), Tris-HCl (pH 8.8, 0.375 M), SDS (0.1% *w/v*), APS (0.1% *w/v*), TEMED (0.04% *v/v*)) overlaid by staking gel (acrylamide (5% *w/v*), *N,N'*-methylene-bis-acrylamide (0.17% *w/v*), Tris-HCl (pH 6.8, 0.125 M), SDS (0.1% *w/v*), APS (0.1% *w/v*), TEMED (0.1% *v/v*)).

2. Mix the cell lysates containing protein (10 mg) and the same volume of sample buffer (SDS (2% *w/v*), dithiothreitol (100 mM), Tris-HCl (pH 6.8, 60 mM), bromophenol blue (0.01% *w/v*)).
3. Heat at 100 °C for 5 min.
4. Load the samples into the bottom of the wells.
5. Start electrophoresis at 100 V.
6. After the dye front has moved into the separating gel, increase voltage to 200 V.
7. When the dye reaches the bottom of the gel, turn off the power supply.

Electrophoretic transfer using a semi-dry method

1. Cut six sheets of absorbent paper (150 mm × 150 mm; Atto) and one sheet of PVDF membrane (Immobilon®; Millipore) to the size of the gel.
2. Soak the sheet of PVDF membrane in methanol.
3. Submerge the sheet in transfer buffer (Tris base (25 mM), glycine (192 mM), methanol (20% *v/v*)) for 15 min.
4. Assemble, in this order, the bottom electrode (the anode), three sheets of absorbent paper soaked in transfer buffer, the PVDF membrane, the polyacrylamide gel, and three sheets of absorbent paper soaked in transfer buffer.
5. Run the transfer for 1.5 h at 1.5 mA cm^{-2}.

Detection of ADF/Trx protein

1. Add blocking solution (skimmed milk (Difco; 10% *w/v*) containing BSA (Sigma Fraction V; 2% *w/v*)) and incubate at room temperature for 1 h.
2. Wash the sheet with TPBS (Tween 20 (0.05% *v/v*) in PBS) for 5 min, shaking tree times.
3. Add the monoclonal antibody solution (5 mL per 8 × 8 cm blot with 0.2 μg mL^{-1} antibody) in TPBS.
4. Incubate at room temperature with shaking for 1.5 h.
5. Wash the sheet with TPBS (3 × 5 min).
6. Add 1:500 diluted horseradish-labelled antibody against mouse immunoglobulin (5 mL).
7. Incubate at room temperature with shaking for 30 min.
8. Wash the sheet with PBS (3 × 5 min).
9. Perform detection by means of ECLTM (Amersham code RPN 2106).

Comments

The quantity of samples can be estimated by simultaneous electrophoresis and blotting of a dilution series of recombinant ADF/Trx (Fig. 1).

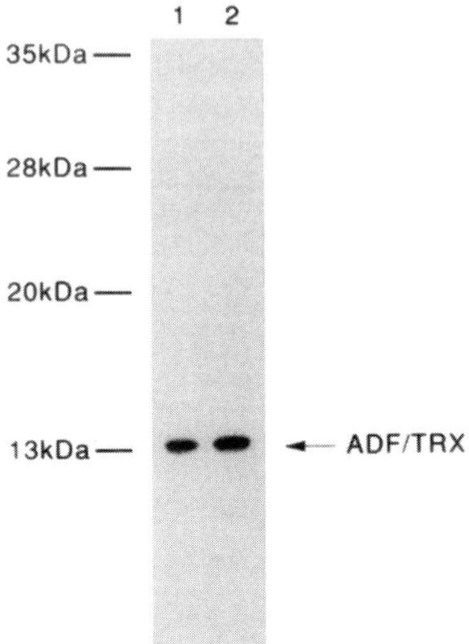

Figure 1. Detection of ADF/Trx protein by Western blotting. Lane 1, cell lysate of Hut101 cells; lane 2, recombinant ADF (10 ng).

Reference

1. Bradford, M. M. (1976). A rapid and sensitive method for the quantification of microgram quantities of protein utilizing the principle of protein-dye binding. *Anal Biochem.*, **72**, 248–54.

60

Peroxiredoxin

SHIRO BANNAI

Introduction

The peroxiredoxin family is a widely distributed class of enzymes that directly reduce H_2O_2 and other hydroperoxides, with hydrogen derived from NAD(P)H, by different routes (1). Two highly conserved cysteine residues are revealed by alignment of the peroxiredoxin family members. Both are necessary for the maintenance of the active dimeric structure of the peroxiredoxins, and the cysteine residue of the *N*-terminal side, which is conserved in all family members, constitutes the site of oxidation by a substrate peroxide. In *Salmonella typhimurium* a family member of peroxyiredoxin is a small subunit of alkyl hydroperoxide reductase (AhpC) and NADH-dependent peroxide reduction by Ahp C is catalysed by a large submit Ahp F. Ahp F is an FAD-containing NAD(P)H dehydrogenase and shows considerable homology to thioredoxin reductase. Similar peroxidase activity has been found in eukaryote peroxiredoxin. In yeast the biologically active hydrogen donor for peroxiredoxin is probably thioredoxin, thioredoxin reductase, and NADPH (2, 3). Salmonella Ahp C is inducible by H_2O_2 and the induction is dependent on oxy R regulatory network. Mammalian peroxiredoxin MSP23 in peritoneal macrophages is also induced by H_2O_2 or by various stress agents such as arsenite and diethyl maleate (4).

Peroxiredoxins protect cellular components against oxidation systems in which a thiol functions as a reducing equivalent. They are thus designated thiol-specific antioxidants (TSA). TSA activity is estimated by protection of glutamine synthetase from inactivation by the metal-catalysed oxidation system dithiothreitol/Fe^{3+}/O_2 (5). The thiol specificity of the TSA is because the oxidized form of TSA can be converted back to its sulfhydryl form by treatment with thiols but not by other reducing agents (e.g. ascorbate).

Protocol

Induction of peroxiredoxin by electrophilic agents

Preparation of peritoneal macrophages

1. Give mice (8 weeks) an intraperitoneal injection of thioglycollate broth (Difco, Brewer's formula; 4% *w*/*v*, 2 mL).

Protocol *Continued*

2. After 4 days collect cells by peritoneal lavage with RPMI 1640 containing heparin (10 U mL^{-1}).
3. Wash twice with RPMI 1640.
4. Plate at 10^6 cells per 35-mm culture dish in RPMI 1640 containing foetal bovine serum (10% *w*/*v*), penicillin (50 units mL^{-1}), and streptomycin (50 μg mL^{-1}).
5. Incubate for 1 h at 37 °C.
6. Change the medium to remove non-adherent cells.

Pulse-labelling

1. Place macrophages in 35-mm dish.
2. Incubate for an appropriate time (h) at 37 °C in the presence and absence of an electrophilic agent (e.g. 100 μM diethyl maleate).
3. Wash twice with PBS.
4. Add methionine- and cysteine-free RPMI 1640 (1 mL) containing dialysed foetal bovine serum (10% *v*/*v*) and ^{35}S-protein labelling mixture (10 μCi; EXPRE^{35}S^{35}S protein mix from Du Pont–New England Nuclear, specific activity >1000 Ci $mmol^{-1}$).
5. Incubate for 15 min at 37 °C.
6. Wash three times with PBS(–) (Ca^{2+} and M_8^{2r} free PBS).
7. Dissolve cells with SDS sample buffer (0.2 mL; containing SDS (2% *w*/*v*), glycerol (10% *v*/*v*), 2-mercaptoethanol (3% *v*/*v*), phenylmethylsulfonyl fluoride (0.1 mM),and Tris-HCl (pH 6.8, 50 mM)).
8. Measure radioactivity in an aliquot of the cell lysate by scintillation counting.

Gel electrophoresis and autoradiography

1. Heat the cell lysate at 100 °C for 3 min.
2. Apply the cell lysate containing approx. 50 000 counts min^{-1} (cpm) on SDS (12.5% *w*/*v*)–polyacrylamide gel and perform electrophoresis.
3. Detect the radiolabel by fluorography at –70 °C using Kodak XAR-5 X-ray film exposed for 1–2 weeks or by means of an imaging analyser (e.g. BASS 5000, Fuji Film) exposed overnight.

Results

These are illustrated in Fig. 1.

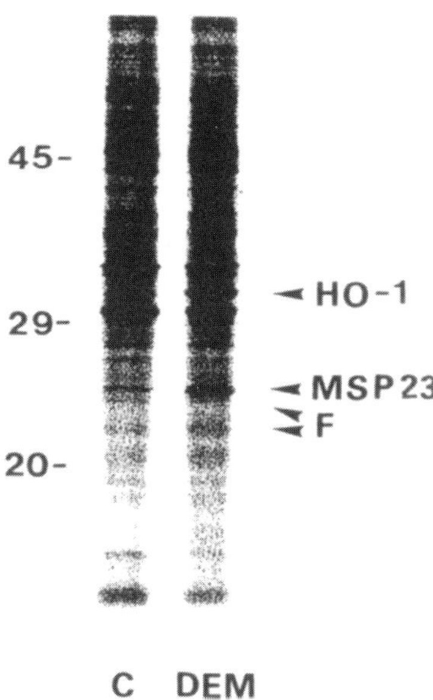

Figure 1. Induction of mouse peroxiredoxin MSP23 in macrophages by diethyl maleate. Peritoneally exuded macrophages were incubated for 12 h with and without diethyl maleate (100 μM). HO-1, haemoxygenase-1; F, heavy and light chains of ferritin; C, control; DEM, with diethyl maleate (100 μM). Molecular weight markers are indicated on the left.

Protocol

Determination of thiol-specific antioxidant activity

1. Prepare a solution of glutamine synthetase[1] (25 μg mL^{-1}) in imidazole-HCl (125 mM, pH 7.0).
2. Add peroxiredoxin solution (10 μL) or vehicle to glutamine synthetase solution (20 μL).
3. Add a freshly prepared solution (20 μL) of dithiothreitol (25 mM)–$FeCl_3$[2] (7.5 mM).
4. Incubate for 10 min at 30 °C.
5. Add γ-glutamyltransferase assay mixture[3] (2 mL) containing ADP (0.4 mM), glutamine (150 mM), potassium arsenate (10 mM), NH_2OH (20 mM), $MnCl_2$ (0.4 mM), and HEPES (pH 7.4, 100 mM).
6. Incubate for 5 min at 30 °C.
7. Add stop-mixture (1 mL) comprising $FeCl_3.6H_2O$ (5.5 g dL^{-1}), trichloroacetic acid (2.0 g dL^{-1}), and concentrated HCl (2.1 mL dL^{-1}).

Protocol *Continued*

8. Remove precipitate, if any, by centrifugation.
9. Measure absorbance at 540 nm.

[1] Purified from *E. coli* or from yeast. The activity of commercially available products should be checked before.
[2] Controls without dithiothreitol and $FeCl_3$ should be run in every experiment. These give the activity of the glutamine synthetase not exposed to the mixed-function oxidation.
[3] The activity of glutamine synthetase can be monitored by use of the γ-glutamyltransferase reaction in which glutamine synthetase catalyses the reaction:

$$\text{L-Glutamine} + NH_2OH \xrightarrow{\text{ADP, Asi, } Mn^{2+}} \text{L-}\gamma\text{-glutamylhydroxamate} + NH_3$$

References

1. Chae, H. Z., Robison, K., Poole, L. B., Church, G., Storz, G., Rhee, S. G. (1994). Cloning and sequencing of thiol-specific antioxidant from mammalian brain: Alkyl hydroperoxide reductase and thiol-specific antioxidant define a large family of antioxidant enzymes. *Proc. Natl. Acad. Sci. USA*, **91**, 7017–7021.
2. Chae, H. Z., Chung, S. J., Rhee, S. G. (1994). Thioredoxin-dependent peroxide reductase from yeast. *J. Biol. Chem.*, **269**, 27670–8.
3. Netto, L. E. S., Chae, H. Z., Kang, S. W., Rhee, S. G., Stadtman, E. R. (1996). Removal of hydrogen peroxide by thiol-specific antioxidant enzyme (TSA) is involved with its antioxidant properties. *J. Biol. Chem.*, **271**, 15315–21.
4. Sato, H., Ishii, T., Sugita, Y., Tateishi, N., Bannai, S. (1993). Induction of a 23 kDa stress protein by oxidative and sulfhydryl-reactive agents in mouse peritoneal macrophages. *Biochim Biophys Acta*, **1148**, 127–32.
5. Kim, K., Kim, I. H., Lee, K. Y., Rhee, S. G., Stadtman, E. R. (1988). The isolation and purification of a specific 'protector' protein which inhibits enzyme inactivation by a thiol/Fe(III)/O_2 mixed-function oxidation system. *J. Biol. Chem.*, **263**, 4704–11.

61

Expression of *bcl-2*

YUTAKA EGUCHI and YOSHIHIDE TSUJIMOTO

Introduction

Human *bcl-2* gene was isolated from the analysis of t(14;18) (q32;q21) chromosomal translocation frequently found in follicular lymphoma (1). Over-expression of Bcl-2 prevents apoptosis induced by a variety of stimuli in various cell lineage (2). Apoptosis is often inhibited by expression of SOD or glutathione peroxidase, suggesting involvement of ROS in apoptosis, although recent analysis of anti-apoptotic activity of Bcl-2 under very low oxygen conditions (3) provided some evidence against the sole function of Bcl-2 as an anti-oxidant (4). Here we describe the over-expression system for Bcl-2, and the quantitation of apoptosis.

Protocol

Overexpression of human Bcl-2 in IL-7 dependent mouse B cell line GB11

Several ways are available for overproduction of Bcl-2 protein*. Here we describe the method of preparing an oligoclonal cell line by use of recombinant viruses.

Preparation of cells producing recombinant viruses carrying human *bcl-2* gene

To prepare recombinant viruses carrying human *bcl-2* gene, plasmid pBC140B4 (Fig. 1) is introduced in packaging cell line ψ2. For control experiments, vector (pBC140) transfectants should always be prepared similarly. We usually use the calcium phosphate precipitation method to introduce DNA in ψ2 cells.

1. Seed 5×10^5 ψ2 cells in DMEM/10% (*v*/*v*) FBS in T25 flask and cultivate for 1 day under 10% (*v*/*v*) CO_2 at 37 °C.
2. Mix DNA (pBC140B4 or pBC140) (5 μg/0.5 mL) in isotonic phosphate solution (NaCl (137 mM), KCl (5 mM), Na_2HPO_4 (0.7 mM), dextrose (6 mM), HEPES (21 mM) pH 7.1) with $CaCl_2$[†] (2 M, 32 μL), and stand for 30 min at room temperature to enable formation of the co-precipitate of plasmid and $CaHPO_4$.

Protocol *Continued*

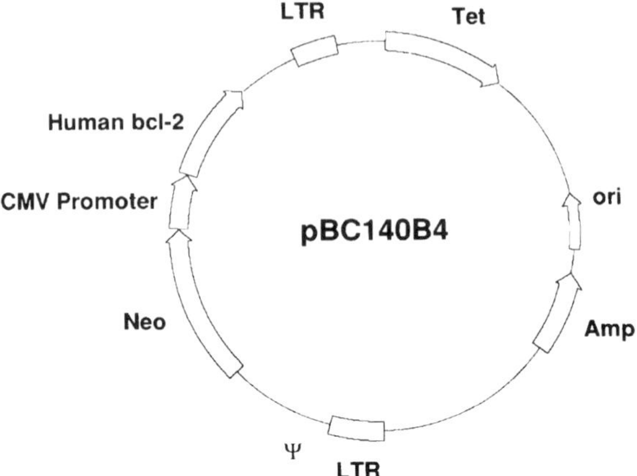

Figure 1. Structure of Recombinant Virus DNA pBC140B4. pBC140B4 was constructed by inserting human *bcl-2* cDNA (B4 fragment) downstream of cytomegalovirus promoter on retrovirus vector pBC140. pBC140 carries LTR and ψ site for packaging in virus particle, neomycin resistant gene, ampicillin resistant gene, and replication origin to expand in *E. coli*.

3. Transfer the suspension of the resulting $CaHPO_4$–DNA co-precipitates into medium (5 mL) above the cell monolayer in a T25 flask.
4. After 24-h incubation, re-plate the cells into two T75 flasks.
5. After 24-h incubation, change medium to DMEM/10% (*v*/*v*) FBS containing G418 (Geneticin, Gibco; 800 μg mL^{-1}) to start selection.
6. Change medium every 3 days.
7. Pool surviving cells and examine the production of human Bcl-2 protein by Western blot analysis.

Western blot analysis to examine the production of ectopically expressed human Bcl-2[‡]

1. Harvest cells and lyse in cell lysis buffer (Tris-HCl (pH 8.0, 50 mM), EDTA (10 mM), NaCl (100 mM), SDS (1% *w*/*v*)).
2. Analyse total protein (10 μg) on SDS–polyacrylamide gel (12.5% *w*/*v*).
3. Transfer proteins in gel to Immobilon P membrane (Millipore).
4. After blocking the filter with skimmed milk in PBS (3% *w*/*v*) for 30 min, incubate the filter with anti-human Bcl-2 monoclonal antibody (Dako) in the same solution at 4°C overnight.
5. Wash the filter at room temperature with PBS containing 0.3% (*v*/*v*) Tween 20 for 5 min. Repeat four times.

6. Incubate the filter with anti-mouse IgG antibody conjugated with HRP (Promega) in PBS containing skimmed milk (3% *w/v*) at room temperature for 1 h.
7. Wash the filter as in Step 5.
8. After washing the filter twice with PBS for 5 min, soak the filter in ECL Western Blotting Detection Reagents (Amersham), and apply to an X-ray film for a period from 1 s to 10 min to detect signals for human Bcl-2.

Infection of GB11 cells with the recombinant viruses

Because growth of GB11 cells depends on IL-7, the culture should be conducted in the presence of IL-7. IL-7 producing cell line T220 is used as feeder cells, or conditioned medium is prepared using T220.

1. Cultivate ψ2/pBC140B4 (and ψ2/pBC140) cells in DMEM/10% (*w/v*) FBS for 3 days after 10-fold dilution. Filter the culture supernatant through a 0.22-μm filter to generate virus solution.
2. Infect GB11 cells with the virus by mixing the virus solution (0.5 mL) and GB11 cells (0.5 mL) grown in RPMI 1640/10% (*v/v*) FBS/50 nM β-mercaptoethanol containing IL-7 to 5×10^5 cells mL^{-1}, and incubate overnight in a 24-well dish under 5% (*v/v*) CO_2 at 37 °C.
3. Change half the medium and incubate for 2 more days.
4. Dilute 5-fold and infect the cells again with the virus as described above.
5. After incubation for 24 h, start selection with G418 at 800 μg mL^{-1}.
6. Change half the medium every three days, leaving all cells behind.
7. Pool surviving cells and determine the amount of the human Bcl-2 protein by Western blotting (as described above).

Analysis of cell survival after triggering apoptosis

To examine the effects of Bcl-2 on cell survival, it is important to determine appropriate death-inducing conditions**. Although several methods are available for estimation of cell death**, we here describe the Hoechst-PI staining method (5) of examining the apoptosis of GB11 cells induced by IL-7 depletion.

1. Culture virus-infected GB11 cells to 1×10^6 mL^{-1} in RPMI 1640 medium containing FCS (10% *w/v*), β-mercaptoethanol (50 nM) and IL-7.
2. Wash cells with the medium containing no IL-7: Centrifuge cells at 1300 rpm for 5 min, remove supernatant, and re-suspend cells in the same volume of the medium. Repeat twice.
3. Finally adjust cell concentration to 1×10^6 mL^{-1}, and transfer suspension (1 mL) to 24-well dish.

Protocol *Continued*

4. Withdraw an aliquot every 24 h, and stain recovered cells with Hoechst 33342 (10 μM, Calbiochem., USA; 10 μL) and propidium iodide (PI, 10 μM, Sigma, USA) in PBS for 3 min and analyse under a non-confocal fluorescent microscope (Olympus BX-50, Japan) with excitation at 360 nm.
5. Quantitate cell death by counting more than 1000 cells under the fluorescent microscope after staining with Hoechst 33342 and PI.

*To achieve the over-expression of Bcl-2, Bcl-2-expressing DNA should be introduced into cells by one of the available methods such as the calcium phosphate precipitation method, virus infection, lipofection, electroporation, and particle gun. These methods can be used for both transient expression and isolation of stable transfectants. When above methods other than virus infection were used for isolation of stable transfectants, monoclonal colonies should be isolated, because some transfectants express Bcl-2, but others do not. For isolation of stable polyclonal or oligoclonal Bcl-2-expressing transfectants, recombinant virus should be used, because each recombinant virus has the same construct and its DNA integrates in the host genome in a way such that Bcl-2 is expressed.
†$CaCl_2$ solution should be added to DNA solution very slowly with gentle mixing over a period of about 30 s to avoid the rapid formation of coarse precipitates that result in reduced efficiency of transformation.
‡The method described here is applicable for most antibodies. When using other antibodies, appropriate second antibodies must be chosen. Because each antibody has a distinct affinity against its epitope, concentrations of the antibody, skimmed milk for blocking (1–5% *w*/*v*), and Tween 20 in PBS for washing (0.01%, 0.05%, 0.1%, or 0.3%*w*/*v*) should be controlled properly. High background levels usually arise from the high concentration of antibodies used, or from incubation for too long with antibodies. For the best results with ECL Western Blotting Detection Reagents (Amersham), detergent used to wash the filter should be removed completely.
**To examine the effects of various gene products on cell survival, it is important to determine appropriate death-inducing conditions, such as depletion of growth factors, genotoxic treatment, Fas treatment, oxygen treatment, heat shock, and so on. Because each cell line has distinct sensitivity, preliminary experiments are inevitable. Various methods are available for estimating cell survival; these include the Hoechst-PI staining method (described here),

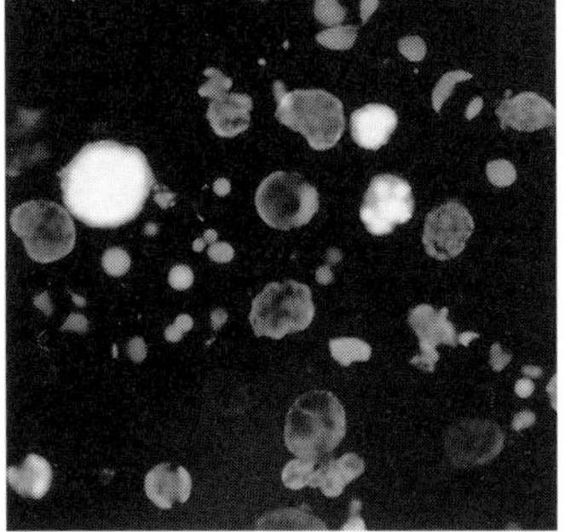

	PI (pink) nuclei	Hoechst (blue) nuclei
viable	not stained	round
necrosis	round	round
early apoptosis	not stained	fragmented
terminal apoptosis	fragmented	fragmented

Figure 2. Detection of apoptosis and necrosis. GB11 cells were treated under IL-7-depleted conditions for two days and examined under a non-confocal fluorescent microscope after staining with PI (pink) and Hoechst (blue). Cells with blue round nuclei, red round nuclei, blue fragmented nuclei and red fragmented nuclei correspond, respectively, to viable, necrotic, 'early' apoptotic, and 'terminal' apoptotic cells.

Trypan-Blue exclusion, crystal violet staining method, MTT assay, PI staining followed by FACScan analysis, annexin V staining method, and TUNNEL staining. It is also important to choose an appropriate method for estimation of cell survival. Because apoptosis is morphologically characterized by the generation of fragmented nuclei with highly condensed chromatin, protrusion of cytoplasm and formation of apoptotic bodies, we usually use the Hoechst-PI staining method. Because Hoechst 33342 stains all nuclei and PI stains nuclei of cells with a disrupted plasma membrane, nuclei of viable, necrotic and apoptotic cells were observed as blue round nuclei, pink round nuclei, and fragmented blue or pink nuclei, respectively (Fig. 2).

Results

Overexpression of human Bcl-2 in IL-7 dependent mouse B cell line GB11

GB11 cells were infected with recombinant viruses carrying *bcl-2* gene or with empty vector viruses to prepare oligoclonal cell lines, as described above. Total cell lysate prepared from each transfected GB11 cell line was subjected to Western blot analysis (Fig. 3), and Bcl-2 protein was found to be expressed in GB11 cells transfected with viruses carrying *bcl-2* gene (lanes 3–6), but not in those transfected with empty vector viruses (lanes 1, 2).

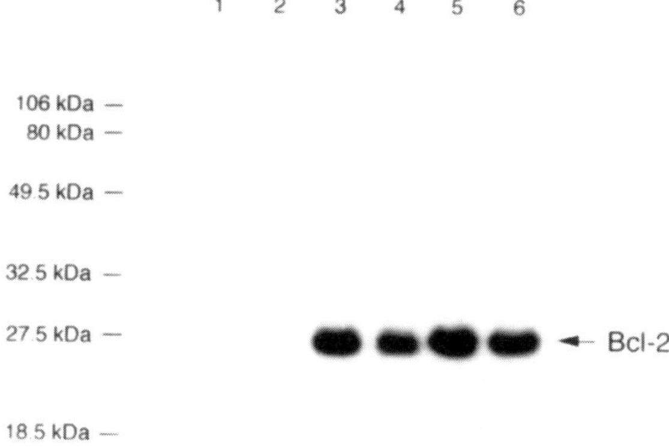

Figure 3. Detection of ectopically expressed Bcl-2 protein by Western blot analysis. Total cell lysates extracted from GB11 cells transformed by recombinant viruses were subjected to Western blot analysis. Bcl-2 protein was detected in GB11 cells infected with viruses carrying *bcl-2* gene (lanes 3–6), but not in those infected with empty vector viruses (lanes 1, 2).

Analysis of cell survival after apoptosis inducing treatment

To test the anti-apoptotic activity of Bcl-2, GB11 cell lines obtained as described above were treated under IL-7-depleted conditions for different

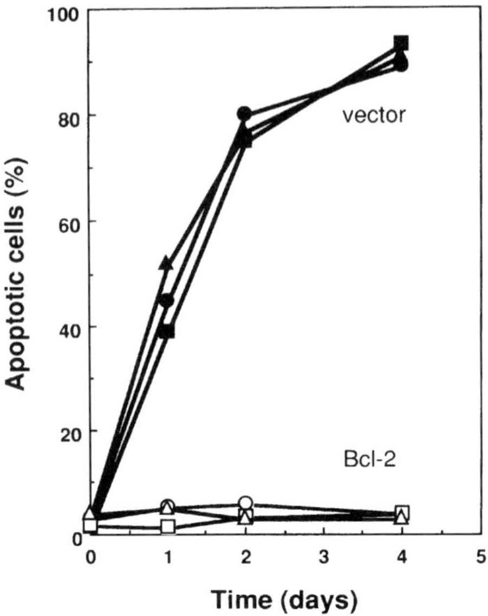

Figure 4. Prevention of apoptosis by Bcl-2. GB11 cells were treated under IL-7-depleted conditions as indicated and examined under a non-confocal fluorescent microscope after staining with PI (pink) and Hoechst (blue), and more than 1000 cells were counted. Open symbols indicate three independent GB11 cell lines transfected with viruses carrying *bcl-2* gene; closed symbols indicate three independent GB11 cell lines transfected with empty vector viruses. Vector transfectants underwent apoptosis upon IL-7 depletion, and the apoptosis was prevented by over-expression of Bcl-2 in *bcl-2*-transfectants.

periods and examined under a non-confocal fluorescent microscope after stained with PI (pink) and Hoechst 33342 (blue), and more than 1000 cells were counted. Apoptotic and necrotic cells can be distinguished by the shape and colour of their nuclei (Fig. 2, also see footnote 4, above). The results shown in Fig. 4 indicate that vector transfectants underwent apoptosis upon IL-7 depletion, and the apoptosis was prevented by over-expression of Bcl-2 in *bcl-2*-transfectants.

References

1. Tsujimoto, Y., Cossman, J., Jaffe, E., and Croce, C. M. (1985). Involvement of the *bcl-2* gene in human follicular lymphoma. *Science*, **228**, 1440–3.
2. Thompson, C. B. (1995). Apoptosis in the pathogenesis and treatment of disease. *Science*, **267**, 1456–62.
3. Shimizu, S., Eguchi, Y., Kosaka, H., Kamiike, W., Matsuda, H., and Tsujimoto, Y. (1995). Prevention of hypoxia-induced cell death by Bcl-2 and Bcl-xL. *Nature*, **374**, 811–13; Jacobson, M. D. and Raff, M. C. (1995).Programmed cell death and Bcl-2 protection in very low oxygen. *Nature*, **374**, 814–16.

4. Hockenbery, D. M., Oltvai, Z. N., Yin, X. M., Milliman, C. L., and Korsmeyer, S. J. (1993). Bcl-2 functions in an antioxidant pathway to prevent apoptosis. *Cell*, **75**, 241–51.
5. Kane, D. J., Sarafian, T. A., Anton, R., Hahn, H., Gralla, E. B., Valentine, J. S., Ord, T., and Bredesen, D. E. (1993). Bcl-2 inhibition of neural death: decreased generation of reactive oxygen species. *Science*, **262**, 1274–7.
6. Shimizu, S., Eguchi, Y., Kamiike, W., Itoh, Y., Hasegawa, J., Yamabe, K., Otsuki, Y., Matsuda, H., and Tsujimoto, Y. (1996). Induction of apoptosis as well as necrosis by hypoxia and predominant prevention of apoptosis by Bcl-2 and Bcl-XL. *Cancer Res.*, **56**, 2161–6.

62

Glutathione peroxidase: preparation of haemolysate and cell homogenates/fixation of tissue/RNA extraction

TETSUYA MORIUCHI

Introduction

Glutathione peroxidases remove H_2O_2 by coupling its reduction to H_2O with oxidation of GSH (see also Chapter 20). The enzyme can also act on peroxides (LOOH) including fatty acid hydroperoxide, cholesterol 7β-hydroperoxide and various synthetic hydroperoxides such as cumene and *t*-butyl hydroperoxides. In all cases the peroxide group is reduced to an alcohol. LOOH + 2GSH → GSSG + H_2O + LOH. There are at least four isozymes and the cytosolic enzyme is the major one and immunochemically distinct.

Protocol

This protocol is used to prepare a variety of samples for comprehensive analysis of cellular glutathione peroxidase (GSH-Px) in the rat. To obtain reliable results, the following steps should be taken before preparation of samples.

1. Allow the rats a good rest for at least 3 days to eliminate and effects of the stress of long-distance transportation.
2. Assemble a team of collaborators to prepare samples promptly.
3. Treat isolated organs or tissues in Petri dishes on ice.
4. Perfuse the organs to avoid contamination of erythrocytes when GSH-Px activity and amount are measured in tissue homogenate.

Preparation of sera and haemolysate

1. Cut open abdomen under ether anaesthesia.
2. Draw blood (5 mL) from abdominal aorta.

3. Expel the blood into a centrifuge tube, and leave the blood to clot at room temperature for 30 min to 2 h.
4. Centrifuge at 3000 g for 20 min at room temperature.
5. Collect serum by means of a Pasteur pipette. Store at –80 °C.
6. Remove the clot[1].
7. Add PBS (10 mL) to the centrifuge tube
8. Centrifuge at 3000 g for 5 min at room temperature.
9. Discard supernatant.
10. Wash erythrocytes with PBS.
11. Add erythrocytes[2] (50 μL) to H_2O (950 μL) in a 1.5-mL microcentrifuge tube.
12. Store haemolysate at –80 °C[3].
13. Perform enzyme assay, ELISA, immunoblot analysis, and measurement of haemoglobin concentration.

[1] After removal of the clot, uncoagulated red blood cells remain in the centrifuge tube.
[2] Cut the end of the micro pipette tip with scissors.
[3] A high concentration of rat haemoglobin might aggregate on freezing. Store it at a low concentration.

Fixation of samples for light microscopy and preparation of tissue

Homogenate for chemical analysis

1. Perfuse organs with PBS from inferior vena cava using peristaltic pump (flow rate 14 mL min^{-1}).
2. Cut the tissue in Petri dish wit scissors[1]. Prepare 5 mm × 5 mm × 2 mm blocks with a razor and fix them with 2–4% paraformaldehyde in 0.1 M phosphate buffer for morphological study
3. Transfer dissected tissue (1 g) into Potter–Elvehjem homogenizer.
4. Add 9 vols ice-cold MS buffer[2] (9 mL).
5. Homogenize the tissue with Teflon pestle (800 rpm, move homogenizer up and down 10 times)
6. Centrifuge at 700 g for 10 min at 4 °C[3].
7. Remove supernatant to fresh tubes.
8. Store at –80 °C.
9. Perform enzyme assay, ELISA, and immunoblot analysis[4].

[1] MS buffer is prepared by mixing Tris-HCl (pH 7.6, 1 M, 5.0 mL), EDTA.3Na (0.5 M, 0.2 mL), D-mannitol (38.3 g), sucrose (24.0 g), and diluting with H_2O to 1 L (store at 4 °C).
[2] Hard tissues such as heart and kidneys should be minced with scissors before homogenization.
[3] Collect necessary fractions of the cell. The nuclei are removed by this centrifugation.

Protocol *Continued*

[4] Anti-GSH-Px monoclonal antibody is available from MBL International Corporation (MIC), 440 Arsenal Street 2nd floor, Watertown, MA, USA. 02172, tel. 617–926–6964, fax. 617–926–6954, or from MBL Japan, fax. 052–971–2337, email: mbl@mxa.meshnet.or.jp

Preparation of RNA from tissues

1. Place tissue fragments (5 g) in 50-mL centrifuge tube (Corning 25331) on ice[1].
2. Add 7 vols GIT buffer[2] (35 mL).
3. Homogenize the tissue in a Polytron on ice at medium speed for 15 s. Samples may be stored at –80 °C at this point.
4. Centrifuge at 3500 g for 20 min at 20 °C.
5. Add 4 mL of CsCl buffer[3] in 10-mL tubes (autoclaved SW40.1 tube) then carefully layer 5 mL of supernatant on top.
6. Fill each tube with GIT buffer to 5 mm below the top and keep balance.
7. Centrifuge overnight (21 h) at 32 000 rpm (174 000 g) in SW40 (Beckman) at 20 °C.
8. [See other protocol for subsequent steps. (3)]
9. Dry pellet in vacuum centrifuge.
10. Re-suspend each pellet in TE-DEPC[4] (50 μL).
11. Perform Northern blot analysis and RT-PCR.

[1] RNA purification kits are commercially available.
[2] GIT buffer is prepared by mixing guanidine isothiocyanate (94.53 g; 4 M final) and sodium acetate (pH 6.0, 3 M, 1.67 mL), diluting with H_2O to 200 mL, filtering through a sterile filter (0.22-μm filter; Falcon 105 bottle top filter), and adding 2-mercaptoethanol (1.67 mL).
[3] CsCl buffer is prepared by mixing CsCl (95.97 g, 5.7 M final), sodium acetate (pH 6.0, 3 M, 0.83 mL), and EDTA.3Na (0.5 M, 1.0 mL), diluting with H_2O to 100 mL, and filtering through a sterile filter (0.22-μm filter)
[4] TE–DEPC is prepared by mixing H_2O–DEPC (98.8 mL), Tris-HCl (pH 7.6, 1 M, 1.0 mL), EDTA.3Na (0.5 M, 0.2 mL) and autoclaving, to sterilize. H_2O–DEPC is prepared by mixing H_2O (1 L), DEPC (diethyl pyrocarbonate; Sigma D-5758; 1 mL) and autoclaving, to sterilize.

References

1. Suemizu, H. *et al.* (1992). Production and characterization of two monoclonal antibodies to human glutathione peroxidase. *Hybridoma*, **11**, 795–801.
2. Suemizu, H. *et al.* (1994). Decreased expression of liver glutathione peroxidase in Long–Evans Cinnamon mutant rats predisposed to hepatitis and hepatoma. *Hepatology*, **19**, 694–700.
3. Sambrook, J., Fritsch, E. F. and Maniatis, T. *Molecular cloning*, 2nd Ed. 1, 1989.

63

Expression of catalase gene

KENZO SATO

Introduction

Catalase (EC 1.11.1.6) is a peroxisomal enzyme catalysing the decomposition of hydrogen peroxide to oxygen and water. The enzyme serves a central function in oxidant defence, together with other enzymes such as superoxide dismutase and glutathione peroxidase. In mammalian tissues, the highest levels of catalase activity are found in the liver, kidney and erythrocytes and the lowest levels in connective tissues (1). Expression of catalase is known to be markedly reduced in the liver of tumour-bearing animals and in some cultured hepatoma cell lines (2).

It is, therefore, necessary to analyse the mechanism of gene expression. In this chapter we describe non-isotopic CAT assay for analysis of promoter activity (3), and the mini preparation of nuclear extracts for gel mobility shift assay (4).

Protocol

Non-isotopic CAT assay

Preparation of cell extracts

1. Transfect monolayer cells (6-cm dish) with CAT plasmid by general method.
2. Wash the dish with PBS(–) (3 × 2 mL).
3. Place TEN (Tris-HCl (pH 7.8, 10 mM), NaCl (25 mM), EDTA (1 mM); 1 mL) in the dish.
4. Scratch the cells off with policeman, and transfer into microtube. To avoid deviation among dishes, careful transfer is necessary.
5. Centrifuge at 12 000 rpm for 30 s at 4 °C, and remove supernatant.
6. Suspend cell pellet in Tris-HCl (pH 7.8, 0.25 M, 50 mL).
7. Disrupt the cell by ultrasonicator (Cellruptor; Cosmo Bio Co.) at 200W for 3 min at 0 °C.
8. Remove cell debris by centrifugation at 12 000 rpm for 5 min at 4 °C.
9. Take supernatant into a microtube, and heat at 65 °C for 10 min, to inactivate chloramphenicol deacetylase.

Protocol *Continued*

10. Remove aggregate by centrifugation, and keep the supernatant for cell extract at −20 °C.

CAT reaction

1. Mix the following reaction mixture on ice;

   5 μl of 10 mM acetyl CoA (suspended in water, and stocked at −20 °C)
   0.2 μl of 100 mM chloramphenicol (suspended in water, and stocked at −20 °C)
   25 μl of 0.25 M Tris–HCl (pH 7.8)
   20 μl of cell extract[1]
2. Incubate the mixture at 37 °C for 30 min to 4 hrs[1] .
3. Terminate the reaction by addition of 200 μl of cold ethyl acetate and vortex mixing.
4. Centrifuge at 12 000 rpm for 10 min at 4 °C.
5. Take upper layer into a glass spits tube.[2]
6. Evaporate ethylacetate in a vacuum-centrifuge concentrator.

[1] Amount of cell extract in the mixture and reaction times are dependent on the level of the gene expression.
[2] Some chemicals are extracted from plastic tube by chloroform.

Detection of acetylated chloramphenicol by HPLC

1. Set up HPLC condition as follows

   column: 4.5 mm diameter × 25 cm length
   resin: TSK gel silica 60
   solvent: 95% chloroform: 5% methanol (freshly prepared)
   flow rate: 1.5 ml/min
   average pressure: 60–80 kgf/cm^2

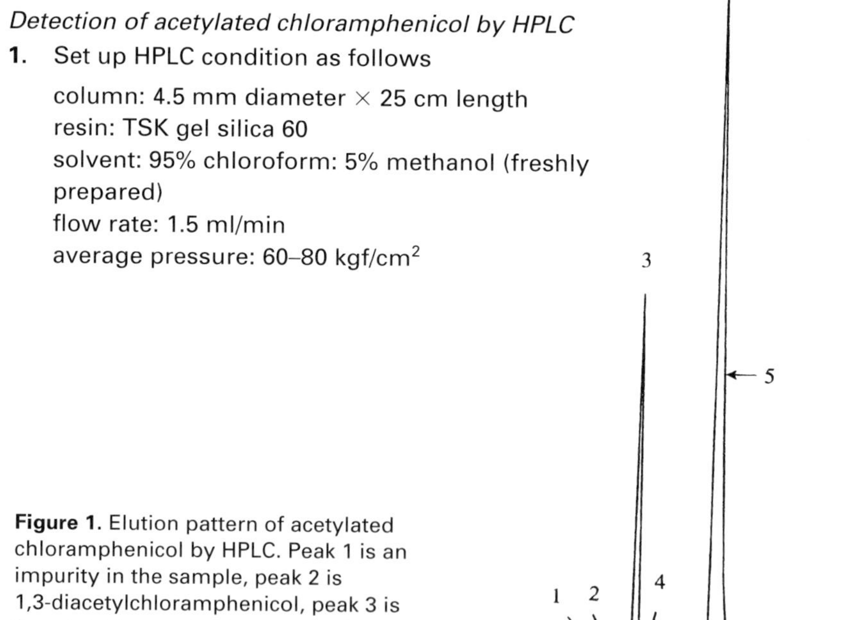

Figure 1. Elution pattern of acetylated chloramphenicol by HPLC. Peak 1 is an impurity in the sample, peak 2 is 1,3-diacetylchloramphenicol, peak 3 is 3-acetylchloramphenicol, peak 4 is 1-acetylchloramphenicol, and peak 5 is non-acetylated chloramphenicol.

2. Pre-run HPLC with solvent for 10 min.
3. Suspend dried sample in 50 μl of chloroform: methanol (95:5).
4. Inject 20 μl of sample suspension into HPLC, and run for 10 min.
5. Calculate the CAT activity by the ratio of acetylated chloramphenicol.

Mini-preparation of nuclear extract from culture cell.

1. Culture cells in a 15 cm dish.
2. Wash the cell with cold PBS(–).
3. Harvest the cell into a microtube with a rubber policeman.
4. Centrifuge at 12 000 rpm for 15 sec. and remove supernatant.
5. Suspend the cell pellet in a cold buffer A (Hepes (pH7.6, 10 mM), KCl (10 mM), EDTA (0.1 mM), PMSF (0.5 mM) and DTT (1 mM) 400 μL) by pipetting.
6. Keep on ice for 15 min.
7. Add NP40 (10% *w/v*, 25 μL) and vortex-mix vigorously for 10 s.
8. Centrifuge at 15 000 rpm for 30 s.
9. Suspend nuclear pellet in buffer B (Hepes (pH 7.6, 20 mM), NaCl (0.4 M), EDTA (1 mM), PMSF (1 mM) and DTT (1 mM); 100 μL).
10. Keep on ice for 15 min, and vortex-mix occasionally.
11. Centrifuge at 15 000 rpm for 15 min at 4 °C.
12. Take supernatant and adjust glycerol concentration to 20% *v/v*.
13. Determine protein concentration[1], freeze the extract in liquid nitrogen and store in –80 °C.

[3]Protein concentration of nuclear extract is given by: $(187 \times A_{230\ nm} - 81.7 \times A_{260\ nm}) \times$ dilution $= \mu g\ mL^{-1}$

References

1. Aebi, H. E. and Wyss, S. R. (1978). Acatalasemia. In *The Metabolic Basis of Inherited Disease*, (ed. J. B. Stanbury, J. B. Wyngarden, and D. S. Fredrickson), pp. 1792–807. McGraw–Hill, New York.
2. Potter, V. R. (1977). Biochemistry of cancer. In *Cancer Medicine*, (ed. J. F. Holland and E. Frei), p. 178. Lea and Febiger, Philadelphia.
3. Sato, K., Ito, R., Beak, K., and Agarwal, K. (1986). A specific DNA sequence controls termination of transcription in the gastrin gene. *Mol. Cell. Biol.* **6**, 1032–43.
4. Lichtsteiner, S., Wuarin, J., and Schibler, U. (1987). The interplay of DNA-binding proteins on the promoter of the mouse albumin gene. *Cell*, **51**, 963–73.

64

L-Gulono-γ-lactone oxidase

MORIMITSU NISHIKIMI

L-Gulono-γ-lactone oxidase (GLO), terminal enzyme in the pathway of L-ascorbic acid synthesis, is expressed in the liver of most mammals, some exceptions being man, monkey, and guinea pig. Most mammalian cell lines have no such enzymic activity, but they are normally cultured in media without supplementation with this vitamin. Because of the importance of L-ascorbic acid as a scavenger of reactive oxygen species, we attempted to express GLO in mammalian cells. Procedures for transfection of the guinea pig cell line 104C1 with a GLO-expressing plasmid and for micro-determination of GLO activity are given below.

Preparation of plasmid-entrapping liposomes (1)

Protocol

1. Prepare solutions (10 mM) of *N*-(α-trimethylammoniomacetyl)didodecyl-D-glutamate (TMAG; Sogo Pharmaceutical, Tokyo), dilauroyl phosphatidylcholine (DLPC; Sigma), and dioleoyl phosphatidylethanolamine (DOPE; Sigma) in $CHCl_3$ and mix in the ratio 1:2:2 (*v*/*v*), respectively. The mixture can be stored at −20 °C.
2. Dispense the TMAG–DLPC–DOPE mixture (100 μL; 1 μmol lipids) into a siliconized conical tube and evaporate the solvent by rotary evaporation at room temperature.
3. Add Dulbecco's modified phosphate-buffered saline without Mg and Ca (PBS(−); 300 μL) containing GLO expression plasmid[1] (20 μg DNA).
4. Vortex-mix the tube vigorously for 1 min.
5. Transfer the resulting GLO expression plasmid-entrapping liposomes into a 1.5-mL microtube, then rinse the conical tube with 200 μL PBS(−) and transfer it into the tube.

[1] The expression plasmid was constructed by inserting the ~1.6 kpb *Bcl*I–*Bam*HI fragment from a rat GLO cDNA (2) into the *Hin*dIII site of pRc/CMV or pRc/RSV (Invitrogen, San Diego, CA, USA) after ligation of a *Hin*dIII linker.

Transfection of cells

Protocol

1. Culture 104C1 cells[1] in RPMI 1640 medium supplemented with FBS (10% *v*/*v*), penicillin (100 units mL^{-1}), and streptomycin (100 μg mL^{-1}).
2. Re-plate the cells at a density of 5×10^5 cells in a 25-cm^2 flask (medium volume, 5 mL) and incubate for 20 h.
3. Replace the medium with fresh and incubate for 4 h.
4. Add GLO expression plasmid-entrapping liposomes (50 μL; 2 μg DNA) and incubate for 24 h.
5. Replace the medium with fresh and incubate for 24 h.
6. Add G418 (100 mg mL^{-1}, 20 μL).
7. Replace the medium every 3 days with a fresh medium containing G418 (400 μg mL^{-1}) until untransfected cells are detached from the flask.
8. Replace the medium with fresh medium containing G418 (200 μg mL^{-1}) and incubate until colonies of cells grow to cover one third to one-half of the surface.
9. Re-plate the cells in a 75-cm^2 flask and incubate the cells to near confluency.
10. Re-plate the cells in two 75-cm^2 flasks and incubate the cells to near confluency.
11. Make frozen stock from one flask, and test the expression of GLO using the cells from the other flask.

[1] The cell line 104C1 is the guinea pig cell that was obtained through transformation of embryonic cells with benzo(*a*)pyrene. It is available from Health Science Research Resources Bank, Osaka, Japan.

Measurement of GLO activity (3)

Protocol

1. Add PBS(–) (20–40 μL) to cell pellet obtained from cells that have been cultured to near confluency in a 75-cm^2 flask and sonicate for several seconds.
2. Dispense potassium phosphate buffer (pH 7.0, 0.1 M, 250 μL, containing 0.1 M sodium citrate), cell extract (20 μL), and water (205 μL) into a 1.5-mL microtube and add L-gulono-γ-lactone (50 mM, 25 μL). (L-Gulono-γ-lactone solution should be used immediately after dissolution.)

Protocol *Continued*

3. Incubate at 37 °C for 15 min.
4. Add metaphosphoric acid (35% *w/v*, 84 μL) and place the tube on ice for 5 min.
5. Centrifuge at 14 000 rpm for 5 min in a microcentrifuge.
6. Transfer the supernatant (330 μL) into a 1.5-mL microtube.
7. Add dichlorophenol indophenol (0.2%, 20 μL) and leave the tube to stand for ~1 min.
8. Add metaphosphoric acid (5% *w/v*, 300 μL; containing 2% *w/v* thiourea) and dinitrophenylhydrazine (2% in 4.5 M H_2SO_4; 60 μL) and incubate at 50°C for 90 min.
9. Add ethyl acetate (400 μL), vortex-mix for ~10 s, and centrifuge at 14 000 rpm for several seconds.
10. Transfer the yellow-coloured upper phase into a new tube.
11. Add sodium sulfate (anhydrous; ~50 mg) and vortex-mix.
12. Pass the solution through a 0.2-μm filter (Millipore Columngard LCR).
13. Analyse 50 μL of the solution by HPLC with a 4.6 mm × 250 mm Develosil silica column, *n*-hexane–ethyl acetate–acetic acid, 4:5:1, as mobile phase at a flow rate of 1 mL min^{-1}, and detection at 495 nm.

Comments

The GLO activity in whole cell extract from the cells after selection with G418 was approx. 0.1 nmol min^{-1}/mg protein (4). After isolation of clones by limited dilution, a clone was obtained with activity as high as 1.3 nmol min^{-1}/mg protein. For production of L-ascorbic acid in cultured cells, add L-gulono-γ-lactone or D-gluculono-γ-lactone (which gives D-glucuronic acid, the intermediate two steps before the substrate L-gulono-γ-lactone) to the culture medium, initially 1/50 volume and subsequently 1/100 volume of 2 M solution (freshly prepared) a few times every 30 min.

References

1. Yagi, K., Noda, H., Kurono, M., and Ohishi, N. (1993). Efficient gene transfer with less cytotoxicity by means of cationic multilamellar liposomes. *Biochem. Biophys. Res. Commun.*, **196**, 1042–8.
2. Koshizaka, T., Nishikimi, M., Ozawa, T., and Yagi, K. (1988). Isolation and sequence analysis of a complementary DNA encoding rat liver L-gulono-γ-lactone oxidase, a key enzyme for L-ascorbic acid biosynthesis. *J. Biol. Chem.*, **263**, 1619–21.

3. Kito, M., Ohishi, N., and Yagi, K. (1991). Micro-determination of L-gulono-γ-lactone oxidase activity. *Biochem. Int.*, **24**, 131–5.
4. Keanov, A., Reinisalo, M., Pitkanen, T. I., Nishikimi, M., and Molsa, H. (1998). Expression of rat gene for L-gulono-γ-lactone oxidase, the key enzyme of L-ascorbic acid biosynthesis, in guinea pig cells and in teleost fish rainbow trout (*Oncorhynchus mykiss*). *Biochim. Biophys. Acta*, **1381**, 241–8.

65

8-OH-dG: extraction/enzyme treatment /measurement of 8-OH-dG

SHINYA ASAMI and HIROSHI KASAI

Introduction

Reactive oxygen species (ROS) are produced by many carcinogenic agents and cause DNA damage in vitro and in vivo (1). ROS are also produced during the normal cellular respiration and oxygen metabolism. Therefore, studies on DNA damage induced by oxygen radicals should be helpful in elucidating the mechanisms of cancer induction by environmental agents and by endogenous processes. We first observed the formation of 8-hydroxy-deoxyguanosine (8-OH-dG, 7,8-dihydro-8-oxodeoxyguanosine) by oxygen radicals in 1984 (2) and there is now an increasing volume of data providing evidence that 8-OH-dG is a key biomarker relevant to carcinogenesis both in animal models and in human studies (3–5). In the following section, we describe our method for measuring 8-OH-dG levels.

Protocol

DNA extraction[1]

Tissue samples

1. Gently homogenize tissue (100–200 mg)[2] with lysis solution (1 mL; kit) on ice.
2. After brief centrifugation (10 000 rpm for 20 s at 4 °C) carefully remove the supernatant.
3. Add lysis solution (1 mL; kit) and place the tube on a microtube mixer for 30 s at moderate speed.
4. Repeat operations 2–3–2.
5. Suspend the resulting pellet in enzyme reaction solution (180 μL; kit) and add RNAse (Sigma, Type I-A, 10 mg mL^{-1}, 2 μL; heated at 100 °C for 15 min).
6. Incubate at 50 °C for 10 min.

7. Add proteinase K (10 μL) and incubate at 50°C for 1 h (mix the solution several times by inversion).
8. Centrifuge at 10 000 rpm for 10 min at room temperature and transfer supernatant (200 μL) to a new tube.
9. Add NaI solution (0.3 mL; kit) and mix well by inversion.
10. Add isopropyl alcohol (0.5 mL) to the mixture and mix well by inversion.
11. Centrifuge at 10 000 rpm for 10 min at room temperature and remove resulting supernatant.
12. Rinse the pellet by adding washing solution A (1 mL; kit) and centrifuge at 10 000 rpm for 5 min at room temperature and remove the supernatant. Repeat the same operation with 1 mL washing solution B (kit) instead of A (kit).
13. Vacuum-dry the pellet for approx. 3 min.
14. Suspend in EDTA solution (1 mM, 100 μL).

White blood cells

1. Centrifuge heparinized blood (5 10 mL)[2] at 1500 rpm for 15 min at 4°C.
2. Remove the buffy coat fractions (ca 1 mL; Fig. 1) and transfer to 2.0-mL microcentrifuge tube.

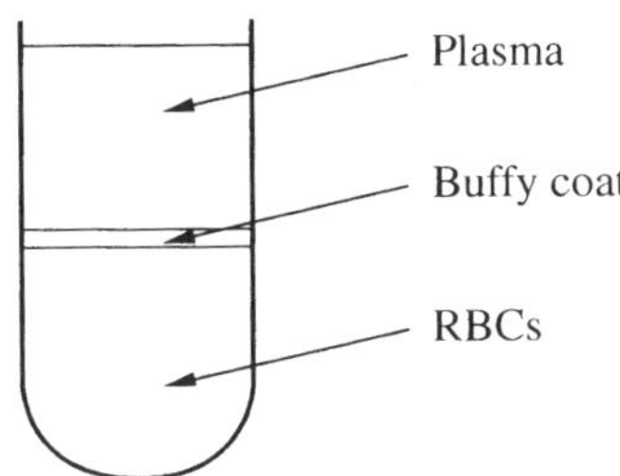

Figure 1. Buffy coat. Collect the white material which resembles a membrane.

3. Add lysis solution (0.7 mL; kit) and place the tube on a microtube mixer for 30 s at moderate speed.
4. After brief centrifugation (10 000 rpm for 20 s at 4°C) carefully remove the supernatant.
5. Add lysis solution (0.5 mL; kit) and place the tube on a microtube mixer for 30 s at moderate speed.
6. Repeat operations 4–5–4.
7. Suspend the resulting pellet in enzyme reaction solution (180 μL; kit) and add RNAse (2 μL).

Protocol *Continued*

8. Incubate at 37 °C for 10 min.
9. Add proteinase K (10 μL) and incubate at 37 °C for 1 h.
10. Add NaI solution (0.3 mL; kit) and mix well by inversion.
11. Add isopropyl alcohol (0.5 mL) to the mixture and mix well by inversion.
12. Centrifuge at 10 000 rpm for 10 min at room temperature and remove resulting supernatant.
13. Rinse the pellet by adding washing solution A (1 mL; kit) and centrifuge at 10 000 rpm for 5 min at room temperature and remove the supernatant. Repeat the same operation with 1 mL washing solution B (kit) instead of A (kit).
14. Vacuum-dry the pellet for approx. 3 min.
15. Suspend in EDTA solution (1 mM, 100 μL).

Enzyme treatment

1. Add sodium acetate (pH 4.5, 2 M, 1 μL; Applied Biosystems), nuclease P1 (5 mg mL^{-1}, 4 μL; Yamasa, Japan, YA7801), and acid phosphatase

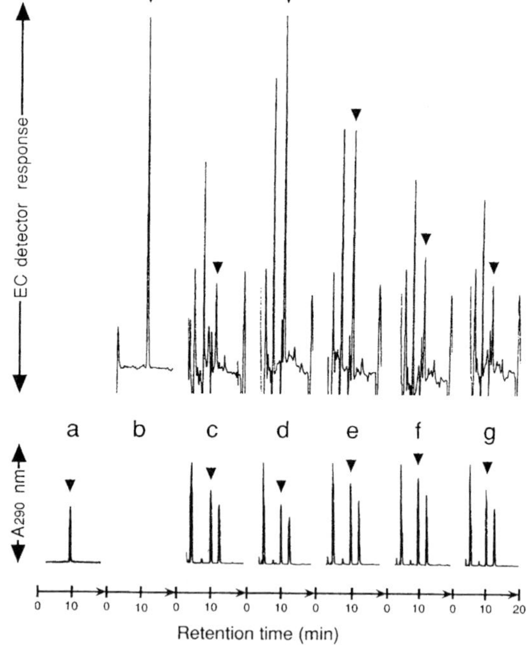

Figure 2. Analysis of 8-OH-dG in kidney at different times after administration of Fe-NTA to rats (4): (a) dG standard; (b) 8-OH-dG standard; (c) control (time 0); (d) 1 h; (e) 6 h; (f) 24 h; (g) 72 h.

(47 mg mL^{-1} suspension in 1.8 M $(NH_4)_2SO_4$; Sigma, P-1435; 2 μL) to the extracted nuclear DNA solution.

2. Incubate at 37 °C for 30 min.
3. Add Muromac ion exchange resin (50 mg mL^{-1}, 10 μL; Muromachi Kagaku Kogyo, Japan) to remove I^-.
4. Centrifuge at 15 000 rpm for 5 min at room temperature to remove the resin and transfer the supernatant to Ultrafree-Probind Filters (Millipore).
5. Centrifuge at 10 000 rpm for 1 min at room temperature.
6. Take sample for HPLC analysis.

Measurement of 8-OH-dG

HPLC analysis is performed by injecting the sample of DNA hydrolysate on to a 4.6 mm × 250 mm, 5 μm, Beckman Ultrasphere-ODS column with NaH_2PO_4 (10 mM) containing 8% *v/v* methanol as mobile phase. The chromatograph is equipped with an ECD (Coulochem II; ESA, USA), model 5020 guard cell (0.35 V), and model 5011 analytical cell (electrode 1, 0.15 V; electrode 2, 0.3V). Deoxyguanosine (dG; 0.5 mg mL, 20 μL) and 8-hydroxydeoxyguanosine (5 ng mL^{-1}, 20 μL) are injected as a standard samples. The amount of 8-OH-dG was calculated as the number of residues per 10^5 dG. Typical results are presented in Fig. 2.

[1] The nuclear DNA was extracted with a DNA Extractor WB Kit (Wako, Japan). The kit includes lysis solution, enzyme reaction solution, sodium iodide solution, washing solutions A and B, and protease. This kit can be used to isolate purified DNA from small samples.
[2] For analysis of formation of 8-OH-dG, investigators are advised to use fresh samples. If this is not possible tissue samples or buffy coat fractions can be stored at—70 °C until the time of analysis of 8-OH-dG in the DNA.

Comments

Several precautions must be taken if reliable 8-OH-dG data are required.

1. Avoid phenol and other organic solvents which usually contain small amounts of photosensitizer;
2. Avoid light, particularly during lysis or during contact with organic solvents;
3. Flush all tubes or buffer solutions with nitrogen gas;
4. Use an appropriate amount of tissue every time. The analysis of DNA from too little tissue results in a higher ratio of 8-OH-dG to dG;
5. Isolate the DNA as rapidly as possible, using the DNA extractor kit;
6. Analyse the DNA digest by the HPLC–ECD as soon as possible; and

7. Analyse samples from the control and treated groups within the same day.

References

1. Ames, B. N. (1983). Dietary carcinogens and anticarcinogens. *Science*, **221**, 1256–64.
2. Kasai, H. and Nishimura, S. (1984). Hydroxylation of deoxyguanosine at the C-8 position by ascorbic acid and other reducing agents. *Nucl. Acids Res.*, **12**, 2127–36.
3. Nakae, D., Mizumoto, Y., Kobayashi, E., Noguchi, O., and Konishi, Y. (1995). Improved genomic/nuclear DNA extraction for 8-hydroxydeoxyguanosine analysis of small amounts of rat liver tissue. *Cancer Lett.*, **97**, 233–9.
4. Yamaguchi, R., Hirano, T., Asami, S., Chung, M. H., Sugita, A., and Kasai, H. (1996). Increased 8-hydroxyguanine levels in DNA and its repair activity in rat kidney after administration of a renal carcinogen, ferric nitrilotriacetate. *Carcinogenesis*, **17**, 2419–22.
5. Asami, S., Hirano, T., Yamaguchi, R., Tomioka, Y., Itoh, H., and Kasai, H. (1996). Increase of a type of oxidative DNA damage, 8-hydroxyguanine, and its repair activity in human leukocytes by cigarette smoking. *Cancer Res.*, **56**, 1279–82.

66

Detection of DNA damage and analysis of its site-specificity

SHINJI OIKAWA, and SHOSUKE KAWANISHI

Introduction

Superoxide ($O_2 \cdot^-$) and hydrogen peroxide (H_2O_2) are constantly formed in the human body, and although they can serve useful physiological functions it is now believed that they are toxic when generated in excess. Certain carcinogens and chemicals produce excess $O_2 \cdot^-$ and H_2O_2, and subsequently more reactive oxygen species, in the presence of endogenous metal ions. These reactive oxygen species cause site-specific DNA damage.

Protocol

Labelling the 5′ ends of DNA fragment

DNA fragments were prepared from plasmid pbcNI which carries a 6.6 kb *Bam*HI chromosomal DNA segment containing the human c-Ha-*ras*-1 protooncogene and from plasmid pUC 18 containing the human *p53* tumour suppressor gene (1, 2).

1. Prepare reaction mixture containing 50 pmol DNA fragment, 10× protruding buffer[1] (5 μL), [γ-^{32}P] dATP (DuPont NEN, NEG-002Z, 9.25 MBq; 10 μL), T_4 polynucleotide kinase (10 units μL^{-1}, 2 μL), and H_2O to 50 μL.
2. Incubate at 37 °C for 60 min.

Digestion with restriction enzyme

1. Mix 5′-end-labelled DNA fragment (50 μL) , 10× digestion buffer (5 μL), and restriction enzyme (5 μL).
2. Incubate at 37 °C for 60 min.
3. Add loading buffer (glycerol (30% *v*/*v*), xylene cyanol FF (0.25% *w*/*v*), and bromophenol blue (0.25% *w*/*v*); 12 μL).
4. Load the DNA sample in the well (1 cm long and 2 cm wide) of a 6% *w*/*v* polyacrylamide gel in 1× TBE buffer[2].

Protocol *Continued*

5. Run at 50 mA for 90–120 min.
6. Remove the upper glass plate and wrap the gel attached to the lower plate in Saran Wrap.
7. Apply an X-ray film on the gel and expose for 1 min at room temperature in dark room.
8. Locate the two DNA bands of interest and remove these bands by cutting with a sharp razor blade.
9. Cut a band in half and transfer pieces to different 2-mL Eppendorf tubes.
10. Crush the piece with a pipette tip and add Gilbert's buffer (ammonium acetate (0.5 M), magnesium acetate (10 mM), EDTA (1 mM), and SDS (0.1% *w*/*v*); 800 μL).
11. Cap the tube and wrap in aluminium foil.
12. Rotate the tube overnight by means of a rotating wheel.
13. Centrifuge the sample at 15 000 rpm for 10 min at 4 °C.
14. Recover the supernatant. (Be careful to avoid transferring acrylamide gel.)
15. Remove any remaining acrylamide gel by passing the supernatant through a small column of glass wool contained in the tip of a 1000-μL pipette.
16. Divide the supernatant into equal parts over 30 kcpm to Eppendorf tubes.
17. Precipitate the DNA with 10 vol. 99.5% *v*/*v* ethanol and keep at –80 °C overnight.
18. Recover the DNA by centrifugation (15 000 rpm for 5 min at 4 °C).
19. Rinse the DNA pellet once with 70% *v*/*v* ethanol.
20. Re-suspend the pellet in H_2O (170 μL).
21. Add sodium acetate (pH 5.2, 3 M, 30 μL) and precipitate the DNA again with 99.5% *v*/*v* ethanol (1000 μL).
22. Store at –80 °C until use (stock solution).

Detection of DNA damage induced by reactive oxygen species (small-scale experiment)

Prepare ^{32}P-labelled DNA fragment

1. Centrifuge the stock solution (over 30 kcpm) at 15 000 rpm for 5 min at 4°C and remove the supernatant.
2. Rinse with 70% *v*/*v* ethanol (400 μL).
3. Dry quickly under vacuum.

4. Dissolve the labelled DNA pellet in H_2O (approx. 500 μL).
5. Divide the labelled DNA solution into equal parts of 5–10 kcpm in Eppendorf tubes.
6. Add sodium acetate (pH 5.2, 3 M, 0.1 vols) and 99.5% *v/v* ethanol (3 vols) to each tube, and vortex-mix.
7. Store at –80 °C for 5 min (or until use).
8. Centrifuge at 15 000 rpm for 5 min at 4°C and remove the supernatant.
9. Rinse with 70% *v/v* ethanol (400 μL).
10. Dry quickly under vacuum.
11. Redissolve the labelled DNA pellet in the desired volume of phosphate buffer (pH 7.8, 20 mM) containing DTPA (10 μM).

Reaction and electrophoresis

1. Prepare reaction mixture containing (per tube) labelled DNA solution (ca. 600 cpm, 5 μL), calf thymus DNA (Sigma, deoxyribonucleic acid, type I: sodium salt, product no. D1505; 0.8 mM, 5 μL), phosphate buffer (pH 7.8, 20 mM, containing 10 μM DTPA; 95 μL), reagents (metal, etc.), and H_2O (to 200 μL).
2. Incubate at 37 °C for 60 min.
3. Add DTPA (10 mM, 15 μL) and vortex-mix.
4. Add tRNA (10 mg mL^{-1}, 5 μL), sodium acetate (pH 5.2, 3 M, 20 μL), and 99.5% *v/v* ethanol (600 μL), and vortex-mix.
5. Precipitate at –80°C for 5 min.
6. Centrifuge at 15 000 rpm for 5 min at 4 °C and remove the supernatant.
7. Rinse with 70% *v/v* ethanol (400 μL).
8. Dry quickly under vacuum.
9. Re-dissolve the pellet in piperidine (10% *v/v*, 100 μL).
10. Incubate at 90 °C for 20 min.
11. Place on ice for 2 min and centrifuge.
12. Add sodium acetate (pH 5.2, 3 M, 10 μL) and 99.5% *v/v* ethanol (300 μL), and vortex-mix.
13. Precipitate at –80 °C for 5 min.
14. Centrifuge at 15 000 rpm for 5 min at 4 °C and remove the supernatant.
15. Rinse with 70% *v/v* ethanol (400 μL).
16. Dry quickly under vacuum.
17. Re-dissolve the pellet in loading buffer (deionized formamide (80% *v/v*), Tris-borate (pH 8.3, 50 mM), xylene cyanol FF (0.1% *w/v*), bromophenol blue (0.1% *w/v*), and EDTA (1 mM); 10–15 μL).

Protocol *Continued*

18. Heat for 2 min at 90 °C then place on ice.
19. Load all of each sample per well on an 8% polyacrylamide–8 M urea gel[3] (18.3 cm × 16.8 cm plate) in 1× TBE buffer (rinse by immersing pipette tip twice in the lower reservoir after dispensing from each reaction tube).
20. Run gel at 50 mA constant current for 30–40 min.
21. Dry the gel thoroughly at 80 °C for 60 to 90 min.
22. Place dried gel in a cassette with an X-ray film and an intensifying screen.
23. Autoradiograph at –80 °C for 1 to 4 days.

Analysis of site-specificity (sequencing)

Prepare ^{32}P labelled DNA fragment

1. Centrifuge the stock solution (over 30 kcpm) at 15 000 rpm for 5 min at 4 °C and remove the supernatant.
2. Rinse with 70% *v*/*v* ethanol (400 μL).
3. Dry quickly under vacuum.
4. Re-dissolve the labelled DNA pellet in the desired volume of phosphate buffer (pH 7.8, 20 mM) containing DTPA (10 μM).

Reaction and electrophoresis

1. Prepare reaction mixture containing (per tube) labelled DNA solution (ca 5 kcpm, 5 μL), calf thymus DNA (0.8 mM, 5 μL), phosphate buffer (pH 7.8, 20 mM, containing 10 μM DTPA; 95 μL), reagents (metal, etc.), and H_2O (to 200 μL).
2. Incubate reaction mixture at 37 °C for 60 min.
3. Add DTPA (10 mM, 15 μL) and vortex-mix.
4. Add tRNA (10 mg mL^{-1}, 5 μL), sodium acetate (pH 5.2, 3 M, 20 μL), and 99.5% *v*/*v* ethanol (600 μL), and vortex-mix.
5. Precipitate at –80 °C for 5 min.
6. Centrifuge at 15 000 rpm for 5 min at 4 °C and remove the supernatant.
7. Rinse with 70% *v*/*v* ethanol (400 μL).
8. Add H_2O (200 μL), sodium acetate (pH 5.2, 3 M, 20 μL) and 99.5% ethanol (600 μL), and vortex-mix.
9. Precipitate at –80 °C for 5 min.
10. Centrifuge at 15 000 rpm for 5 min at 4 °C and remove the supernatant.
11. Rinse with 70% *v*/*v* ethanol (400 μL).

12. Dry quickly under vacuum.
13. Re-dissolve the pellet in piperidine (10% *v/v*, 100 μL).
14. Incubate at 90 °C for 20 min.
15. Place on ice for 2 min and centrifuge.
16. Add sodium acetate (pH 5.2, 3 M, 10 μL) and 99.5% *v/v* ethanol (300 μL), and vortex-mix.
17. Precipitate at –80 °C for 5 min.
18. Centrifuge at 15 000 rpm for 5 min at 4 °C and remove the supernatant.
19. Rinse with 70% *v/v* ethanol (400 μL).
20. Dry quickly under vacuum.
21. Re-dissolve the pellet in loading buffer (deionized formamide (80% *v/v*), Tris-borate (pH 8.3, 50 mM), xylene cyanol FF (0.1% *w/v*), bromophenol blue (0.1% *w/v*), and EDTA (1 mM); 1 kcpm μL^{-1}).
22. Heat for 2 min at 90 °C then place on ice.
23. Load each sample (1 μL per well) on an 8% polyacrylamide–8 M urea gel in 1× TBE buffer.
24. Run gel at 40 mA constant current for 2 h, using Pharmacia-LKB 2010 Macrophor Sequencing System (Fig. 1).

Figure 1.

Protocol *Continued*

25. Dry the gel thoroughly for 30 to 60 min at 80 °C.
26. Place dried gel in a cassette with an X-ray film and an intensifying screen.
27. Autoradiograph at –80 °C for 1 day.

Prepare G+A and T+C standards (3)

1. Centrifuge the stock solution (over 30 kcpm) at 15 000 rpm for 5 min at 4 °C and remove the supernatant.
2. Rinse with 70% *v/v* ethanol (400 μL).
3. Dry quickly under vacuum.
4. Re-dissolve the labelled DNA pellet in H_2O (35 μL).
5. Prepare G + A standard by mixing the labelled DNA solution (15 μL), H_2O (5 μL), and formic acid (50 μL).
6. Incubate at room temperature for 6 min.
7. Prepare T + C standard by mixing the labelled DNA solution (20 μL) and hydrazine (30 μL).
8. Incubate at room temperature for 7 min.
9. Add Hz stop buffer (sodium acetate (pH 5.2, 0.3 M), EDTA (0.1 mM), tRNA (25 μg mL^{-1}); 200 μL) and 99.5% *v/v* ethanol (750 μL), and vortex-mix.
10. Store overnight at –80 °C.
11. Centrifuge at 15 000 rpm for 10 min at 4 °C and remove the supernatant.
12. Rinse with 70% *v/v* ethanol (400 μL).
13. Dry quickly under vacuum.
14. Re-dissolve the pellet in piperidine (10% *v/v*, 100 μL).
15. Incubate at 90 °C for 20 min.
16. Place on ice for 2 min and centrifuge.
17. Add sodium acetate (pH 5.2, 3 M, 20 μL) and 99.5% *v/v* ethanol (500 μL), and vortex-mix.
18. Store overnight at –80 °C.
19. Centrifuge at 15 000 rpm for 10 min at 4 °C and remove the supernatant.
20. Rinse with 70% *v/v* ethanol (400 μL).
21. Dry quickly under vacuum.
22. Re-dissolve the pellets in the desired volume (1–2 kcpm μL^{-1}) of loading buffer (deionized formamide (80% *v/v*), Tris-borate (pH 8.3, 50 mM), xylene cyanol FF (0.1% *w/v*), bromophenol blue (0.1% *w/v*), and EDTA (1 mM)).

23. Heat for 2 min at 90 °C, then place on ice.

24. Load each standard (1 μL per well) on an 8% polyacrylamide–8 M urea gel in 1× TBE buffer.

[1] 10× Protruding buffer is Tris-HCl (500 mM) containing $MgCl_2$ (100 mM), DTT (50 mM), spermidine (1 mM), and EDTA (1 mM). It should be stored at –20 °C

[2] 6% Acrylamide sol is prepared by mixing acrylamide (58 g), bisacrylamide (2 g), 10× TBE buffer (100 mL), and H_2O to 1 L. 10× TBE buffer is prepared by mixing Tris (108 g), boric acid (55 g), EDTA.2Na (9.3 g) and H_2O to 1 L.

[3] 8% acrylamide–8 M urea sol is prepared by mixing acrylamide (77.6 g), bisacrylamide (2.4 g), 10× TBE (100 mL), urea (480 g), and H_2O to 1 L.

References

1. Yamamoto, K. and Kawanishi, S. (1989). Hydroxyl free radical is not the main active species in site-specific DNA damage induced by copper(II) ion and hydrogen peroxide. *J. Biol. Chem.*, **264**, 15435–40.
2. Yamashita, N., Murata, M., Inoue, S., Hiraku, Y., Yoshinaga, T., and Kawanishi, S. (1998). Superoxide formation and DNA damage induced by a fragrant furanone in the presence of copper (II). *Mutat. Res.*, **397**, 191–201.
3. Maxam, A. M. and Gilbert, W. (1980) Sequencing end-labelled DNA with base-specific chemical cleavages. *Methods Enzymol.*, **65**, 499–560.

67

Detection of DNA/protein cleavages by glycation

HIDEAKI KANETO and NAOYUKI TANIGUCHI

Introduction

Reactive oxygen species, which are known to be produced by the glycation reaction seen under diabetic conditions, trigger DNA cleavage (1) as well as protein cleavage. In this chapter we describe a method of detecting DNA and protein cleavage (2) induced by the glycation reaction.

Protocol

Detection of DNA cleavage by glycation

1. Mix digested DNA fragment (4 pmol; 500–1000 bp fragments of human Mn-SOD DNA were used in our experiment.), CIP (calf intestine alkaline phosphatase; 2 μL), 10× CIP buffer ($ZnCl_2$ (10 mM), $MgCl_2$ (10 mM), Tris-HCl (pH 8.3, 100 mM, 20 μL), and H_2O (178 μL)).
2. Incubate at 37 °C for 60 min.
3. Add EDTA (100 mM, 10 μL).
4. Incubate at 75 °C for 10 min.
5. Add TE saturated phenol (100 μL) and chloroform–isoamylalcohol (24:1, 100 μL).
6. Collect the aqueous layer.
7. Add NaCl (5 M, 10 μL) and 100% EtOH (500 μL).
8. Incubate at –35 °C for more than 1 h.
9. Centrifuge at 5000 g for 10 min at 4 °C.
10. Wash with 70% *v/v* EtOH.
11. Suspend in H_2O (12 μL).
12. Add [γ-^{32}P] ATP (4 μL), 10× kinase buffer (2 μL), and T4PNK (polynucleotide kinase; 2 μL).
13. Incubate at 37 °C for 1 h.

14. Add labelled DNA (1 μL), KPB (pH 7.4, 50 mM, 11 μL), protein (1 mg mL^{-1}; recombinant human Cu, Zn-SOD was used in our experiments), reducing sugar (10–500 mM; glucose, fructose, and ribose were used in our experiments.)
15. Incubate at 37 °C for an appropriate time (1–10 days).
16. Add 8 μL stop solution (formamide (95%), EDTA (20 mM), bromophenol blue (0.05% *w*/*v*), xylene cyanol (0.05% *w*/*v*)).
17. Boil for 2 min and cool on ice.
18. Perform electrophoresis at 100V in 5% (*w*/*v*) acrylamide gel in the presence of 50% (*w*/*v*) urea. (acrylamide (30% *w*/*v*, 1.67 mL), 10× TBE (1 mL), ammonium persulfate (10 % *w*/*v*, 50 μL), TEMED (20 μL), and urea (5 g) in H_2O (10 mL))
19. Dry and expose to X-ray film at –80°C overnight.

Detection of protein cleavage by glycation

1. Mix protein (1 mg mL^{-1}; recombinant human Cu, Zn-SOD was used in our experiments), reducing sugar (10–500 mM; glucose, fructose, and ribose were used in our experiments), KPB (pH 7.4, 50 mM; phosphate buffer is suitable for glycation reaction). NaCl (150 mM), and NaN_3 (0.025% *w*/*v*).
2. Incubate at 37 °C for an appropriate time (1–10 days).
3. Add sample buffer (SDS (1% *w*/*v*), bromophenol blue (0.1%, *w*/*v*), glycerol (30% *v*/*v*), mercaptoethanol (10% *v*/*v*); 5 μL for each 10 μL of sample).
4. Heat at 100°C for 1 min.
5. Perform electrophoresis at 20 mA in 10% (*w*/*v*) acrylamide gel in the presence of SDS. (Separation gel: Tris-HCl (pH 8.9, 1 M, 7.5 mL), acrylamide (30% *w*/*v*, 6.7 mL), H_2O (5.4 mL), ammonium persulfate (10 % *w*/*v*, 100 μL), TEMED (10 μL), SDS (10 % *w*/*v*, 200 μL). Upper gel: Tris-HCl (pH 6.7, 1 M, 1.25 mL), acrylamide (30% *w*/*v*, 1 mL), H_2O (7.5 mL), ammonium persulfate (10 % *w*/*v*,100 μL), TEMED (5 μL), SDS (10 % *w*/*v*,100 μL). Running buffer: Tris (3 g), glycine (14.4 g), SDS (1 g), H_2O (1000 mL).
6. Agitate the gel gently for 30 min in staining solution (coomassie brilliant blue R-250 (1.25 g), MeOH (227 mL), AcOH (46 mL), H_2O (500 mL)).
7. Agitate the gel gently overnight in destaining solution (MeOH (75 mL), AcOH (50 mL), H_2O (500 mL)).

References

1. Ookawara, T., Kwamura, N., Kitagawa, Y., and Taniguchi, N. (1992). Site-specific and random fragmentation of Cu,Zn-superoxide dismutase by glycation reaction. Implication of reactive oxygen species. *J. Biol. Chem.*, **267**, 18505–10.
2. Kaneto, H., Fujii, J., Suzuki, K., Kasai, H., Kawamori, R., Kamada, T., and Taniguchi, N. (1994). DNA cleavage induced by glycation of Cu,Zn-superoxide dismutase. *Biochem. J.*, **304**, 219–25.

68

4-hydroxy-2-nonenal protein adducts

KOJI UCHIDA

Introduction

Among the aldehydes which originate from the peroxidation of cellular membrane lipids, 4-hydroxy-2-nonenal (HNE) is believed to be largely responsible for the cytopathological effects of oxidative stress in vivo (1). HNE has a wide range of biological activity including inhibition of protein and DNA synthesis, inactivation of enzymes, stimulation of phospholipase C, reduction recently of gap–junction communication, and stimulation of neutrophil migration (1). Many of these in vitro effects, which are observed at low micromolar or even submicromolar concentrations of HNE, have been attributed to the modification of cellular proteins with HNE.

We have recently developed procedures for the detection of HNE adducts generated in HNE-modified proteins and shown that, in addition to cysteine (2), HNE forms stable Michael addition-type adducts with histidine and lysine residues of proteins (Fig. 1) (3,4). A sensitive immunochemical procedure that uses an antibody specific to the HNE adducts was developed for detection of these adducts (5). The selectivity and sensitivity of this procedure were established by results of studies in which the HNE-specific antibody was used to detect HNE-modified proteins in a variety of human tissue samples.

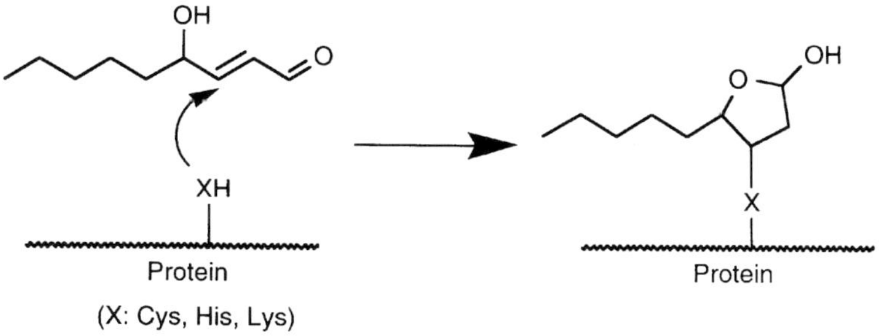

Figure 1. The Michael-type addition of HNE to protein. X represents amino acid side-chains such as cysteine, histidine, and lysine.

Protocol

Preparation of Mono-specific Antibodies.

Preparation of antigen

1. Incubate keyhole limpet haemocyanin (KLH; 1 mg) with HNE (2 mM) in sodium phosphate buffer (pH 7.2, 50 mM, 1 mL) for 2 h at 37 °C.
2. Add *N*-acetylcysteine (100 mM, 0.1 mL) to terminate the reaction.

Immunization

1. Emulsify the HNE-modified KLH prepared above with the same volume of Freund's complete adjuvant.
2. Inject intradermally into several sites in New Zealand White rabbits.
3. After 7–14 days, take samples of the serum and check antibody production, (Antibody response is monitored by immunoblotting or by ELISA using HNE-modified BSA as the antigen).
4. When a good titre has developed, repeat booster injections of Freund's incomplete adjuvant and collect samples of sera (20–40 mL) 10 days after each boost.

Purification of mono-specific antibody

Preparation of ligand

1. Incubate the peptide (NH_2–Ala_3–His–Ala_3–CO–NH_2; 5 mg) with HNE (10 mM) in sodium phosphate buffer (pH 7.2, 50 mM, 1 mL) for 20 h at 37 °C.
2. Purify the peptide containing the HNE–histidine adduct by HPLC on a 0.46 cm × 25 cm TSK-GEL ODS-80® column (TOSO HAAS) with a linear gradient from 0.05% trifluoroacetic acid in water at time 0 to 100% acetonitrile after 20 min at a flow rate of 1 mL min^{-1}.

Immunoaffinity purification

1. Bind the ligand to the affinity resin by incubation of a gel slurry of Affi-Gel 10 with the HNE-derivatized Ala_3–His–Ala_3 amide in Hepes (pH 8.0, 0.1 M) for 20 h at 4 °C.
2. Dilute the antiserum (2 mL) to 20 mL with Hepes buffer (pH 8.0, 0.1 M).
3. Pass the serum at 10 mL h^{-1} through an affinity column containing affinity resin (0.1 M, 1 mL).
4. Repeat step 3 three times.
5. Remove unbound proteins by washing with Hepes buffer (pH 8.0, 0.1 M, 5 mL) then NaCl (100 mM, 5 mL).
6. Elute bound antibodies with glycine (pH 2.5, 0.1 M, 5 mL as 1 mL × 5).

7. Immediately add Tris-HCl (pH 8.0, 14 M, 0.5 mL) to the eluate (1 mL).
8. Store the mono-specific antibody at –80 °C.

Immunohistochemistry (6)

1. Cut small blocks of tissue (approximately 1 $cm^2 \times 0.4$ cm).
2. Place in Bouin's fixative (picric acid–acetic acid–formaldehyde, 15:5:1) overnight and immerse sequentially for 24 h in 50% and 70% picric acid.
3. Follow standard paraffin embedding procedures.
4. Mount 4-mm sections on clean glass slides.
5. Dewax in xylene (3 × 5 min).
6. Dehydrate by passing through graded alcohols.
7. Rinse in water.
8. Incubate with hydrogen peroxide (0.09 %) in 30 % methanol for 30 min to inhibit endogenous peroxidase.
9. Incubate with normal goat serum (diluted 1:75) to inhibit non-specific binding of the secondary antibody.
10. Add rabbit mono-specific antibody solution (0.5 mg mL^{-1}).
11. Apply the labelled secondary antibody (biotin-labelled goat anti-rabbit IgG serum).
12. Detection (avidin–biotin complex).

References

1. Esterbauer, H., Schauer, R. J., and Zoller, H. (1991). Chemistry and biochemistry of 4-hydroxynonenal, malonaldehyde and related aldehydes. *Free Radicals Biol. Med.*, **11**, 81–128.
2. Uchida, K. and Stadtman, E. R. (1992). Modification of histidine residues in proteins by reaction with 4-hydroxynonenal. *Proc. Natl. Acad. Sci. USA*, **89**, 4544–8.
3. Uchida, K. and Stadtman, E. R. (1992). Selective cleavage of thioether linkage in protein modified with 4-hydroxynonenal. *Proc. Natl. Acad. Sci. USA*, **89**, 5611–15.
4. Uchida, K. and Stadman, E. R. (1993). Covalent attachment of 4-hydroxynonenal to glyceraldehyde-3-phosphate dehydrogenase: a possible involvement of intra-molecular and intermolecular cross-linking reactions. *J. Biol. Chem.* **268**, 6388–93.
5. Uchida, K., Szweda, L. I., Chae, H. Z., and Staddtman, E. R. (1993). Immunochemical detection of 4-hydroxy-2-nonenal-modified proteins in oxidized hepatocytes. *Proc. Natl. Acad. Sci. USA*, **90**, 8742–6.
6. Toyokuni, S., Uchida, K., Okamoto, K., Hattori-Nakakuki,Y., Hiai, H., and Stadtman, E. R. (1994). Formation of 4-hydroxy-2-nonenal-modified proteins in the renal proximal tubules of rats treated with a renal carcinogen, ferric nitrilotriacetate. *Proc. Natl. Acad. Sci. USA*, **91**, 2616–20.

69

Assay for 8-oxodGTP hydrolase

HISASHI MAKI

Introduction

Mutagenesis caused by reactive oxygen species is a consequence of either DNA or dNTP damage. 8-oxodG is produced by oxidation of guanine nucleotides, known to be the most potent mutagenic damage inducing specific base substitutions. Organisms have cellular functions which avoid such mutageneses. 8-oxodG in DNA can be repaired by DNA glycosylase specific to 8-oxodG, and 8-oxodGTP is hydrolysed by 8-oxodGTP hydrolase (1–3). When these enzymes are hampered, increased spontaneous mutation is observed in cells. These enzymes, therefore, suppress spontaneous mutagenesis and carcinogenesis in normally growing cells.

Protocol

Preparation of 8-oxodGTP[1]

1. Mix, at 0 °C[2], dGTP (12 mM in potassium phosphate buffer (pH 6.8, 0.2 M); 0.5 mL), distilled water (0.1 mL), ascorbic acid (0.1 M, 0.3 mL), and H_2O_2 (1 M, 0.1 mL).
2. Incubate at 37 °C for 2 h in the dark.
3. Inject 100 μL into an FPLC MonoQ column equilibrated with triethylammoniumbicarbonate (TEAB; 10 mM)[3].
4. Elute 8-OxodGTP in 20 mL of a gradient from 10 mM to 1 M TEAB[4].
5. Collect peak fractions, lyophilize, and store at –20 °C[5].
   The results are illustrated in Fig. 1

[1] Kasai and Nishimura (4) reported methods for synthesis of 8-oxo derivatives by oxidation of guanine nucleoside and nucleotide. The most efficient conditions are those using ascorbic acids and H_2O_2 at neutral pH . Free dGTP is readily oxidized to produce 8-oxodGTP in an amount corresponding to approx. 30% of the starting material under these conditions. One of the oxidized products, 8-oxodGTP, is, however, also subject to further oxidation and disappears when the reaction goes for too long. It is not easy to purify 8-oxodGTP after the synthesis because it is hardly separated from dGTP. 8-oxodGTP is determined by measuring its absorbance at 245 nm. The molar absorption coefficient of 8-oxodGTP at pH 7.0 is $1.23 \times 10^4\ M^{-1}\ cm^{-1}$

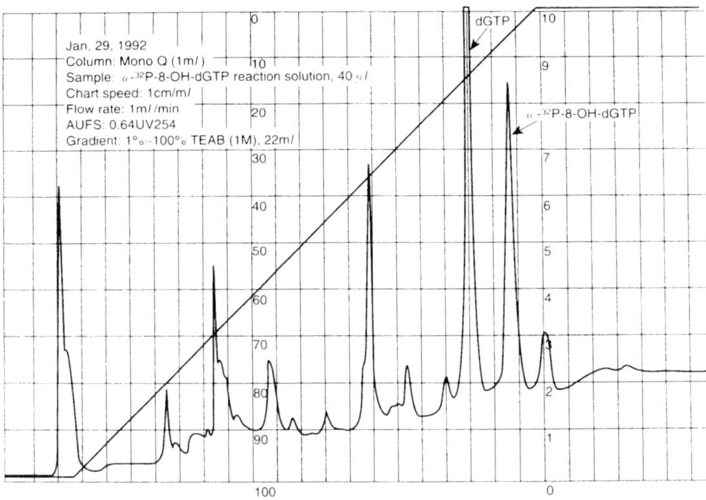

Figure 1. Purification of 8–0H-dGTP on a mono Q column.

[2] Radioisotope-labelled 8-oxodGTP should be used for assay of 8-oxodGTPase. The reaction should, therefore, contain approx. 50 μCi of α-^{32}P-dGTP. Prepare solutions of ascorbic acid and H_2O_2 just before use.
[3] 50–100 μL sample can be injected to 1-mL MonoQ column. TEAB (1 M, pH 7.0) can be purchased from Wako Pure Chemicals.
[4] The MonoQ column should be equilibrated with 10 mM buffer. Purified and lyophilized materials should be stable and kept at −20 °C. The solution should be prepared at neutral pH and at 1 mM or higher concentration.

Assay of 8-oxodGTPase activity[1]

1. Prepare the 8-oxodGTPase reaction mixture by mixing, at 0 °C, Tris-HCl buffer (pH 7.5, 20 mM), glycerol (4% *v*/*v*), bovine serum albumin (100 μg mL^{-1}), $MgCl_2$ (8 mM), dithiothreitol (5 mM), and 8-oxodGTP (10 μM).
2. Add cell crude extract (0.5–1.0 mL) to the 8-OH-dGTPase reaction mixture (10 μL).
3. Incubate at 30 °C for 5 min.
4. Stop reaction by addition of EDTA (10 mM , 10 μL).
5. Apply 2 μL to PEI cellulose sheet.
6. After drying spot, develop against 1 M LiCl for 1 h.[2]
7. Dry PEI cellulose sheet
8. Measure radioactivity of 8-oxodGMP produced by autoradiography.

The results are illustrated in Fig. 2.

[1] Assay should be performed with α-^{32}P-8-oxodGTP. 8-oxodGMP produced by 8-oxodGTpase-catalysed hydrolysis of 8-oxodGTP can be separated by thin layer chromatography on a Figure

Protocol *Continued*

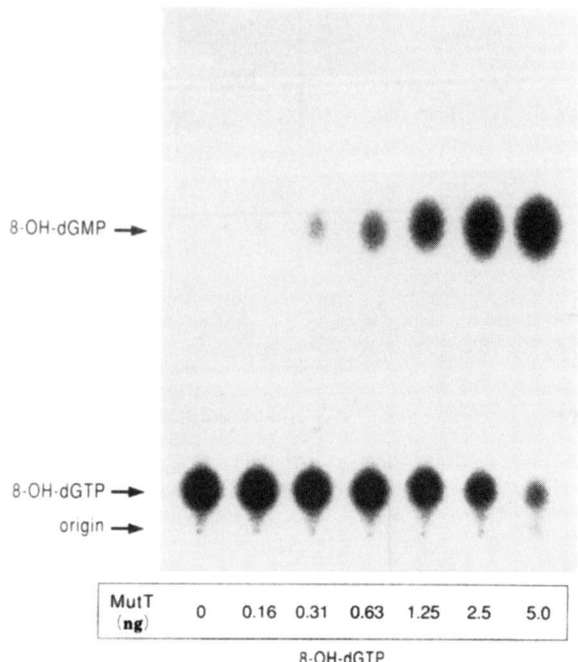

Figure 2. TLC assay of 8-OH-dGTP.

[1] Assay should be performed with α-^{32}P-8-oxodGTP. 8-oxodGMP produced by 8-oxodGTpase-catalysed hydrolisis of 8-oxodGTP can be seperated by thin layer chromotography on a polyethyleneimine (PEI)–cellulose plate (Merck). The amount of product can be determined by measuring the radioactivity of the spot corresponding to 8-oxodGMP. 1 unit of the activity is the amount of enzyme that produce 1 pmol 8-oxodGMP in 1 min.
[2] 1 M LiCl should be fresh.

References

1. Maki, H. and Sekiguchi, M. (1992). MutT protein specifically hydrolyses 8-oxodGTP, a potent mutagenic substrate for DNA synthesis. *Nature*, **355**, 273.
2. Akiyama, M., Maki, H., Sekiguchi, M., and Horiuchi, T. (1989). A specific role of MutT protein: To prevent dG:dA mispairing in DNA replication. *Proc. Natl. Acad. Sci. USA*, **86**, 3949.
3. Mo, J.-Y., Maki, H., and Sekiguchi, M. (1992). Hydrolytic elimination of a mutagenic nucleotide, 8-oxodGTP, by human 18-kD protein: Sanitization of nucleotide pool. *Proc. Natl. Acad. Sci. USA*, **89**, 11021.
4. Kasai, H. and Nishimura, S. (1984). Hydroxylation of deoxyguanosine at the C-8 position by ascorbic acid and other reducing agents. *Nucl. Acids Res.*, **12**, 2137.

70

Mitochondrial DNA damage

MIKA HAYAKAWA and TAKAYUKI OZAWA

Introduction

Mitochondrial DNA (mt DNA) is constantly exposed to reactive oxygen species generated in the mitochondrial respiratory chain and it is, therefore, damaged much more rapidly than nuclear DNA. Reactive oxygen species convert deoxyguanosine in mt DNA to 8-hydroxydeoxyguanosine (8-OH-dG) (Fig. 1).

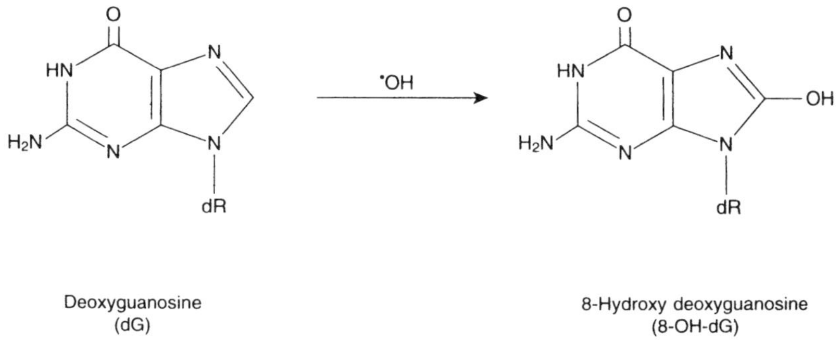

$\cdot$OH

Deoxyguanosine
(dG)

8-Hydroxy deoxyguanosine
(8-OH-dG)

Figure 1. Conversion of dG to 8-OH-dG by hydroxyl radical.

The method is useful for the direct assay of mt DNA 8-OH-dG by micro HPLC–mass spectrometry (Figs. 2 and 3).

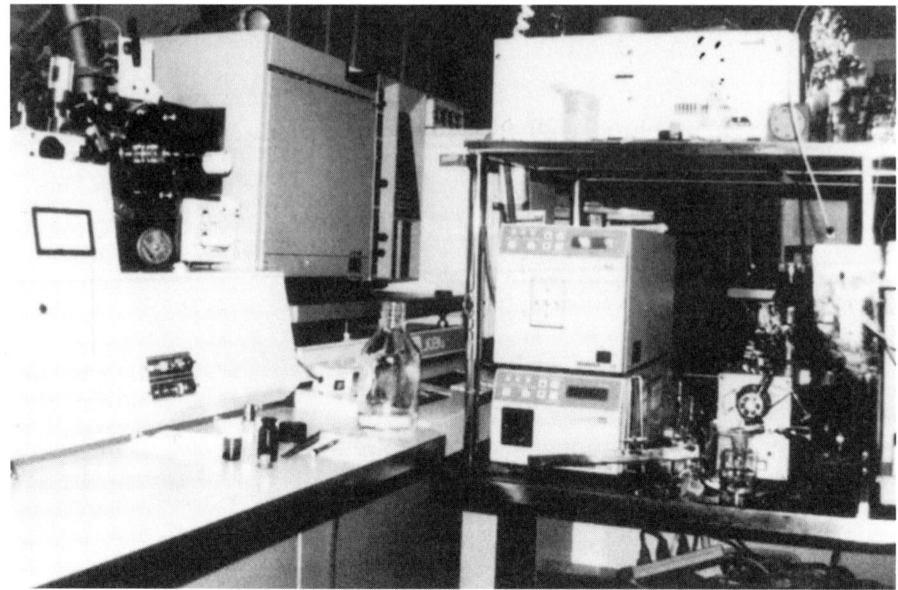

Figure 2. The micro HPLC–MS system

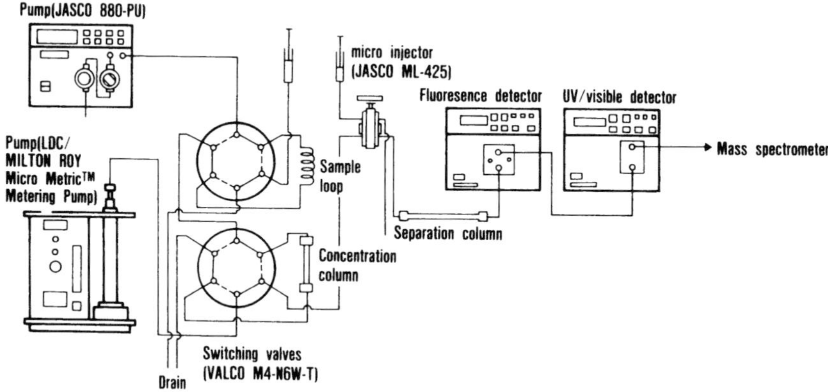

Figure 3. Schematic diagram of the micro HPLC–MS system.

Protocol

Preparation of mitochondrial fraction from liver

This was performed by a modification of the method of Hogeboon, by means of discontinuous sucrose gradient fractionation (1).

1. Weigh liver samples and homogenize for 2 min at 0–4 °C in 10 vol. Tris-

HCl (pH 7.4, 5 mM) containing sucrose (0.25 M) and EDTA (40 μM), by means of a Potter–Elvehjem-type homogenizer.

2. Centrifuge the homogenate at 80 g for 7 min.
3. Add the supernatant to an equal volume of sucrose solution (0.35 M) containing EDTA (40 μM) and Tris-HCl (pH 7.4, 5 mM).
4. Centrifuge at 700 g for 10 min.
5. Carefully collect approx. two-thirds of the supernatant and centrifuge the remaining fluid at 7000 g for 10 min.
6. Discard the opalescent supernatant and partially sedimented pink layer.
7. Wash the tightly packed precipitate and mitochondrial fraction twice with the homogenizing buffer.
8. Measure the protein concentration by the Biuret method.

Extraction of mt DNA

1. Suspend freshly isolated mitochondria in EDTA (0.1 M)–Tris-HCl (pH 7.4, 10 mM).
2. Incubate in a shaking water bath for 4 h at 37 °C in the presence of sodium dodecyl sulfate (0.5%) and proteinase K (100 mg mL^{-1}).
3. Extract twice with phenol–chloroform–isoamyl alcohol (25:24:1 *v*/*v*).
4. Extract with diethyl ether.
5. Add one-tenth vol. sodium acetate (pH 7.4, 3 M) and 2 vol. of absolute ethanol.
6. Precipitate at –20 °C overnight.
7. Centrifuge at 13,000 X g for 10 min
8. Dry the precipitate in vacuo.
9. Add Tris-HCl (pH 7.4, 10 mM) containing EDTA (1 mM).
10. Incubate with pancreatic ribonuclease (10 mM) for 3 min at 37 °C.
11. Extract with phenol–chloroform–isoamyl alcohol (25:24:1 *v*/*v*).
12. Add one-tenth vol. sodium acetate (pH 7.4, 3 M) and 2 vol. of absolute ethanol.
13. Precipitate at –20 °C overnight.

Comments

Phenol should be neutralized by addition of 1 M Tris, pH 8.0. Because dG is readily converted to 8-OH-dG by the extraction procedures, the extraction solvent should be as pure as possible. The absence of nuclear DNA in the mt DNA preparation should be confirmed by 1% agarose gel electrophoresis and ethidium staining after restriction endonuclease digestion.

Protocol *Continued*

Enzymatic hydrolysis of mt DNA to nucleosides

1. Add DNAaseI (200 units/mg DNA) to mt DNA (0.2–1.0 mg) in Tris-HCl (pH 8.5, 40 mM, 100 μL) containing $MgCl_2$ (10 mM).
2. Add spleen exonuclease (0.01 units/mg DNA).
3. Add snake venom exonuclease (0.5 units/mg DNA).
4. Incubate for 2 h at 37 °C to give DNA hydrolysate. Completion of hydrolysis should be confirmed by agarose-gel electrophoresis and ethidium bromide staining.

Analysis by liquid chromatography–mass spectrometry (2)

Analysis is performed by means of two chromatography pumps, switching valves, a concentration column (Develosil ODS), a separation microcolumn (0.3 mm i.d. × 150 mm, 5 μm particle, Develosil ODS column (Nomura Kagaku), total bed volume is 7 μL), UV-detector, fluorescence detector, and mass spectrometer. The mobile phase is ammonium formate (50 mM) containing 7% methanol and 0.5% glycerol, at a flow rate of 4 μL min^{-1}.

There are two methods of injection—direct injection for microcolumn analysis and injection after sample concentration. In the direct injection method, the amount of sample applied is only 0.02 μL, less than one-hundredth the volume of the microcolumn bed. In the other method up to 500 μL is applied to the concentration column and up to 25 000-fold concentration is possible. The concentrate is directed to the analytical column by switching a valve.

Eluate from the microcolumn is monitored by fluorescence and ultraviolet detection and directed to the mass spectrometer without any split. Compounds are ionized by FAB (fast atom bombardment) and the chromatogram and mass spectra are recorded.

Procedure

1. Apply DNA hydrolysate (50–100 μL) to the precolumn for concentration.
2. Transfer to the micro column by means of a Valco M4-N6W-T valve.
3. Monitor UV absorbance of column eluent at 260 nm.
4. Direct all column eluent directly into the mass spectrometer by means of a connecting capillary tube.
5. The sample is analysed by both selected ion traces and acquisition of full-scan mass spectra.

Comments

The micro HPLC column is very thin (i.d. 30–50 μm; Fig. 4) and the contaminants in the sample or mobile phase readily contaminate the column.

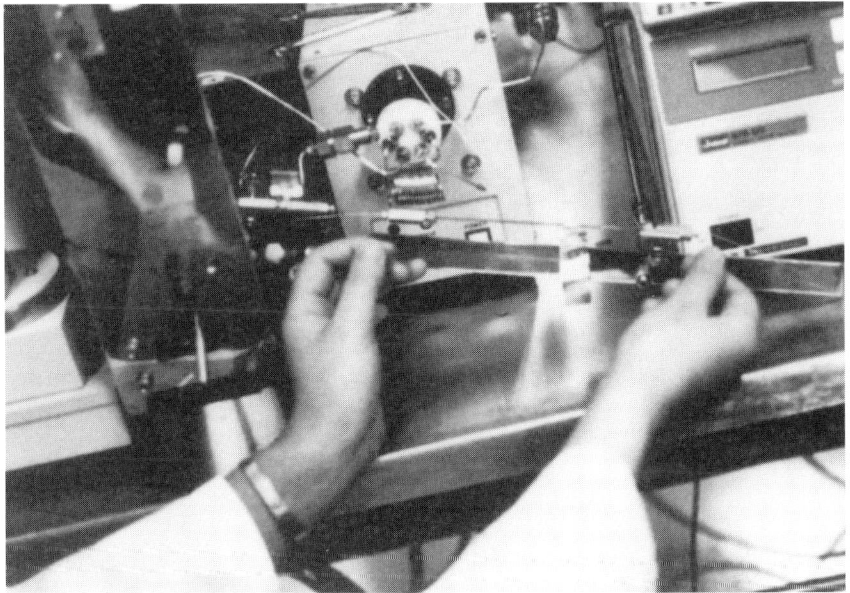

Figure 4. The analytical microcolumn. The tiny capillary-like column which the operator holds in his hands is the microcolumn.

It should also be noted that the flow rate is very low and it is difficult to find leaks from the tube which might affect the analytical data.

Results

The results obtained are illustrated in Figs. 5 and 6.

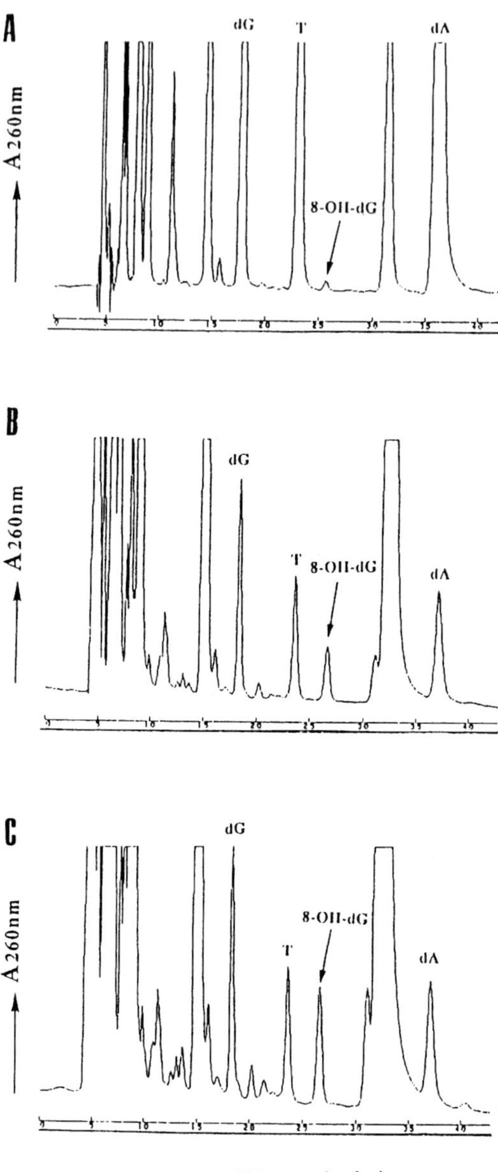

Figure 5. HPLC chromatograms obtained from for the hydrolysates of liver mt DNA from mice after administration of AZT (a drug used to treat AIDS). A, control group; B, group treated with AZT (1 mg kg^{-1}); C, group treated with AZT (5 mg kg^{-1}). Mice treated with 1 mg AZT for 4 weeks per os were used for preparation of live mt DNA. After extraction and hydrolysis, the samples were analysed by micro HPLC–MS. Eluent was monitored by absorbance at 260 nm. 8-OH-dG peak-size increased in a dose dependent manner with increasing AZT dose.

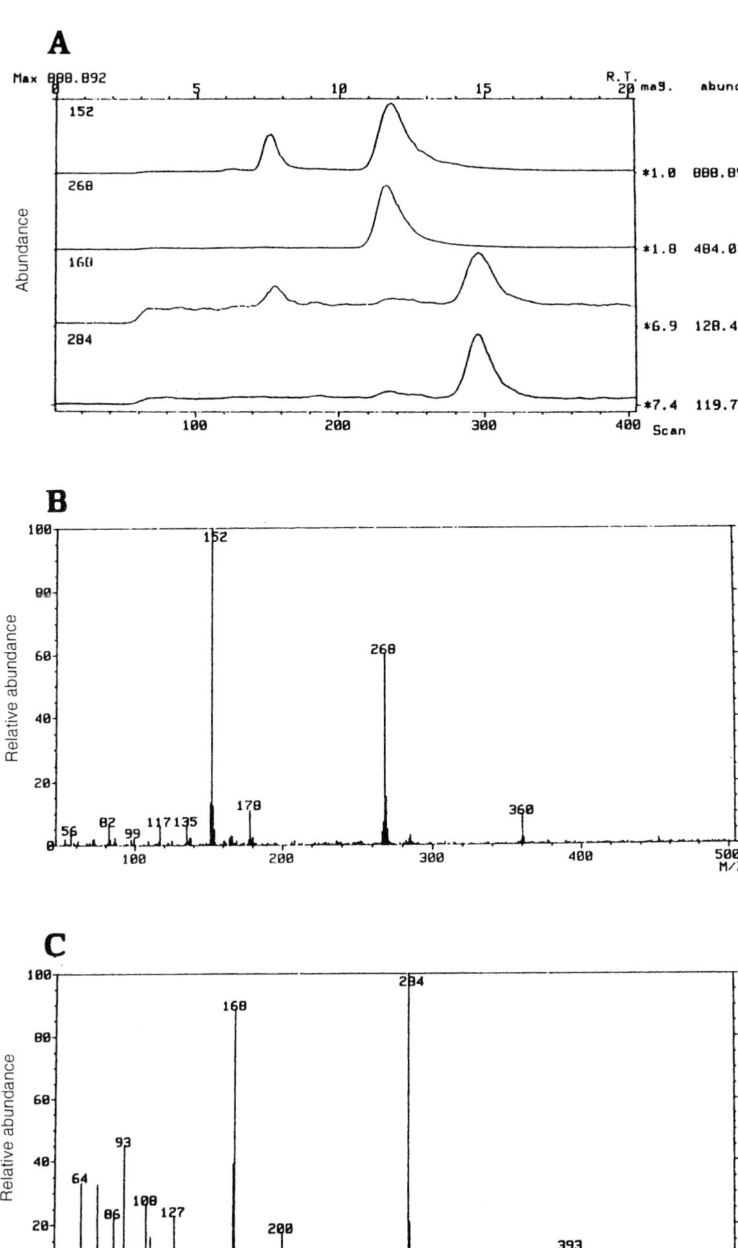

Figure 6. Mass spectrometric analysis of liver mt DNA hydrolysates from mice treated with AZT. The hydrolysates of mt DNA were separated on a microcolumn and then subjected to mass spectrometry.

References

1. Utsumi, K. (1963). Relation between mitochondrial swelling induced by inorganic phosphate and accumulation of P32 in mitochondrial pi fraction. *Acta Medicinae Okayama*, **17**, 259.
2. Hayakawa, M., Ogawa, T., Sugiyama, S., and Ozawa, T. (1991). Massive conversion of guanosine to 8-hydroxyguanosine in mouse liver mitochondrial DNA by administration of azidothymidine. *Biochem. Biophys. Res. Commun.* **176**, 87–93.

71

Hydroxyl radical damage to DNA bases detected by GC–MS

G. J. QUINLAN and J. M. C. GUTTERIDGE

Introduction

DNA bases are readily damaged by ·OH radicals, producing patterns that are recognized as highly characteristic of the ·OH radical. GC–MS techniques, with selective ion monitoring, have been pioneered by Miral Dizdaroglu and his colleagues (1, 2).

Protocol

DNA hydrolysis and derivatization of bases

1. Into clean hydrolysis tubes place DNA sample followed by 6-azathymine (10 mg) as internal standard and formic acid (98%, 0.9 mL).
2. Evacuate air from tube before sealing and heating at 130 °C for 3 min.
3. Cool and freeze-dry hydrolysis product.
4. Derivatize bases by adding bis(trimethylsilyl)trifluoroacetamide (BSTFA)–acetonitrile (1:1, 0.2 mL).
5. Seal tubes and heat for 20 min at 120 °C.
6. Cool and store at –20 °C before GC–MS.

GC–MS separation of oxidised bases

The derivatized sample (1.0 μL) is injected with a split ratio of 12:1 into a Perkin–Elmer 8420 capillary gas chromatograph coupled to a Perkin–Elmer Ion Trap detector and fitted with a wall-coated open-tubular fused-silica 25 m × 0.25 mm i.d. capillary column coated with cross linked and bonded 5% phenylmethylsiloxane (Chrompack, CP Sil 8 CB); the column head pressure is 18 psig. The oven temperature is held at 100 °C for 1 min then increased at 10° min^{-1} to 250 °C. Data are collected between 6–19 min between the mass range 100 to 480 amu. Total ion chromatograms are individually analysed for selective ions by ion extraction.

Calculations

A total ion chromatogram of DNA bases after treatment with an ·OH generating system (antibiotic and copper salt) is shown in Fig. 1. The ions used to identify the peaks are listed in Table 1.

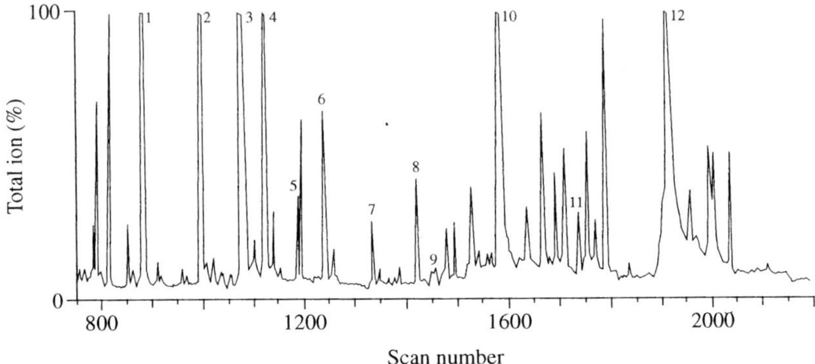

Figure 1. Total ion chromatogram of DNA bases after treatment with rifamycin SV and copper salt.

TABLE 1. Peak identities and characteristic ions used for peak identification in single-ion monitoring.

Peak no.	Compound	Base peak	Molecular/significant ion
1	Thymine	255	270
2	6-Azathymine	256	271
3	Cytosine	240	255
4	Methylcytosine	254	269
5	5-Hydroxyuracil	329	344
6	Cytosine	312	327
7	5-Hydroxycytosine	328	343
8	*cis*-Thymine glycol	259	433
9	*trans*-Thymine glycol	259	433
10	Adenine	264	279
11	8-Hydroxyadenine	352	367
12	Guanine	352	367

Except for peaks 8 and 9 the base peaks corresponded to the *m/z* value of the $(M - CH_3)^+$ ion. Total ion chromatograms obtained from DNA, DNA + Cu^{2+}, DNA + H_2O_2, and DNA + antibiotic alone were as in Fig. 2 minus peaks 5, 7, 8, 9, and 11.

Comment

Because 8-hydroxyguanine is extremely labile and is probably lost during the assay procedure (3), it was not detected as a major base modification by use of this technique.

References

1. Dizdaroglu, M. (1985). Application of capillary gas chromatography–mass spectrometry to chemical characterisation of radiation-induced base damage to DNA: Implications for assessing DNA repair processes. *Anal. Biochem.*, **144**, 593–603.
2. Dizdaroglu, M. (1991).Chemical determination of free radical-induced damage to DNA. *Free Rad. Biol. Med.*, **10**, 225–42.
3. Quinlan, G. J. and Gutteridge, J. M. C. (1991). DNA base damage by β-lactam tetracycline, bacitracin, and rifamysin antibacterial antibiotics. *Biochem. Pharmacol.*, **42**, 1595–9.

72

Peroxynitrite damage to DNA

J. M. C. GUTTERIDGE

Introduction

Peroxynitrite ($ONOO^-$) is a powerful oxidant formed by the reaction of superoxide with nitric oxide. Its decomposition at acid pH also gives rise to an aggressive oxidant with properties similar to those of the hydroxyl radical. When peroxynitrite oxidises a DNA base (2, 3) it leaves behind a nitrated footprint, such as 8-nitroguanine (1). The use of HPLC and GC–MS to profile base damage to DNA has been reviewed elsewhere (4).

References

1. Yermilov, V., Rubio, J., and Ohshime, H. (1995). Formation of 8-nitroguanine in DNA treated with peroxynitrite in vitro and its rapid removal from DNA by depurination. *FEBS Lett.*, **371**, 207–210.
2. Inoue, S. and Kawanishi, S. (1995). Oxidative DNA damage induced by simultaneous generation of nitric oxide and superoxide. *FEBS Lett.*, **371**, 86–8.
3. Salgo M. G., Stone, K., Squadrito, G. L. *et al.* (1995). Peroxynitrite carries DNA risks in plasmid pBR 322. *Biochem. Biophys. Res. Commun.*, **210**, 1025–30.
4. Szabo, C. (1996). DNA strand breakage and activation of poly-ADP ribosyltransferase: a cytotoxic pathway triggered by peroxynitrite. *Free Rad. Biol. Med.*, **21**, 855–69.

73

Carbonyl assay for oxidative damage to proteins

G. J. QUINLAN and J. M. C. GUTTERIDGE

Introduction

Free radical damage to proteins (radiation or Fenton chemistry) was shown in the 1950s to leave carbonyl functions that reacted with 2,4-dinitrophenylhydrazine (1). The technique has been widely introduced into free radical biology and medicine by Professor Stadtman and his colleagues (2, 3). Techniques for cells and fluids are given elsewhere (3, 4). Here we describe a spectrophotometric method for use with plasma.

Protocol

1. In two clean glass tubes (one to serve as a blank) place freshly taken plasma (0.20 mL).
2. To the blank tube add HCl (2 M, 1.0 mL).
3. Add carbonyl reagent (2,4-dinitrophenyl hydrazine (10 mM in 2 M HCl);1.0 mL) to the test.
4. Leave tubes at 37 °C for 90 min.
5. Cool, add trichloroacetic acid (28% *w*/*v*, 1.0 mL)
6. Vortex-mix for 3 min.
7. Centrifuge at 6000 rpm for 3 min and discard upper layer.
8. Suspend deposit in ethanol–ethyl acetate (1:1, 1.0 mL).
9. Vortex-mix for 2 min.
10. Centrifuge at , 6000 rpm 6 min and discard upper phase.
11. Repeat above procedure twice.
12. Suspend deposit in guanidinium chloride (6 M, 1.0 mL) and vortex-mix for 1 min.
13. Transfer to Eppendorf tubes and centrifuge at 6000 rpm for 3 min.
14. Read absorbance at 360 nm.

Calculation

Subtract blank reading from the test, and use a molar absorption coefficient of 21 000 M^{-1} cm^{-1} for the hydrazones. Convert value to mg protein.

Comments

The normal plasma value is less than 1 nmol/mg protein.

References

1. Bennett, W. and Garrison, W. M. (1959). Production of amide groups and ammonia in the radiolysis of aqueous solutions of protein. *Nature*, **183**, 889.
2. Oliver, C. N., Ahn, B. W., Moerman, E. J., Goldstein, S., and Stadtman, E. R. (1987). Age-related changes in oxidised proteins. *J. Biol. Chem.*, **262**, 5488–91.
3. Punchard, N. A. and Kelly, F. J. (ed.) (1996). *Free Radicals. A practical Approach*, pp. 160–170. IRL Press, Oxford.
4. Rice-Evans, C. A., Diplock, A. T., and Symons, M. C. R. (1991). *Techniques in Free Radical Research*, pp. 223–34. Elsevier, Amsterdam.

74

ROS damage to protein

KOJI UCHIDA

Introduction

Several lines of evidence indicate that protein oxidation by free radicals and the subsequent accumulation of oxidatively modified proteins, which could be an early indication of oxygen radical-mediated tissue damage, have been found in cells during ageing, after oxidative stress, and in various pathological states including premature ageing diseases, muscular dystrophy, rheumatoid arthritis, and atherosclerosis (1). Furthermore, free radical-mediated oxidative inactivation of enzymes, especially by metal-catalysed oxidation systems, have been postulated as marking steps in enzyme turnover.

The types of protein modification that have been observed and the kinds of 'active-oxygen' generating systems employed are so varied that detailed descriptions of the numerous methods used in their study cannot be achieved within the space allocated. This review therefore focuses on the products generated by the acid-hydrolysis of oxidized amino acids.

Protocol

Assays for oxidized proteins as monitored by increase in the acidic amino acids[1]

Metal-catalysed oxidation of protein

1. Add protein solution (100 μL) and of $CuSO_4$ (100 μL) to sodium phosphate buffer (pH 7.2, 50 mM, 700 μL) and mix well.
2. Initiate the free-radical reaction by addition of ascorbate solution (100 μL; final concentrations: protein, 1 mg mL^{-1}; $CuSO_4$, 50 μM; ascorbate, 5 mM).
3. Incubate at 37 °C for 2–24 h.
4. Stop the reaction by the addition of EDTA solution (1 mM, 100 μL).

Acid-hydrolysis and amino acid analysis

1. Add trichloroacetic acid (20% *w/v*, 100 μL) to an equal volume of the reaction mixture containing the oxidized protein.

Protocol *Continued*

2. Centrifuge at 11 000 g for 3 min and discard the supernatant.
3. Rinse twice with diethyl ether–ethyl acetate (1:1).
4. Hydrolyse *in vacuo* with HCl (6 M) for 24 h at 105 °C.
5. Concentrate the hydrolysate and dissolve in distilled water (100 μL).
6. Analyse by amino acid analysis (2, 3).

Detection of 2-oxo-histidine in oxidized protein (4,5)

Preparation of authentic *N*-benzoyl-2-oxo-histidine

1. Add *N*-benzoylhistidine (51.8 mg) and $CuSO_4$ (5 mg) to sodium phosphate buffer (pH 7.2, 50 mM, 100 mL) and mix well.
2. Initiate the free radical reaction by the addition of ascorbate (875 mg; final concentrations: *N*-benzoylhistidine, 2 mM; $CuSO_4$, 0.5 mM; ascorbate, 50 mM).
3. Incubate at 37 °C for 48 h (bubbling oxygen gas is effective).
4. Stop the reaction by the addition of EDTA (37.2 mg; final concentration 1 mM).
5. Concentrate the reaction mixture.
6. Analyse by HPLC (Fig. 1A)[2].

HPLC is performed on an ODS column with 25% *w/v* methanol in 0.1% *w/v* trifluoroacetic acid as mobile phase at a flow rate of 5 mL min^{-1}, and with both UV and electrochemical detection. (A Jasco HPLC system equipped with a PU-980 HPLC pump, 807-IT Integrator, both UV-970 UV and 840-EC electrochemical detectors, and a 0.46 mm i.d. × 25 cm TSK-GEL ODS-80® column (TOSOH) were used throughout this study.) Elution profiles are monitored at 210 nm on the UV detector and at an applied oxidation potential of 0.85 V on the electrochemical detector.

Acid-hydrolysis of *N*-benzoyl-2-oxo-histidine

1. Hydrolyse the purified *N*-benzoyl-2-oxo-histidine with HCl (6 M) *in vacuo* at 105 °C for 24 h.
2. Concentrate the hydrolysate and dissolve in distilled water.
3. Analyse by HPLC–ECD (Fig. 1B).

HPLC is performed with the same column with 5% methanol in 0.1% *w/v* heptafluorobutyric acid as mobile phase at a flow rate of 0.5 mL min^{-1}. Elution is monitored at an applied oxidation potential of 0.85 V on the electrochemical detector.

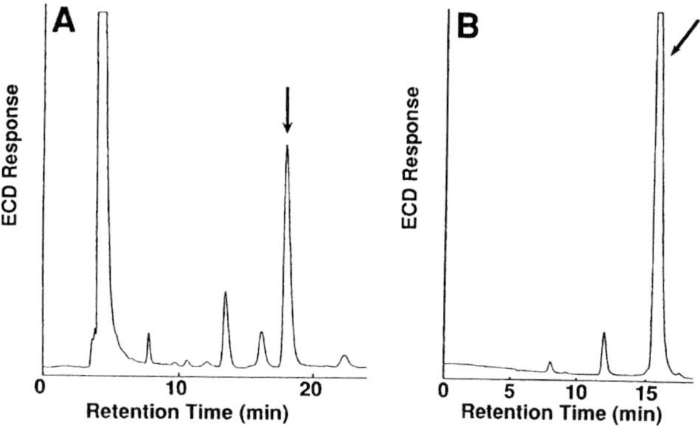

Figure 1. Detection of *N*-benzoyl-2-oxo-histidine (A) and its hydrolysed product 2-oxo-histidine (B) by HPLC–ECD. (A) *N*-Benzoylhistidine (1 mM) was incubated with $CuSO_4$ (50 mM) and ascorbate (5 mM) in sodium phosphate buffer (pH 7.2, 50 mM, 1 mL) for 1 h at 37 °C. The main peak indicated by the arrow at 18.2 min represents *N*-benzoyl-2-oxo-histidine. The starting material (*N*-benzoylhistidine), for which there was no ECD (electrochemical detection) response, eluted after 13 min. (B) Authentic *N*-benzoyl-2-oxo-histidine (0.1 mg) was hydrolysed with HCl (6 M) for 24 h at 105 °C. The hydrolysate was concentrated, dissolved in sodium phosphate buffer (pH 7.2, 50 mM, 0.5 mL), and an aliquot (10 μL) was analysed by HPLC–ECD. The main peak indicated by the arrow at 15.8 min represents 2-oxo-histidine.

Detection of 2-oxo-histidine in Cu,Zn-superoxide dismutase exposed to hydrogen peroxide *(6)*

1. Incubate Cu,Zn-SOD (0.5 mg) with H_2O_2 (5 mM) in sodium phosphate buffer (pH 7.2, 50 mM, 1 mL) at 37 °C.
2. Add EDTA (10 mM, 10 μL) and catalase solution (10 μL) to the reaction mixture (0.1 mL).
3. Treat with an equal volume of trichloroacetic acid solution (20% *w/v*).

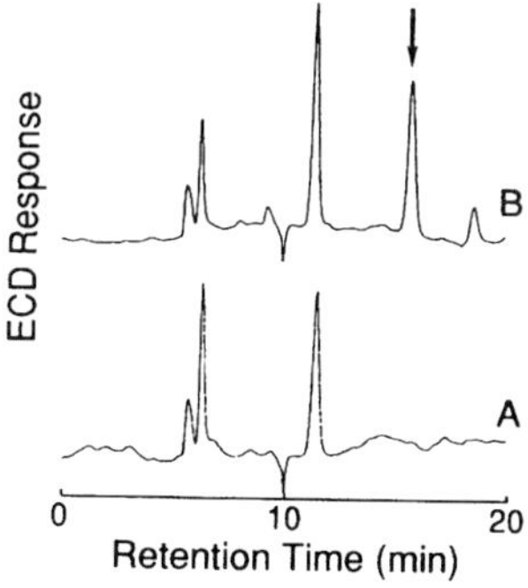

Figure 2. Detection of 2-oxo-histidine in H_2O_2-treated Cu,Zn-SOD by HPLC–ECD. The H_2O_2-treated Cu,Zn-SOD was prepared by incubating the enzyme (0.5 mg mL^{-1}) with H_2O_2 (5 mM) for 30 min at 37 °C. Control (A) and H_2O_2-treated (B) Cu,Zn-SOD were hydrolysed and analysed by reversed-phase HPLC with electrochemical detection. The peak indicated by the arrow represents 2-oxo-histidine.

Protocol *Continued*

4. Centrifuge at 10 000 g for 3 min.
5. Hydrolyse *in vacuo* with HCl (6 M) for 24 h at 105 °C.
6. Concentrate the hydrolysate and dissolve in distilled water (100 μL).
7. Analyse by HPLC–ECD as described above (Fig. 2).

[1] Oxidative modification of protein is usually associated with an increase in the amount of acidic amino acids, such as aspartate and glutamate; this has been ascribed to the oxidation of histidine and proline residues, respectively.
[2] Once hydrolysed, 2-oxo-histidine is unstable. It is therefore recommended that the hydrolysed sample is analysed as soon as possible. Otherwise, the sample should be stored at −80 °C.

References

1. Stadtman, E. R. (1992). Protein oxidation and ageing. *Science*, **257**, 1220–4.
2. Uchida, K., Kato, Y., and Kawakishi, S. (1990). A novel mechanism for oxidative cleavage of prolyl peptides induced by hydroxyl radical. *Biochem. Biophys. Res. Commun.*, **169**, 265–71.
3. Kato, Y., Uchida, K., and Kawasaki, S. (1992). Oxidative fragmentation of collagen and prolyl peptide by Cu (II)/H_2O_2. *J. Biol. Chem.*, **267**, 23646–51.
4. Uchida, K. and Kawakishi, S. (1986). Selective oxidation of imidazole ring in histidine residues by the ascorbic acid–copper ion system. *Biochem. Biophys. Res. Commun.*, **138**, 659–65.
5. Uchida, K. and Kawakishi, S. (1986). Ascorbate-mediated specific oxidation of imidazole ring in a histidine derivative. *Bioorg. Chem.*, **17**, 334–43.
6. Uchida, K. and Kawakishi. S. (1992). Identification of oxidized histidine generated at the active site of Cu,Zn-superoxide dismutase exposed to H_2O_2. Selective generation of 2-oxo-histidine at the histidine-118. *J. Biol. Chem.*, **269**, 2405–10.

75

Damage to proteins by reactive oxygen and reactive nitrogen species

G. J. QUINLAN and J. M. C. GUTTERIDGE

Measurement of *ortho*-tyrosine, chlorotyrosine, and nitrotyrosine by HPLC

Introduction

The aromatic amino acids of proteins, e.g. phenylalanine and tyrosine, are particularly susceptible to oxidative modification and, therefore, provide useful evidence of oxidative damage in biological systems. Activated neutrophils produce $O_2\cdot^-$, H_2O_2, HOC1, and ·NO which can hydroxylate, chlorinate and nitrate protein amino acid residues by a variety of complex interactions between these species and iron.

Protocol

Protein hydrolysis and amino acid purification for biological samples such as plasma

The proteins in the sample are concentrated by ultrafiltration under pressure or by centrifugal techniques.

1. Place protein concentrate (0.150 mL), mercaptoethanol (0.005 mL), and HCl (6 M, 0.500 mL) in a hydrolysis tube and remove the air under vacuum for 10 min.
2. Flush tubes four times with oxygen-free nitrogen.
3. Seal and heat at 110°C for 18 h.
4. Cool and apply to a cation-exchange solid-phase column (benzene-sulfonic acid) previously conditioned with methanol (5 mL), HCl (0.1M, 5 mL) in methanol, and HCl (0.1 M, 5 mL).
5. Wash columns with distilled water (3 × 2 mL).
6. Amino acids produced by hydrolysis of the protein are selectively eluted with ammonia (0.1 M, 2 mL).

Protocol *Continued*

7. Freeze-dry solutions overnight and suspend in potassium dihydrogen phosphate (5 mM, 1 mL).
8. Apply freeze-dried samples to preconditioned ion exchange columns (see Step 4) and wash and elute as described in Step 5.
9. After freeze drying, the samples are suspended in running buffer (0.2 mL) for HPLC analysis.

HPLC analysis of o*-tyrosine, chlorotyrosine, and nitrotyrosine*

Sample or standard (60 μL) is injected on to a 150 mm × 4.6 mm i.d., 5 μm particle, Varian Res-Elut C1890A column with 1:9 (%*v*/*v*) methanol–mixed potassium dihydrogen phosphate (5 mM) phosphoric acid buffer (pH 3.0) as isocratic mobile phase at a flow rate of 1 mL min^{-1}. Detection is by UV absorbance at 274 nm. (Diode-array detection is strongly recommended.) Compound identification is by comparison of retention times with those of pure standards

Calculation

Peak areas are measured and compared with calibration curves prepared from pure standards (range 0.625–20 mM). Results can be corrected to the total protein value and expressed as nmol/mg protein.

Measurement of *ortho*-tyrosine, *meta*-tyrosine, chlorotyrosine, and nitrotyrosine by GC–MS

Protocol

Protein hydrolysis

Perform protein hydrolysis as described above.

Derivatisation

1. Mix purified amino acids (50 mL) and 3-(2-thienyl)-L-alanine (1 mM, 5 mL; internal standard).
2. Freeze dry overnight.
3. Add propan-2-ol (1 mL) and sulfuric acid (85 mL).
4. Seal and heat at 100°C for 2 h.
5. Concentrate the propyl ester formed under a stream of N_2 at 40°C.
6. Add trifluoroacetic anhydride (0.5 mL), seal and leave 30 min at room temperature.

7. Concentrate the trifluoroacetic propyl amino acid derivatives formed, by means of a stream of N_2 at room temperature.
8. Re-suspend in 50 mL dry acetone.

Gas chromatography–mass spectrometry

The sample (2 μL) is injected (splitless) on to a gas chromatograph containing a WCOT column and the temperature is ramped from 60 to 200 °C. The modified tyrosines are detected by positive-ion selective-ion monitoring mass spectrometry. The ions used are: *m*/*z* 203 and 147 for *ortho*-tyrosine, *m*/*z* 203 and 260 for *meta*-tyrosine, *m*/*z* 141 and 294 for chloro-tyrosine (2 derivatives formed), and *m*/*z* 152 and 209 for nitrotyrosine.

Calculation

Quantitation is based on comparison of the intensities of ions of *m*/*z* 147, 260, 294, and 152 with standard curves prepared over appropriate ranges (Fig. 1).

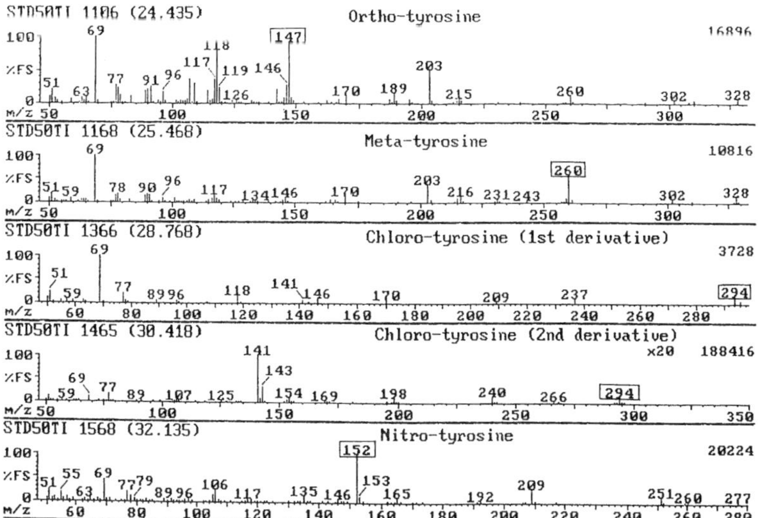

Figure 1. Quantitation of *ortho*-tyrosine, *meta*-tyrosine, chlorotyrosine, and nitrotyrosine by GC–MS.

References

1. van der Vliet, A., O'Neil, C. A., Halliwell, B. *et al.* (1994). Aromatic hydroxylation and nitration of phenylalanine and tyrosine by peroxynitrite. Evidence for hydroxyl radical production from peroxynitrate. *FEBS Lett.*, **339**, 89–92.
2. Lamb, N. J., Gutteridge, J. M. C., and Baker, C. *et al.* (1999). Oxidative damage to proteins of bronchoalveolar lavage fluid in patents with seas. Evidence for neutrophil-mediated hydroxylation, nitration and chlorination. *Crit. Cave Med.*, **27**, 1738–44.

Part 3

Experimental Pathophysiological Models

76

Isolated, perfused heart systems (Langendorf's model)

HIDEO KUSUOKA and TSUNEHIKO KUZUYA

Introduction

There are two different types of perfusion in isolated, perfused heart systems —constant-pressure mode and constant-flow mode. In perfusion with constant pressure, the perfusion pressure in coronary arteries is equal to the difference between the perfusion pressure and the end-diastolic pressure of the left ventricle (EDP). One of the demerits of constant pressure perfusion is that effective coronary perfusion pressure decreases when EDP increases. It is, however, advantageous that the flow rate can be matched with the oxygen demand by coronary autoregulation. In contrast, in constant flow perfusion the flow rate is maintained even when the EDP increases significantly, whereas the beneficial effects of coronary autoregulation are lost.

Protocol

Heart preparation

1. Remove the heart from a heparinized (500–1000 units, i.v.) and anaesthetized (sodium pentobarbital; 80 mg kg^{-1}, i.p.) rat.
2. Mount the heart on a cannula inserted into the aorta above the aortic valve and tied securely.
3. Insert a latex balloon tied to the end of a tube into the left ventricle through the mitral valve.
4. Start perfusion and stabilize the function.

Ischaemia and reperfusion

1. Stop the flow by cross-clamping[1] the perfusion line.
2. Turn off the pump, if used.
3. Induce global ischaemia[2].

Protocol *Continued*

4. Reperfuse the heart. If ventricular fibrillation occurs, try defibrillation[3] by massage, electrical shock, or drugs.

[1] Stop the perfusion by closing a three-way stopcock placed above a cannula. If the constant flow method is used, it is necessary not only to turn off the pump but also to clamp the line near the heart. The perfusate in the line might be sucked into the coronary arteries when line clamping is not accompanied by turning-off of the pump.
[2] It is important to keep the temperature of the heart constant. The myocardial temperature might decrease owing to non-transport of heat by warm perfusate, resulting in cardiac protection. When warmed water is circulated outside the heart to maintain the temperature, it is necessary to use de-oxygenated water; otherwise, oxygen might be supplied to the myocardium by sucking the water through the mitral valve. It is also better to pace the heart at least during the early phase of ischaemia. Heart rate may be decreased by ischaemia, resulting in cardiac protection. The volume of the balloon in the left ventricle should be kept constant throughout the ischaemic period for the same reason.
[3] Defibrillation can be achieved by mild massage of the heart. Electrical shock with a direct current is also effective. Drugs such as lidocaine are not suitable for defibrillation when cardiac function is being assessed, because these drugs reduce heart function.

77

Myocardial ischaemia–reperfusion models *in vivo*

SHIRO HOSHIDA

Introduction

To determine the role of active oxygen species on the progression of myocardial injury, many investigators use a coronary occlusion–reperfusion model, but not a permanent occlusion model, *in vivo* in dogs, cats, pigs, rabbits, and rats. A different protocol is needed to clarify a different end point of myocardial injury such as limitation of infarct size, alteration of cardiac function, and occurrence of cardiac arrhythmia. In this section, we focus on the limitation of infarct size by use of a coronary artery occlusion–reperfusion model in dogs. (1, 2)

Coronary artery occlusion–reperfusion model of myocardial infarction

Protocol

1. Anaesthetize the dog with pentobarbital sodium (30 mg kg^{-1}, i.v.).
2. Intubate with a cuffed endotracheal tube.
3. Perform left thoracotomy at the fifth intercostal space[1].
4. Suspend the heart in a pericardial cradle.
5. Isolate the left anterior descending coronary artery[2].
6. Insert catheters into the left carotid artery, for measurement of blood pressure and heart rate[3], into the left jugular vein, for blood sampling and drug administration[4], and into the left atrium, for administration of coloured microspheres.

Protocol *Continued*

7. Perform sustained ligation of the coronary artery followed by reperfusion.

[1] Aseptic surgical technique is needed throughout and antibiotics such as benzylpenicillin (30 000 units kg^{-1}) are injected intramuscularly for infection prophylaxis.
[2] The left anterior descending coronary artery is isolated carefully distal to the first major diagonal branch.
[3] Lead II of the ECG is monitored throughout the experiment to evaluate the incidence of cardiac arrhythmia.
[4] Arterial blood is monitored for evaluation of pH , pO_2, and pCO_2.

Measurement of infarct size

Protocol

1. Re-ligate the coronary artery at the end of the experiment.
2. Insert a catheter into the coronary artery distal to the ligated site.
3. Inject Evans blue dye (0.5%) into the jugular vein and perfuse autologous heparinized blood into the cannulated coronary artery.
4. Remove the hearts and divide into six transverse slices of approximately equal thickness.
5. Measure the area of left ventricle at risk (lack of Evans blue staining).
6. Incubate transverse slices in triphenyltetrazolium chloride solution (phosphate buffered saline pH 7.4, 20 min, 37 °C)
7. Measure the infarcted area of the left ventricle (lack of triphenyltetrazolium chloride staining).
8. Calculate the percentage of the myocardium at risk that is infarcted (index of infarct size)

Comments

It is essential to decide a suitable occlusion time (e.g. dog 90 min–3 h; rabbit 30 min–1 h) for evaluation of limiting infarct size by treatment with pharmacological agents in each experiment.
To quantitate infarct size accurately, it is better to use a longer reperfusion model (1–7 days).

Measurement of regional myocardial blood flow

It is essential to measure coronary blood flow during myocardial ischaemia because of the wide variety of coronary collateral flow in a canine model.

Myocardium beneath the endocardium and epicardium must be discarded when sampling tissue if regional myocardial blood flow is to be measured precisely.

Protocol

1. Inject coloured microspheres (5×10^6; 11.9 ± 1.9 μm) into the left atrium before (baseline), during (ischaemia), and after (reperfusion) coronary artery ligation followed by a saline flush (10 mL).
2. Sample reference arterial blood at a constant rate (10 mL min^{-1}) 10 s before microsphere injection and continuing until 50 s after the injection.
3. Perform myocardial sampling from ischaemic and non-ischaemic regions and divide into equal subendocardial and subepicardial portions at the end of the experiment.

Blood

1. Centrifuge for 30 min at 3000 rpm.
2. Dilute the red cell sediment with Blood Haemolysis Reagent.
3. Centrifuge for 30 min at 3000 rpm.
4. Dilute the sediment with distilled water.
5. Centrifuge for 30 min at 3000 rpm.
6. Add Blood Digest Reagent I to the sediment and incubate in boiling water for 30 min.
7. After cooling, dilute the digest with distilled water.
8. Centrifuge for 30 min at 3000 rpm.
9. Add Blood Microsphere Counting Reagent to the sediment.

Myocardium

1. After weighing, add Tissue Digest Reagent I to myocardial samples (1–2 g).
2. Leave at room temperature for 5–7 days.
3. Incubate the myocardial samples in boiling water for 30 min.
4. After vigorous vortex-mixing, add Tissue Digest Reagent II to the sample.
5. Centrifuge for 30 min at 3000 rpm.
6. Add Tissue Microsphere Counting Reagent to the sediment.

Calculations

The final solution is placed in a haemocytometer and ten chambers are counted for each sample. The total number of microspheres in each reference blood and tissue sample is computed by use of the formula:

$$\text{Total microspheres} = \frac{\text{No. counted}}{\text{No. chambers} \times 0.9\ \text{mm}^3} \times \frac{1.000\ \text{mm}^3}{\text{m}l} \times \text{ml/suspension}$$

where no. counted is the number of coloured microspheres counted, no. chambers is the number of chambers counted, 0.9 mm^3 is the ruled volume of the chamber, and mL suspension is the final dilution volume.

Regional myocardial blood flow (RMBF) is computed by use of the formula:

$$\text{RMBF} = \frac{\text{CT} \times \text{R}}{\text{CR} \times \text{WT}}$$

where CT is the total number of microspheres in the tissue sample, R is the reference flow rate (mL min^{-1}), CR is the total number of microspheres in the reference blood sample, and WT is the weight of the tissue sample in grams. Results are expressed as mL $\text{min}^{-1}\ \text{g}^{-1}$.

References

1. Hoshida, S. *et al.* (1994). Gamma-glutamylcysteine ethyl ester for myocardial protection in dogs during ischaemia and reperfusion. *J. Am. Coll. Cardiol.*, **24**, 1391–7.
2. Hoshida, S. *et al.* (1994). Ebselen protects against ischaemia–reperfusion injury in a canine model of myocardial infarction. *Am. J. Physiol.*, **267**, H2342–7.

78

GI tract ischaemia–reperfusion model

TOSHIKAZU YOSHIKAWA and YUJI NAITO

Introduction

We introduced a new animal model of ischaemia–reperfusion-induced gastric mucosal injury in rats by applying a vascular clamp to the celiac artery, then removing it (1). As a result of the clamping of the celiac artery the gastric mucosal blood flow was reduced to 10% of that measured before clamping; the flow returned to the normal range on subsequent reperfusion. The total area of erosion, a morphological index of gastric injury, increased gradually after clamping of celiac artery and increased significantly on reperfusion after 30 min of gastric ischaemia (Fig. 1). Although thiobarbituric acid (TBA)-reactive substances in the gastric mucosa, an index of lipid peroxidation, did not increase 30 min after ischaemia, however, they increased significantly between 30 and 60 min after reperfusion. The ischaemia–reperfusion-induced gastric injury model was used effectively to determine whether a novel agent can alleviate reperfusion injury and whether it can inhibit lipid peroxidation *in vivo*.

Preparation of rats for ischaemia–reperfusion and assays

Protocol

1. Fast male Sprague–Dawley rats (180–220 g)[1] for 18 h before the experiments, but allow free access to water.
2. Anaesthetize with urethane (1000 mg kg^{-1}, i.p.) or pentobarbital (25 mg kg^{-1}, i.p.)
3. Pre-treat with scavengers[2] (this step is optional).
4. Applying a small vascular clamp to the celiac artery for 30 min[3,4].
5. Remove the clamp for 30 min or 60 min.
6. Sample the stomach and blood.

Protocol *Continued*

Stomach

1. Measure the total area of erosion under a dissecting microscope.
2. Homogenize in potassium phosphate buffer (pH 7.8, 10 mM, 1.5 mL) containing KCl (30 mM).
3. Perform assays for TBA, α-tocopherol, myeloperoxidase, and total protein.

Serum

1. Perform assays for TBA, total cholesterol, and α-tocopherol.
2. Sonicate the sample.
3. Centrifuge at 20 000 g for 20 min at 4 °C
4. Isolate the supernatant.
5. Measure CuZn-SOD, Mn-SOD, and glutathione peroxidase activity and determine total protein.

[1] To study the effect of drugs on ischaemia–reperfusion injury, they administered to rats intravenously, subcutaneously, or intraperitoneally before ischaemia.
[2] We used several scavengers: recombinant human CuZn-SOD (Nippon Kayaku; 50 000 units kg^{-1}, s.c. or i.v.); bovine liver catalase (Sigma; 90 000 units kg^{-1}, s.c.); dimethylsulphoxide (Wako Pure Chemical; 550 mg kg^{-1}, i.p.); allopurinol (Sigma; 50 mg kg^{-1}, orally, 48 and 24 h before ischaemia); anti-neutrophil serum obtained from immunized rabbits (10 mL kg^{-1}, i.p., 18 h before ischaemia)
[3] It is easy to identify the celiac artery. After abdominal median incision, the abdominal aorta is carefully exposed by removal of the soft tissue. The celiac artery is the first branch from the abdominal aorta. The second branch from the aorta is the superior mesenteric artery and should be treated with care.
[4] Use the small vascular clamp with a spring. To prevent dehydration, the abdominal incision should be sutured after both ischaemia and reperfusion.

Measurement of lipid peroxides (TBA assay) (2)

Protocol

1. Obtain mucosal homogenate (0.2 mL).
2. Add sodium dodecyl sulphate (SDS; 8.1% *w/v*, 0.2 mL).
3. Add acetate buffer (pH 3.5, 20% *w/v*, 1.5 mL)[1].
4. Add TBA solution (0.8% *w/v*, 1.5 mL)[2].
5. Add double-distilled water (0.6 mL).
6. Heat at 95°C for 60 min on an oil bath.
7. Cool in cold water.

8. Add double-distilled water (1.0 mL).
9. Add a mixture of *n*-butanol and pyridine (15:1, *v/v*, 5.0 mL).
10. Shake vigorously by vortex-mixing for 2 min.
11. Centrifuge at 3000 rpm for 10 min at room temperature.
12. Measure absorbance at 532 nm.

[1] The reactions of lipid peroxides with TBA in whole homogenate are affected by the pH of the reaction mixture; it is reported that the optimum pH for the reaction with whole homogenate is 3.5.
[2] TBA solution should be prepared on the day of the experiment.

Measurement of SOD activity

Protocol

1. Obtain mucosal homogenate.
2. Sonicate for 2 min at 60 W on ice.
3. Centrifuge at 78 000 g for 60 min.
4. Take supernatant (50 μL) and add Tris-HCl (pH 7.8, 125 mM, 950 μL).

Total SOD activity[1]

1. Dilute with double-distilled water.
2. Measure Total SOD activity

CuZn-SOD and Mn-SOD activity

1. Mix sample (100 μL), Tris-HCl (260 μL) and SDS (10% *w/v*, 90 μL).
2. Incubate for 2 h at 37 °C.
3. Cool on ice for 10 min.
4. Add KCl (3 M, 50 μL).
5. Shake for 10 min.
6. Cool on ice for 30 min.
7. Centrifuge at 20 000 g for 10 min.
8. Dilute supernatant with double-distilled water.
9. Measure CuZn-SOD activity.
10. Mn-SOD activity = total SOD activity – CuZn-SOD activity.

[1] SOD activity is measured by the cypridina luciferin analogue (CLA)-dependent chemiluminescence assay (3). The reaction mixture is: Tris-HCl buffer (pH 7.8, 125 mM, 800 μL), double-distilled water (850 μL), DETAPAC (50 μL, 4 mM), hypoxanthine (2 mM, 50 μL), sample or standard SOD (100 μL), xanthine oxidase (100 μL), and CLA (50 μL).

Results

The results obtained are illustrated in Fig. 1.

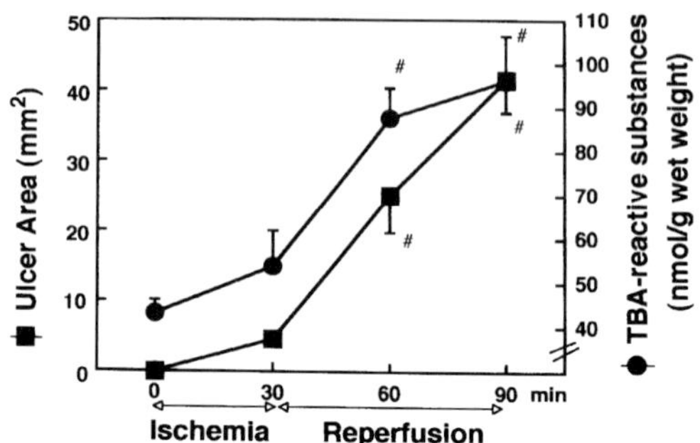

Figure 1. Time-course of changes in total areas of erosion and in thiobarbituric acid-reactive substances (TBA-RS) of the gastric mucosa during ischaemia and reperfusion in rats ($p < 0.05$ compared with before ischaemia).

References

1. Yoshikawa, T., Ueda, S., Naito, Y., Takahashi, S., Oyamada, H., Morita, Y., Yoneta, T., and Kondo, M. (1989). Role of oxygen-derived free radicals in gastric mucosal injury induced by ischaemia or ischaemia–reperfusion in rats. *Free Rad. Res. Commun.*, **7**, 285–91.
2. Ohkawa, H., Ohnishi, N., and Yagi, K. (1986). Assay for lipid peroxides for animal tissues by thiobarbituric acid reaction. *Anal. Biochem.*, **95**, 351–8.
3. Nakano, M., Kimura, H., Hara, M., Kuroiwa, M., Kato, M., Totsune, K., and Yoshikawa, T. (1990). A highly sensitive method for determining both Mn- and CuZn- superoxide dismutase activities in tissue and blood cells. *Anal. Biochem.*, **187**, 277–80.

79

Models of brain ischaemia

MASAYASU MATSUMOTO

Introduction

Brain ischaemia followed by reperfusion generates oxygen-derived free radicals which destroy cellular lipids, proteins, DNA, and mitochondrial respiratory chains, killing neurons. In this section, we describe representative, reproducible, and reliable models of rodent brain ischaemia.

Complete forebrain ischaemia was induced by occluding bilateral common carotid arteries in adult Mongolian gerbils because of a lack of posterior communicating arteries between carotids and vertebro-basilar arteries. Induction of rat forebrain ischaemia by occluding the two carotid and two vertebral arteries, does not completely stop forebrain blood flow but the reduction is sufficient to cause neuronal death in the hippocampus, caudo-putamen, and posterior cerebral cotex. To produce focal cerebral ischaemia, occlusion of proximal middle cerebral artery in rat is sufficiently reproducible for investigation of the pathophysiology of development of cerebral infarction and brain oedema. In the hypoperfused regions around such ischaemic cores, oxygen-derived free radicals are formed, and these kill neurons. Enabling reperfusion of the ischaemic tissue induces production of free radicals. If the temporary middle cerebral artery occlusion is used, these radicals can be observed; this mimics the actual pathophysiology of human cerebral ischaemia.

Gerbil model of transient forebrain ischaemia

Protocol

1. Anaesthetize a Mongolian gerbil (male, 55–75 g, 8–12 weeks) with 2% *v*/*v* halothane and maintain anaesthesia with 0.5–1.0% (*w*/*v*) halothane under spontaneous breathing.
2. Place the animal in the supine position.
3. Make a ventral midline incision in the neck after disinfection of skin.

Protocol *Continued*

4. Expose both common carotid arteries. Avoid mechanical stimulation of the vagal nerve during separation, otherwise the vagotomy which occurs will cause the death of the animal.
5. After complete recovery from anaesthesia, occlude both common carotids simultaneously with miniature aneurysmal clips.
6. Remove both clips to restore cerebral blood flow. During the operation body temperature should be monitored with rectal and temporal muscular thermometers and controlled at 37 °C by means of a heating pad and light. After reperfusion, rectal temperature should be monitored 2–3 h after clip removal.

Results

Ischaemia for 5 min causes neuronal death selectively in the CA1 neurons of the hippocampus, whereas 2-min ischaemia causes no neuronal damage in the brain. Such neuronal degeneration happens 3–7 days after reperfusion; this is denoted 'delayed neuronal death'. Post-ischaemic hyperthermia (over 38.8 °C) is always observed 20 min to 3 h after reperfusion. Animals showing seizure and no post-ischaemic hyperthermia should be excluded from the experiment. A mouse model of transient forebrain ischaemia is also available but its reproducibility is not as high as that of the gerbil model. Transient hind-brain ischaemia can be also reproducibly produced in gerbils by occluding the bilateral vertebral arteries, as described by Hata *et al.*

Rat model of transient forebrain ischaemia, Pulsinelli's four vessel occlusion

Protocol

1. Anaesthetize a Wistar rat (male, 300–400 g,12–16 weeks) with 4% halothane, and maintain anaesthesia with 0.5% halothane in a nitrous oxide–oxygen mixture (70:30% *v/v*).
2. Use the femoral artery and vein to measure blood pressure and to take a blood sample. Make a half-inch incision behind the occipital bone directly overlying the first two cervical vertebrae.
3. Under an operating microscope expose both alar foramina of the first vertebra by separating paraspinal muscles from the midline.
4. Drill around the foramina to expose the vertebral artery directly through larger holes (Fig. 1-1). The rat's vertebral arteries pass within the

vertebral canal and beneath the alar foramen before entering the posterior fossa.

5. Occlude the vertebral arteries permanently, electrocauterize both vertebral arteries three times for 2–3 s by means of a 1.0-mm diameter electrocautery needle through each alar foramen. Between each electrocauterization, cool the coagulated tissues by washing with saline for 30–60 s.
6. Check to ensure occlusion is complete by inserting a 28-gauge needle into the foramina. No bleeding means electrocauterization is complete.
7. Let the animal recover from the first operation for 18–24 h with free access to water but without food.
8. On next day, anaesthetize the animal as described above. Make a ventral midline incision of the neck in the supine position.
9. Isolate both common carotid arteries. Avoid touching the accompanying vagal nerve (Fig. 1-2)

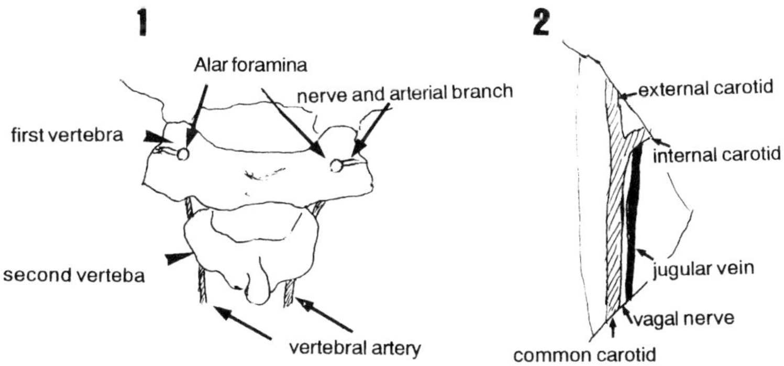

Figure 1.

10. After a recovery from anaesthesia, occlude both carotids with miniature aneurysmal clips simultaneously for 10–20 min. Body temperature should be monitored with rectal and temporal muscular thermometers and controlled at 37.0 °C by means of a heating pad and light.

Results

Carotid occlusion for 1–2 min of causes loss of the righting reflex and 10–20 min carotid occlusion causes delayed selective neuronal death in the hippocampus, striatum, and the 3rd, 5th, and 6th layers of the fronto-parietal cerebral cortex. Cerebral blood flow to the hind-brain is not usually affected in this model. Exclude animals that retain their righting reflex during occlusion experiment.

Administration of atropine sulphate, 1–5 mg kg^{-1}, i.p., might be required to prevent acute circulatory–respiratory insufficiency. The animals subjected to ischaemia do not show post-ischaemic hyperthermia.

Rat model of permanent focal ischaemia

Protocol

1. Induce inhalative anaesthesia in a Wistar rat (male, 250–350 g, 8–10 weeks) with 4% *v*/*v* halothane and maintain it with 1.5–0.8% *v*/*v* halothane under spontaneous breathing.
2. Place the animal in the recumbent position with the left side upward.
3. Make an 1-inch incision in the skin at the midpoint between the left orbit and the external auditory canal.
4. Under surgical microscopic observation diathermize the vascular supply to the parotid gland and soft tissues by electrocoagulation.
5. Partially incise the temporal muscle as Fig. 2-1 and retract it postero-inferiorly.
6. Prepare a burrhole with a saline-cooled dental drill (1-mm diameter) at the mid-position between the foramen ovale and the foramen orbitorotundum (3 mm anterior and 1 mm lateral to the foramen ovale) as shown in Fig. 2–2.
7. Open the dura mater carefully with a fine needle.
8. Confirm the presence of the middle cerebral artery crossing over the olfactory tract (Fig. 2–3).

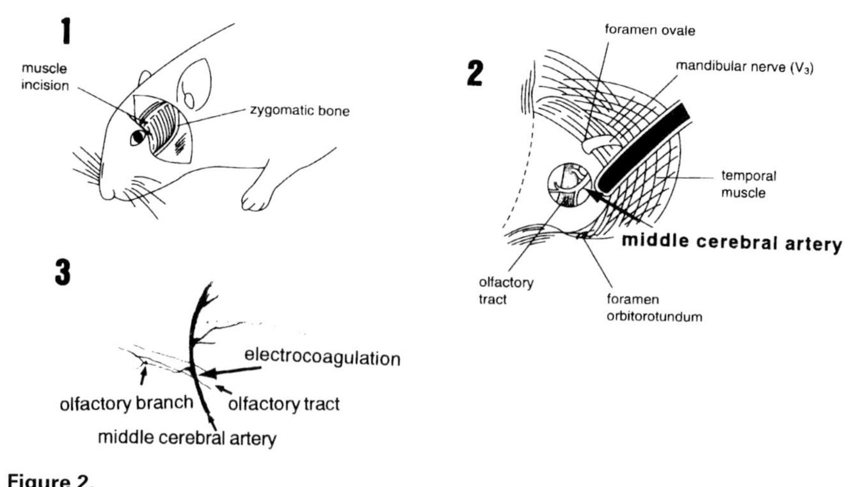

Figure 2.

9. Divide the arachnoid on either side of the artery by means of a fine needle.
10. Electrocauterize the middle cerebral artery which crosses over the olfactory tract as shown in Fig. 2–3, several times at low electrical power. Between each coagulation, the surgical region must be cooled with saline, and the tips of the microbipolar coagulator must be cleaned and wetted to prevent their adhering to the artery.
11. After confirming interruption of the blood flow of middle cerebral artery, cover the craniectomy by a small piece of gel foam and replace the soft tissues before suturing. Body temperature should be monitored with rectal and temporal muscular thermometers and controlled at 37 °C by means of a heating pad and light.

Results

This procedure produces pure cortical infarction over the area perfused by the middle cerebral artery. The same procedure is applicable to a mouse model of permanent middle cerebral artery occlusion.

Rat model of transient focal ischaemia

Protocol

1. Prepare an intraluminal occluder from 4–0 surgical nylon monofilament (either Ethilon (Ethicon) or Dermalon (Davis–Geck)). The tips of the occluders should be rounded by flame heating but not coated with silicone.
2. Fast Wistar or Sprague–Dawley rats (male, 270–340 g) overnight but allow free access to water.
3. Anaesthetize a rat with 4% halothane, and maintain anaesthesia with 0.5% *v*/*v* halothane in a nitrous oxide–oxygen mixture (70:30% *v*/*v*) through a face mask.
4. Expose the left common, external and internal carotid arteries and separate the adjacent vagal nerve from the carotids by midline incision.
5. Ligate the external carotid at its origin with a 6–0 silk suture (Fig. 3–1).
6. Clip the common carotid at its bifurcation and tie a 6–0 silk suture loosely around the common carotid just proximal to clip occlusion (Fig. 3–3).
7. Ligate the common carotid as proximal as possible.

Protocol *Continued*

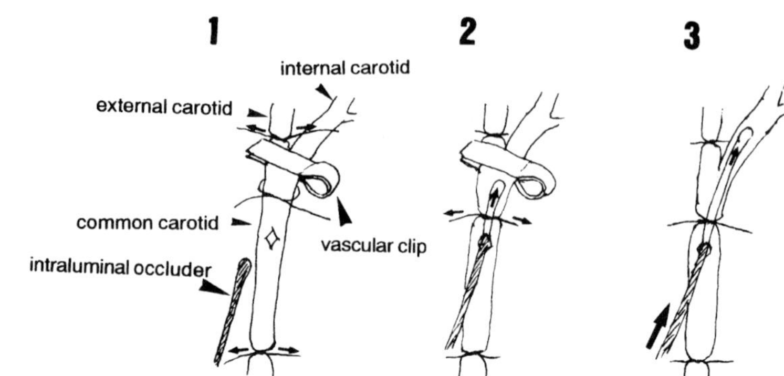

Figure 2.

8. Cut the common carotid with blade (length 0.5 mm or less).
9. From the puncture introduce a nylon suture into the common carotids and then tighten the silk suture at the carotid bifurcation around the intraluminal nylon

References

1. Ohtsuki, T., Ruetzler, C. A., Tasaki, K., and Hallenbeck, J. M. (1996). Interleukin-1 mediates induction of tolerance to global ischaemia in gerbil hippocampal CA1 neurons. *J. Cereb. Blood Flow Metab.*, **16**, 1137–42.
2. Yang, G., Kitagawa, K., Matsushita, K., Mabuchi, T., Yagita, Y., Yanagihara, T., and Matsumoto, M. (1997). C57BL/6 strain is most susceptible to cerebral ischaemia following bilateral common carotid occlusion among seven mouse strains: selective neuronal death in the murine transient ischaemia. *Brain Res.*, **752**, 209–18.
3. Hata, R., Matsumoto, M., Kitagawa, K., Matsuyama, T., Ohtsuki, T., Tagaya, M., Handa, N., Niinobe, M., Mikoshiba, K., and Nishimura, T. (1994). A new model of hind-brain ischaemia by extracranial occlusion of the bilateral arteries. *J. Neurol. Sci.*, **121**, 79–89.
4. Pulsinelli, W. A. and Brierley, J. B. (1979). A new model of bilateral hemispheric ischaemia in the unanaesthetized rat. *Stroke*, **10**, 267–72.
5. Tamura, A., Graham, D. I., McCulloch, J., and Teasdale, G. M. (1981). Focal cerebral ischaemia in the rat: 1. Description of technique and early neuropathological consequences following middle cerebral artery occlusion. *J. Cereb. Blood Flow Metab.*, **1**, 53–60.
6. Mandai, K., Matsumoto, M., Kitagawa, K., Matsushita, K., Ohtsuki, T., Mabuchi, T., Colman, D. R., Kamada, T., and Yanagihara, T. (1997). Ischaemic damage and subsequent proliferation of oligodendrocytes in focal cerebral ischaemia. *Neuroscience*, **77**, 849–61.

80

Isolated rat lung preparation

T. W. EVANS and J. M. C. GUTTERIDGE

Introduction

This model enables assessment of lung injury, and protection by pharmacological agents (Fig. 1).

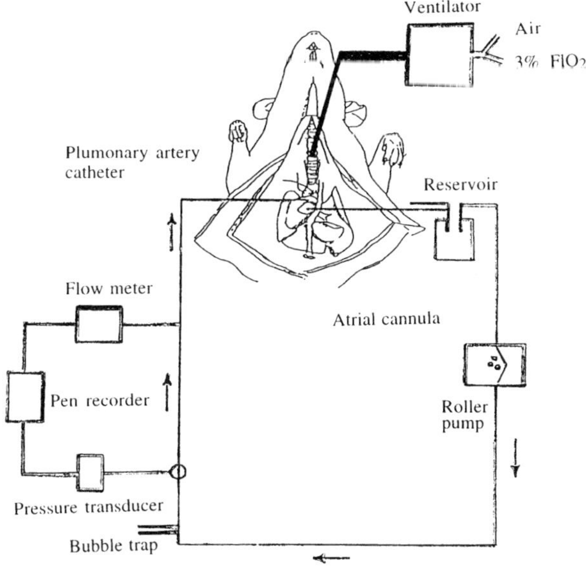

Figure 1. Schematic diagram of rat pulmonary perfusion circuit.

Protocol

Male Wistar rats, 300–350 g, housed in cages at 21 °C are anaesthetised with diazepam (0.6 mg kg^{-1}, intraperitoneal) and hypnorm (fentanyl 0.315 mg mL^{-1} and fluanisone 10 mg mL^{-1}; 0.15 mL intramuscular). Before each experiment one rat (the donor rat) is anticoagulated with 30 mg kg^{-1} heparin, exsanguinated, and the blood is placed in a reservoir maintained at 40 °C. The donor blood is also used to fill a perfusion circuit consisting

Protocol *Continued*

of: left atrial cannula, reservoir, peristaltic pump, connecting tubing, bubble trap, sidearm pressure transducer and pulmonary artery cannula. The pressure transducer is connected to a recorder for recording changes in pulmonary artery pressure (Ppa). After anaesthesia and cannulation of the trachea, the rats are ventilated with normoxic gas (21% O_2, 5% CO_2, 74% N_2), total volume 4 mL, frequency of 16 breaths min^{-1}, before being heparinized (30 mg kg^{-1}) and exsanguinated *via* the abdominal aorta. The blood is added to the reservoir and the lungs are exposed by means of a median sternotomy. A purse-string suture is placed around the left atrium to secure a catheter inserted through an atriotomy, which drains freely into the reservoir. A second cannula is inserted into the main pulmonary artery *via* a right ventriculotomy and tied in place. The lungs are perfused with whole blood at a constant flow of 18 mL min^{-1} and changes in Ppa are recorded. After setting up the perfusion circuit, a 30 min period of equilibration is allowed for establishment of baseline Ppa.

Calculations

Changes in Ppa represent changes in pulmonary vascular resistance (PVR) according to the formula:

$$\text{PVR} = \frac{(\text{Ppa} - \text{Pla})}{\text{Pulmonary blood flow}}$$

where Pla is the left atrial pressure. Because Pla was kept at zero and the pulmonary blood flow remained constant, Ppa was directly proportional to, and influenced only by, changes in PVR.

References

1. Hauge, A. (1968). Role of histamine in hypoxic pulmonary hypertension in the rat. *Circulation Res.*, **22**, 371.
2. Emery, C. J., Bee, D., Barer, G. R. (1981). Mechanical properties and reactivity of vessels in isolated perfused lungs of chronically hypoxic rats. *Clin. Sci.* **61**, 569.
3. Messent, M., Griffiths, M. J. D., and Evans, T. W. (1993). Pulmonary vascular reactivity and ischaemia–reperfusion injury in the rat. *Clin. Sci.*, **85**, 71.

81

Cell culture model for oxidative stress

KEIICHIRO SUZUKI

Introduction

Cells in culture are widely used as experimental models to study pro-oxidant and antioxidant responses. Although the conditions used are usually simple, and the results are obtained easily and rapidly, proteins, e.g. foetal calf serum, can act as radical scavengers, and iron and copper salts can also contaminate culture media during free-radical experiments. Appropriate controls must therefore be included and conditions carefully standardized, for example by using foetal calf serum (FCS)-free media.

NO overload by SIN-1

Protocol

1. Culture cells in appropriate dishes (subconfluent).
2. Aspirate the culture medium, wash dishes with PBS(–) and add PBS(–).
3. Add SIN-1 (3-morpholinosydnonimine *N*-ethylcarbamide) solution[1] (e.g. 0.1–1 mM final conc.).
4. Incubate for an appropriate time.
5. Aspirate PBS(–) and add culture medium. Or harvest cells

[1] SIN-1 is available commercially. This compound will produce superoxide anion and NO (and consequently peroxynitrite). SOD or another superoxide anion scavenger can be added if necessary. Prepare appropriate solutions (e.g. 1 mg/100 μL) in methanol or HCl (0.1 M)

Hypoxia–reoxygenation

Several methods are used for hypoxic experiments. A CO_2 incubator can be used to adjust the O_2 concentration. (Although the O_2 concentration is stable and easily changeable, the machine is more expensive and complete anoxia is impossible to achieve—the minimum level of oxygen is 1%.) Disposable O_2-absorbing reagents within a sealed box are highly effective, and can result in almost complete anoxia; it can, however, be difficult to maintain CO_2 concentration. Other methods which can bring about hypoxia/anoxia are:

sealed glass desiccator equilibrated with a desired gas in a 37 °C oven; (1)

anaerobic incubator; and

sealed bottle with Mg combustion.

Protocol

1. Culture cells in appropriate dishes under normoxic conditions.
2. If necessary use FCS-free media or PBS(–) instead of the usual culture media[1].
3. Incubate under hypoxic conditions.
4. Incubate under normal conditions and harvest culture cells.

[1] If possible, culture medium or PBS used in hypoxic conditions should be equilibrated with N_2.

Reference

1. Yamashita, N., Nishida, M., Hoshida, S., Kuzuya, T., Hori, M., Taniguchi, N., Kamada, T., and Tada, M. (1994). *J. Clin. Invest.* **94**, 2193–9.

82

Atherosclerosis (oxidized LDL): preparation of oxidized LDL/detection of cholesterol accumulation in macrophages

MASAYUKI YOKODE and TORU KITA

Introduction

A characteristic feature of atherosclerotic lesions is macrophage-derived foam cells in the intima filled with cholesteryl esters. Recent studies have shown that macrophages take-up oxidatively-modified low density lipoprotein (LDL) (oxidized LDL) by the scavenger receptor and are converted into foam cells. Thus, it is important to understand how oxidized LDL is produced in the body. Here we describe a method for isolating LDL and inducing its oxidation with a copper salt

Isolation of LDL

Protocol (1, 2)

1. Add EDTA.2Na (pH 7.4, 0.25 M, 7.5 mL) to blood (500 mL).[1]
2. Obtain 80 ml plasma fraction by low-speed centrifugation.
3. Add potassium bromide (density 1.019 g mL^{-1}, 3.81 g)[2].
4. Ultracentrifuge at 100000 g for 20 h at 4°C in a Beckman polyallomer centrifuge tube.[3]
5. Slice tube, discard top fraction, and add KBr (5.27 g) to the bottom fraction (80 mL).
6. Ultracentrifuge at 100000 g for 20 h at 4°C in a Beckman polyallomer centrifuge tube.[3]
7. Slice tube, save top fraction (density 1.040 g mL^{-1}; 10 mL) and add KBr (density 1.063; 0.34 g).
8. Ultracentrifuge at 100000 g for 20 h at 4°C in a Beckman polyallomer centrifuge tube.[3]

Protocol *Continued*

9. Slice tube and save top fraction (10 mL).
10. Dialyse against PBS (12 L; two changes of 6 L each) for 48 h at 4°C.
11. Sterilize by filtration through a 0.45-μm filter.
12. Determine the protein content by the Lowry method.

[1] All procedures should be conducted at 4°C.
[2] The plasma (initial density d_0) is adjusted to a final density (d_f) by addition of potassium bromide according to $X = V \times (d_f - d_0) / (1 - 0.312d_f)$, where X is the amount of potassium bromide (g) and V is the volume of plasma (mL).
[3] Do not use the centrifuge brake during deceleration of the ultracentrifuge.

Preparation of oxidized LDL

Protocol (3)

1. Dissolve LDL (5 mg) in PBS (1 mL).
2. Add solution of PBS containing $CuSO_4$ (10 μM, 1 mL).
3. Transfer to 3.5-cm sterile plastic culture dish and incubate for up to 48 h at 37°C.
4. Dialyse against PBS (12 L; two changes of 6 L each) for 48 h at 4°C.
5. Determine the protein content by the Lowry method.

Comment

Although there is increasing evidence of the presence of oxidized LDL in the body, little is known of the mechanism of this oxidation. This protocol describes a method of preparing oxidatively modified LDL by use of a copper salt ex vivo. Probucol, a lipid-soluble antioxidant, which protects LDL from copper-induced modification, has been shown to inhibit atherosclerosis in hypercholesterolaemic animals. Therefore, the sensitivity of the LDL towards copper-induced modification is likely to correlate with its atherogenicity.

Measurement of LDL thiobarbituric acid-reactive substances (TBARS) in oxidised LDL

Protocol (3)

1. Add LDL (50 μg) to NaCl (150 mM, 1.5 mL) in a 13-cm × 10-cm culture tube.
2. Add TCA (20% (*w/v*), 0.5 mL) and TBA reagent[1] (0.5 mL).

3. Boil at 95 °C for 1 h.
4. Cool with tap water.
5. Add butan-1-ol (2.0 mL) and vortex-mix vigorously.
6. Centrifuge at 4000 g for 15 min.
7. Remove top layer and measure fluorescence (excitation at 515 nm; emission at 550 nm).

[1]A 1:1 (*v/v*) mixture of thiobarbituric acid solution (0.67% *w/v*) and glacial acetic acid.

Primary culture of mouse peritoneal macrophage

Protocol (2)

1. Harvest peritoneal cells from unstimulated mice in phosphate buffered saline (PBS; 40 mL).
2. Collect cells by centrifugation at 400 g and 4 °C for 10 min.
3. Wash cells once in PBS by centrifugation at 400 g and 4 °C for 10 min.
4. Suspend at a density of 3×10^6 cells mL^{-1} in DME containing foetal calf serum (FCS; 10% *v/v*), penicillin, and streptomycin.
5. Disperse the cell suspension (1 mL) on a 3.5×1.0 cm plastic dish and incubate in humidified CO_2 (5%) at 37 °C for 2 h.
6. Wash cells with DME (2×2 mL) and incubate in DME containing FCS (10% *v/v*), penicillin, and streptomycin for 18 h before initiation of experiment.

Measurement of cholesteryl ester content of macrophages

Protocol (2)

1. Add lipoproteins (0–200 μg mL^{-1}) in FCS-free DME (0.6 mL) to the macrophage monolayer[1].
2. To each dish add [^{14}C]oleate–BSA complex[2](12 μL).
3. Incubate at 37 °C for 5 hr.
4. Remove the medium and wash twice with Buffer B[3] and once with Buffer C[4]
5. Add hexane–isopropanol (3:2; 1 mL) and incubate for 30 min.

Protocol *Continued*

6. Transfer the hexane–isopropanol mixture to a glass tube.
7. Add the hexane–isopropanol mixture (1 mL) to the cell monolayer and transfer to the tube.
8. Add [^{3}H]cholesteryl oleate (internal standard) to the tube and vortex-mix vigorously.
9. Evaporate organic solvents with nitrogen gas.
10. Dissolve the pellet in hexane (100 μL), apply to a silica gel TLC plate, and develop with hexane–diethyl ether–acetic acid, 90:30:1.
11. Identify the cholesteryl ester spot with iodine vapour and count the radioactivity.
12. Dissolve the cells in NaOH (0.1 M, 1 mL) and determine protein concentration by the Lowry procedure.

[1] All procedures should be conducted at 4 °C.
[2] Evaporate [1-^{14}C]oleic acid (500 μCi; 55–60 mCi mmol^{-1}) in 25-mL glass tube and re-suspend in a solution (8.7 mL) of sodium oleate (12.7 mM) and bovine albumin (12% *w/v*) in NaCl (pH 7.4, 150 mM). Add bovine albumin (12% *w/v*, 3.2 mL) in NaCl (pH 7.4, 150 mM) and stir gently for 4 to 6 h at room temperature.
[3] NaCl (150 mM) containing Tris-HCl (pH 7.4, 50 mM) and bovine albumin (2 mg mL^{-1}).
[4] NaCl (150 mM) containing Tris-HCl (pH 7.4, 50 mM).

References

1. Goldstein, J. L., Basu, S. K., and Brown, M. S. (1983). Receptor-mediated endocytosis of low-density lipoprotein in cultured cells. *Methods Enzymol.*, **98**, 241.
2. Kita, T., Yokode, M., Watanabe, Y., Narumiya, S., and Kawai, C. (1986). Stimulation of cholesteryl ester synthesis in mouse peritoneal macrophages by cholesterol-rich very low density lipoproteins from the Watanabe heritable hyperlipidaemic rabbit, an animal model of familial hypercholesterolaemia. *J. Clin. Invest.*, **77**(5), 1460.
3. Yokode, M., Kita, T., Kikawa ,Y., Ogorochi, T., Narumiya, S., and Kawai, C. (1988). Stimulated arachidonate metabolism during foam cell transformation of mouse peritoneal macrophages with oxidized low density lipoprotein. *J. Clin. Invest.*, **81**(3), 720.

83

Dicarbonyls and *HB-EGF* gene expression

WENYI CHE, YOUNG HO KOH, and NAOYUKI TANIGUCHI

Introduction

The serum concentration of the reactive dicarbonyls, methylglyoxal and 3-deoxyglucosone, increases under diabetic conditions. Methylglyoxal and 3-deoxyglucosone induce the production of heparin-binding epidermal growth factor-like growth factor (HB-EGF) in primary rat aortic smooth muscle cells by increasing the level of intracellular peroxides (1). HB-EGF is known to be a potent mitogen in smooth muscle cells and is abundant in atherosclerotic plaques (2). Methylglyoxal also activates c-Jun NH_2-terminal kinases.

Commercial methylglyoxal solutions contain many impurities and methylglyoxal readily undergoes polymerization and oxidation reactions. Methylglyoxal can be prepared by hydrolysis of 1,1-dimethoxypropanone.

Preparation and purification of methylglyoxal

Protocol

1. Mix well 1,1-dimethoxypropanone (Sigma; 100 mmol, 12.12 mL), sulphuric acid (5% *v*/*v*, 100 mL), and distilled water (200 mL) and boil for 1 h in a water bath.
2. Take the resulting mixture (or commercial methylglyoxal (Sigma) solution) and distil on a column containing helices under reduced pressure (b.p. 20 mmHg, 26 °C).
3. Prepare methylglyoxal stock solution[1].

[1] The purity of methylglyoxal stock solution can be confirmed by NMR spectroscopy.

Determination of the concentration of methylglyoxal

The concentration of methylglyoxal can be determined by end point enzymatic assay involving conversion to *S*-D-lactoylglutathione with glyoxalase I, and hydrolysis catalysed by glyoxalase II.

Protocol

1. Mix well methylglyoxal (500 μL; appropriately diluted), reduced glutathione (Boehringer Mannheim; 100 mM, 20 μL), glyoxalase I (Sigma; 1 unit mL^{-1}, 5 μL), sodium phosphate buffer (pH 6.6, 500 mM, 100 μL), and H_2O_2 (375 μL).
2. Monitor the absorbance at 240 nm and incubate at room temperature until a steady maximum is achieved.
3. Add glyoxalase II (Sigma; 1 unit mL^{-1}, 1 mL) to the reaction solution, mix well.
4. Monitor the absorbance at 240 nm and incubate at room temperature until a steady minimum is attained.

Calculation

The methylglyoxal concentration can be calculated from the equation:

$$\text{Methylglyoxal concentration (M)} = \Delta A_{240}/(-3.1 \times 10^{-3})$$

(For methylglyoxal solution, $\varepsilon_{240} = -3.1\ \text{mM}^{-1}\ \text{cm}^{-1}$.)

Induction of HB-EGF by methylglyoxal and 3-deoxyglucosne

Protocol

1. Prepare cultured cells (e.g. rat smooth muscle cells in Dulbecco's MEN + 10% (*w/v*) foetal bovine serum).
2. Treat with methylglyoxal or 3-deoxyglucosone[1] for 1 to 12 h.
3. Extract total RNA with acid guanidinium thiocyanate–phenol–chloroform.
4. Analyse by Northern blot with a rat HB-EGF cDNA as a probe.

[1] 3-DG can be synthesized chemically by the method of Khadem *et al.*

Assay of c-Jun NH_2-terminal kinases

Protocol

Preparation of whole-cell extract (WCE)

1. Obtain cultured cells.
2. Harvest cells after washing twice with ice-cold phosphate-buffered saline.
3. Suspend cells in WCE buffer (200 μl) and rotate at 4 °C for 30 min.
4. Centrifuge at 1000 g for 10 min.
5. Collect supernatant WCE.
6. Determine the concentration of protein in the WCE by the Bio-Rad protein assay.

WCE kinase assay

1. Prepare a mixture of WCE (20 μg), 10× reaction buffer (3 μL), [γ-^{32}P]ATP (5 μCi), and GST-c-Jun (1–79) (1 μg).
2. Add to H_2O_2 (30 μL).
3. Incubate at 30 °C for 30 min.
4. Terminate the reaction by boiling.
5. Separate the protein by 10% (*w/v*) SDS–PAGE.
6. Dry and autoradiograph.

[1] 10 X reaction buffer contains HEPES (pH 7.5, 250 mM), magnesium acetate (100 mM), and ATP (500 mM).
[2] WCE buffer contains HEPES (pH 7.7, 20 mM), Triton X-100 (0.05%), EDTA (0.1 mM), NaCl (75 mM), $MgCl_2$ (2.5 mM), DTT (dithiothreitol) (0.5 mM), β-glycerophosphate (20 mM), Na_3VO_4 (0.1 mM), PMSF (phenylmethyl-sulfonyl fluoride) (100 mg mL^{-1}), and leupeptin (2 mg mL^{-1}).

References

1. Che, W.-Y., Asahi, M., Takahashi, M., Kaneto, H., Okado, A., Higashiyama, S., and Taniguchi, N. (1997). Selective induction of heparin-binding EGF-like growth factor by methylglyoxal and 3-deoxyglucosone in rat aortic smooth muscle cells, *J. Biochem. Chem.*, **272**, 18453–9.
2. Miyagawa, J., Higashiyama, S., Kawata, S., Inui, Y., Tamura, S., Yamamoto, K., Nishida, M., Nakamura, T., Yamashita, S., Matsuzawa, Y., and Taniguchi, N. (1995). Location of heparin-binding EGF-like growth factor in the smooth muscle cells and macrophages of human atherosclerotic plaques, *J. Clin. Invest.*, **95**, 404–11.

84

UV damage models

YOSHIKI MIYACHI

Introduction

Because the skin is always in contact with oxygen and occasionally exposed to UV light, different kinds of photobiological cutaneous reactions inevitably occur resulting in photo-oxidative stress. Acute actinic damage, e.g. sunburn, photoallergic reactions, and chronic cumulative UV injuries including photo-ageing and photocarcinogenesis are the targets of investigation in photo-biology (1).

Acute UV-light injuries in animals

Sunburn reaction

Protocol

1. Use Skh-1 hairless mice[1] (2).
2. Put mice in a special cage 25 mm high[2] without anaesthesia[3].
3. Irradiate mice with appropriate sources of UV light[4].
4. Obtain skin samples for biochemical analysis[5] and histological examination[6], or measure the ear thickness[7].

[1] White hairless mice are preferable to eliminate any effect of melanin on photobiochemical reactions as an ROS quencher, and possible skin injuries caused by shaving.
[2] Otherwise mice will climb on each other to the roof.
[3] Anaesthetics can behave as photosensitizers and thus should be avoided.
[4] For example Toshiba FL-20SE (275–410 nm) as a source of UVB light and FL-20BLB (300–420 nm) as a source of UVA light, respectively. Measure the UV doses by means of a (Toshiba UVR-305/365D) UV meter.
[5] Skin homogenates can be obtained as a powder by use of the Spex Industries 6700 Freezer/Mill.
[6] Counting the number of sunburn cells is regarded as a reasonable quantitative indicator of photo-oxidative stress.
[7] Ear swelling responses are measured with a special gauge (Peacock, Tokyo, Japan)

Photoallergic/phototoxic reactions (3)

Protocol

1. Photosensitize mice either by applying a photosensitizer[1] to the skin of the abdominal wall or by intraperitoneal administration with subsequent UVA irradiation.
2. Photochallenge animals by painting the same photosensitizer on the ear lobe followed by UVA irradiation.
3. Measure the increment in ear thickness.

[1] Contact photosensitizers are usually dissolved in acetone. Porphyrins can be administered intraperitoneally. No such procedures are required for evaluation of phototoxic reaction.

Chronic UV light injuries in animals

Photoageing (4, 5) (Fig. 1)

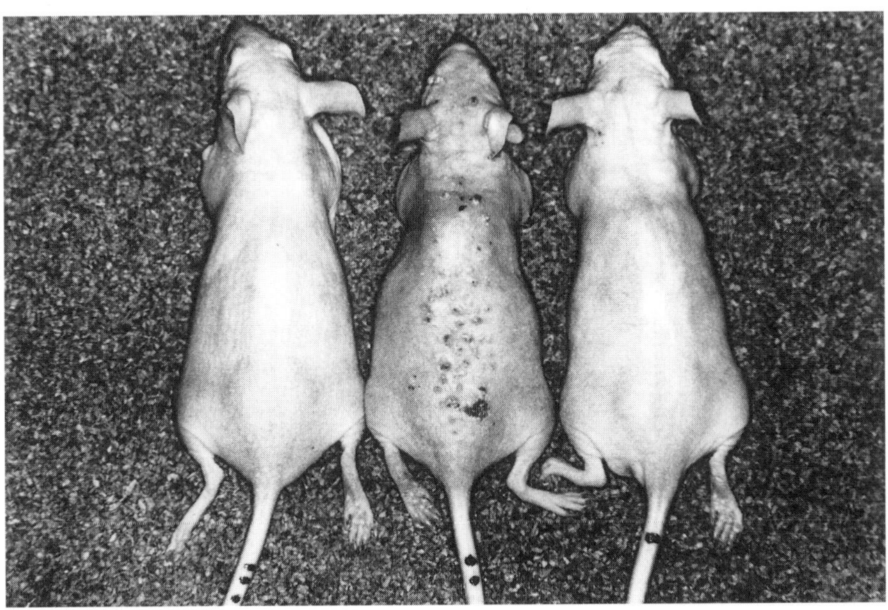

Figure 1. Clinical manifestations of chronic UV-irradiated hairless mice. Left. Mice received 30 J cm^{-2} UVA three times a week for 36 weeks. Middle. Mice received 40 mJ cm^{-2} UVB three times a week for 36 weeks. Right. Control mice.

Protocol

1. Irradiate hairless mice (6–8 weeks) with appropriate UV sources and doses (e.g. UVB 40 mJ cm^{-2}, UVA 30 J cm^{-2}).
2. Repeat irradiation three times a week for 24–36 weeks.
3. Obtain skin samples for biochemical assay or histochemical evaluation.

2.2. Photocarcinogenesis (6)

Protocol

1. Irradiate hairless mice (6–8 weeks) with appropriate doses of UVB[1].
2. Repeat irradiation a few times a week for several months.
3. Obtain skin samples for biochemical or histochemical procedures.

[1] Larger doses and more frequent irradiation result in faster development of skin cancer, although this is a less physiological condition.

UV injuries in cultured cells

Protocol

1. Suspend cells[1] in culture dishes or in three-dimensional cultures (7).
2. Change the medium for colourless and FCS-free medium a few hours before irradiation.
3. Irradiate the cells with the cover open, preferably in an incubator.
4. Harvest the cells for assay.

[1] Fibroblasts, lymphocytes, mast cells, endothelial cells are frequently used.

References

1. Miyachi, Y. (1995). Photoaging from an oxidative standpoint. *J. Dermatol. Sci.*, **9**, 79–86.
2. Takahashi, Y., Okada, K., Ohnishi, K., Ishikawa, O., and Miyachi, Y. (1995). Disaccharide analysis of the skin glycosaminoglycans in chronically ultraviolet-irradiated hairless mice. *J. Dermatol. Sci.*, **10**, 139–44.
3. Takigawa, M., Miyachi, Y., Toda, K. I., Yoshioka, A. (1984). Mechanisms of photosensitivity in mice. *J. Immunol.*, **132**, 1124–9.

4. Okada, K., Nonaka, Y., Ohnishi, K., Ishikawa, O., Miyachi, Y. (1994). Time-dependent effect of chronic UV irradiation on superoxide dismutase and catalase activity in hairless mice skin. *J. Dermatol. Sci.*, **8**, 183–6.
5. Miyachi, Y. and Ishikawa, O. (1998). Dermal connective tissue metabolism in photoaging. *Australas. J. Dermatol.*, **39**, 19–23.
6. Hamanaka, H., Miyachi, Y., Moriwaki, S. I., Nishigori, C., Imamura, S. (1991). Superoxide dismutase activity of murine ultraviolet radiation-induced fibrosarcoma cell strains. *J. Dermatol.*, **18**, 36–8.
7. Ishikawa, O., Kondo, A., Okada, K., Miyachi, Y., and Furumura, M. (1997). Morphological and biochemical analyses on fibroblasts and self-produced collagens in a novel three-dimensional culture. *Br. J. Dermatol.*, **136**, 6–11.

85

Model for ageing: longevity of *Caenorhabditis elegans*

NAOAKI ISHII

Introduction

The free-living nematode *Caenorhabditis elegans* is a small, rapidly growing organism that can easily be raised in the laboratory on the bacterium *Escherichia coli*. *C. elegans* has received much attention as a model for dissecting the genetics and, hence, the mechanisms that underlie ageing. Here we describe a method for measuring the life-span of wild and oxygen-sensitive mutants under different oxygen concentrations.

Protocol

Synchronization of animals

1. Incubate on NG agar plates (NaCl (3 g), peptone (2.5 g), agar (17 g), water (975 mL)).
2. Autoclave. Cool to 55°C, then using sterile technique add, in order, cholesterol (5 mg mL^{-1} in EtOH; 1 mL), $CaCl_2$ (1 M, 1 mL), $MgSO_4$ (1 M, 1 mL), and potassium phosphate (pH 6.0, 1 M, 25 mL). The cholesterol solution need not be sterilized.
3. After 24 h seed the plates with OP50, a uracil-requiring mutant of *Escherichia coli*, and incubate at 20°C.
4. Select a plate that contains many gravid adults and eggs and wash its surface with S buffer[1] (4.5 mL).
5. Transfer the suspension to a 15-mL centrifuge tube.
6. Add NaOCl (5%, 0.5 mL) and NaOH (10 M, 100 μL).
7. Shake at room temperature for 5 to 10 min (until all the developmental stages but eggs are dissolved) and centrifuge briefly (30 s, 1500 rpm).
8. Re-suspend the pellet in S buffer and repeat several times to remove all alkaline bleach.

9. Incubate overnight at 20°C in S buffer (~100 μL)), allowing for completion of embryogenesis and hatching to the first larval stage.

Measurement of longevity

1. Transfer newly hatched animals to a seeded 9-cm NG agar plate.
2. Incubate at 20°C until animals develop into adults.
3. Transfer ten animals to each of ten 3-cm NG agar plates containing 40 mM 5-fluoro-2-deoxyuridine (FUdR; Sigma, St Louis MO, USA)
4. Expose to different oxygen concentrations in an airtight plastic chamber and incubate at 20°C.
5. Examine daily to determine the time of death.

[1] S buffer is prepared by mixing NaCl (5.9 g), potassium phosphate (pH 6.0, 1 M, 50 mL), and KH_2PO_4 (136 g) and diluting with water to 900 mL. Adjust pH to 6.0 with concentrated KOH. Add water to 1 L. Autoclave.

Results

The results obtained are shown in Fig. 1.

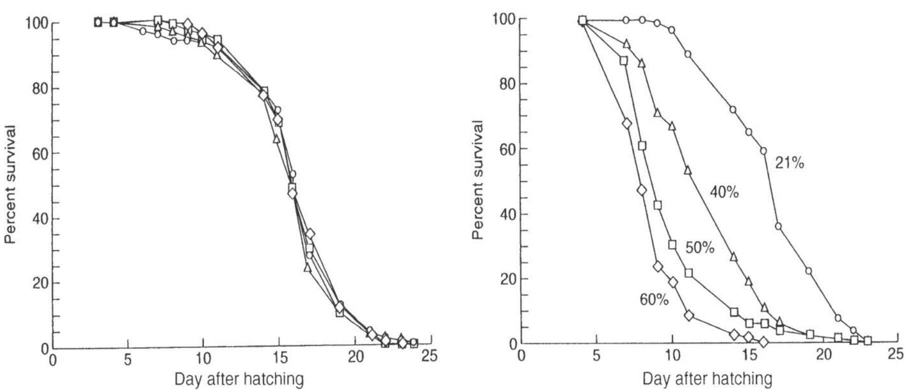

Figure 1. Life-spans of wild-type (upper panel) and *rad-8* (UV and oxygen-sensitive mutant; lower panel) *Caenorhabditis elegans* grown under different oxygen concentrations. Open circles, 21% oxygen; triangles, 40% oxygen; squares, 50% oxygen; diamonds, 60% oxygen (all *v/v*).

Comments

The gas in the chambers is replaced once a day and the oxygen concentration is measured by means of an oxygen analyser (e.g. Iijima Model G-101-Y). In the absence of an analyser, the oxygen concentration in the chamber can be estimated from:

$$\text{Exposure time (s/1000 mL chamber)} = 2.26e0.038R$$

where R (%) is the oxygen concentration from 60 to 90% and the flow rate is 2000 mL min^{-1}.

References

1. Brenner, S. (1974). The genetics of *Caenorhabditis elegans*. Genetics, **77**, 71–94.
2. Ishii, N., Suzuki, N., Hartman, P. S., and Suzuki, K. (1993). The radiation-sensitive mutant *rad-8* of *Caenorhabditis elegans* is hypersensitive to the effects of oxygen on ageing and development. *Mech. Ageing Develop.*, **38**, 1–10.
3. Wood, B. (ed) (1988). *The nematode Caenorhabditis elegans.*, Cold Spring Harbour Laboratory, New York.
4. Riddle D. L., Blumenthal, T., Mayer, B. J., and Priess, J. R. (ed.) (1997). *C. elegans* II. Cole Spring Harbour Laboratory, New York.
5. Epstein, H. H., Shakes, D. C. (ed) (1995). *Caenorhabditis elegans: Modern Biological Analysis of an Organism*, Methods in Cell Biology, Vol. 48. Academic Press.

86

Ovulation: superovulation and inhibition by SOD/*in vitro* fertilization and SOD

MUTSUO ISHIKAWA

Introduction

The phenomenon of follicular rupture is thought to be associated with inflammatory-like changes. Superoxide radicals are generally observed in inflammatory diseases and levels of these radicals also increase during follicular rupture. In addition, the administration of SOD (superoxide dismutase) greatly reduces ovulatory efficiency in rats and perfused rabbit ovaries.

In rat ovaries, SOD activity can protect intrafollicular ova from reactive oxygen species. In this section, we describe the protocol for investigating interactions between the reactive oxygen-scavenger system and ovulatory processes including oocyte maturation and fertilization.

Protocol

Superovulation and inhibition of ovulation by SODs

1. Use immature female rats (28 days), Wistar Imamichi strain, with stable oestrous cycles and normal general characteristics.
2. Inject PMSG (15 Int. units) intraperitoneally 50 h after administration of hCG (10 Int. units).
3. Inject SOD and/or catalase.
4. Anaesthetize with Nembutal.
5. Remove ovaries and oviducts.
6. Rupture the oviduct.
7. Pick out the oocyte–cumulus complex and put into media supplemented with hyaluronidase (300 Int. units mL^{-1}).
8. Count the ova (Fig. 1).

Protocol *Continued*

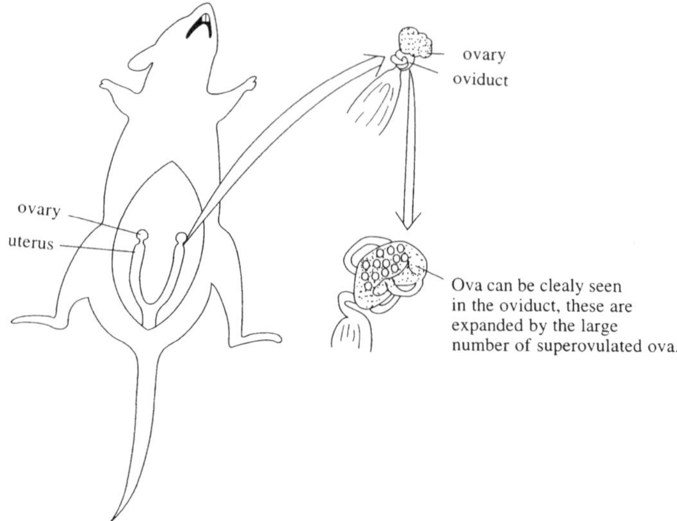

Figure 1. Collection of ova from the oviduct. Ova can be clearly seen in the oviduct, these are expanded by a large number of superovulated ova.

In vitro fertilization and effect of SODs on fertilization rate

1. Use mature female mice (6–8weeks), ICR strain, with stable fertilization rates (80–100%) by artificial insemination methods, despite easy handling and sperm preparation.
2. Inject PMSG (10 Int. units) intraperitoneally 48 h after administration of hCG 10 (10 Int. units).
3. Inject SOD and/or catalase.
4. Anaesthetize with Nembutal.
5. Remove ovaries and oviducts.
6. Rupture the oviduct.
7. Pick out the oocyte–cumulus complex and place in culture medium[1].
8. Inseminate with sperm[2] of concentration $0.25–9 \times 10^6$ mL^{-1}.
9. Count embryos at the two pronuclei stage 6 h after insemination.

[1] We use mKRB or HTF media made with sterile water and left in the culture incubator for at least a few hours before use. (All instruments must also be sterile.) mKRB (modified Krebs Ringer bicarbonate) contains NaCl (5.525 mg mL^{-1}), KCl (336 μg mL^{-1}), $CaCl_2.2H_2O$ (251 μg mL^{-1}), $MgSO_4.7H_2O$ (294 μg mL^{-1}), KH_2PO_4 (162 μg mL^{-1}), $NaHCO_3$ (2.106 mg mL^{-1}), glucose (1.000 mg mL^{-1}), Na pyruvate (60 μg mL^{-1}), Na lactate (3.6 μg mL^{-1}), penicillin G potassium (75 μg mL^{-1}), and BSA (40 μg mL^{-1}).

[2] Epididymal sperm are collected from a mature male mouse (over 12 weeks), placed in mKRB culture medium, and pre-incubated for 30 min to 2 h. (For rats, 5 h pre-incubation is needed.)

[3] Two pronuclei are usually visible 6 h after insemination. If these are not clear, wait until 24–30 h after insemination when the embryo is at the two-cell stage, or confirm there are two pronuclei and two polar bodies, and a sperm tail using whole-mount fixation.

Comments

One important point in the investigation of ovulation and fertilization is the selection of an animal species. Many species have different ovulatory cycles and fertilization phenomenon. To select an animal suitable for the investigation, it is necessary to be familiar with the characteristics of that species.

Recently developed, molecular biological methodology enables the use of very small samples such as oocytes and embryos for PCR. Collecting these samples at very precise phases might be one of the most important points to consider in investigations of reproduction.

References

1. Ishikawa, M., Tamate, K., Nakata, T., Sengoku, K., Suzuki, K., and Taniguchi N. (1993). Superoxide dismutase/corpus luteum: immunohistochemical localization of copper zinc and manganese superoxide anion in ovulation/luteal function. *ARTA*, **4**, 251–9.
2. Tamate, K., Sengoku, K., and Ishikawa, M. (1995). The role of superoxide dismutase in the human ovary and fallopian tube. *J. Obstet. Gynaecol.*, **21**, 401–9.
3. Ishikawa, M., Tamate, K., Sengoku, K., and Taniguchi, N. (1996). The role of superoxide dismutase and ovarian function. *Assisted Reprod. Rev.*, **S6**(1), 16–19.

87

Scavenging NO in experimental animals

TAKAAKI AKAIKE and YOICHI MIYAMOTO

Introduction

It is important to examine the effects of NO scavengers on animal models of diseases to evaluate the pathophysiological significance of NO. In animals with endotoxin-induced shock, excessive production of NO causes hypotension. In this section, we describe methods using PTIO derivatives (1), scavengers of NO (Fig. 1), in animals with endotoxic shock (2).

PTIO derivatives

+ NO → + NO_2

R = H: 2-phenyl-4,4,5,5-tetramethyl-imidazoline-1-oxyl-3-oxide (PTIO)
R = COOH: carboxy-PTIO

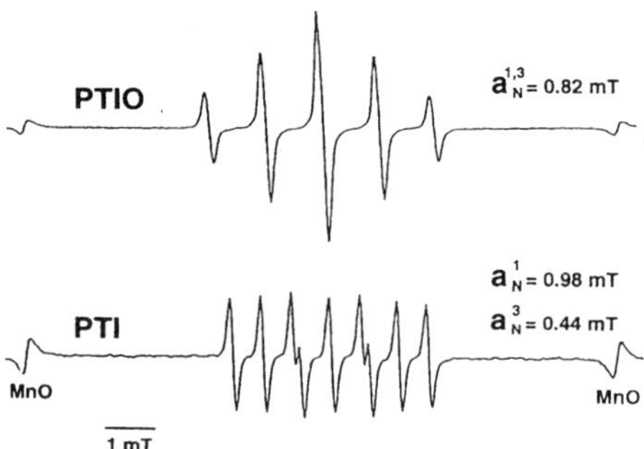

Figure 1. Chemical structures and ESR spectrum of PTIO (PTI) derivatives and their reaction with NO.

Analysis of haemodynamics in the animal model of endotoxin shock

Protocol

1. Anaesthetize a male Wistar rat (250–300 g) with pentobarbital sodium (60 mg kg^{-1}, i.p.).
2. Place the rat in a dorsal position and fix the limbs. Insert PE50 catheters (polyethylene tubes, 0.58 mm i.d. and 0.965 mm o.d.; Becton Dickinson, Sparks, MD, USA) into the left external jugular vein and the abdominal aorta.[1] The tube cannulated into the external jugular vein is used for injection of lipopolysaccharide (LPS) and the other reagents, that into the abdominal aorta for monitoring the blood pressure and heart rate. In some experiments, the carotid artery is cannulated for sampling blood. Insert a catheter into urinary bladder for collection of urine to evaluate renal function (Fig. 2).

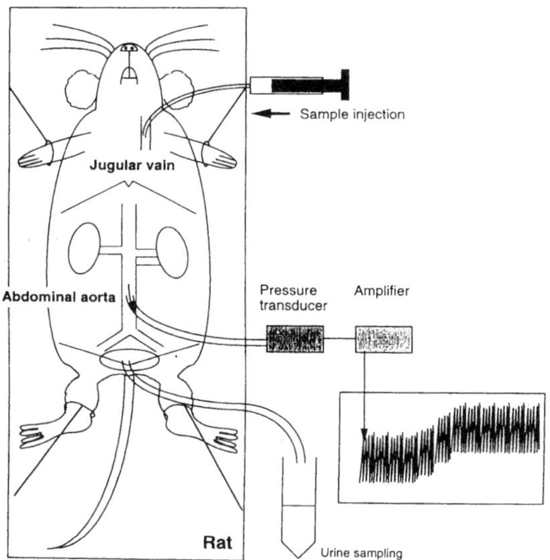

Figure 2. Evaluation of haemodynamics of endotoxin-shock rats.

3. Connect the catheter inserted into the abdominal aorta to a pressure transducer and monitor blood pressure (systolic, diastolic, and mean blood pressures) and heart rate continuously.
4. After blood pressure and heart rate have stabilized, administer LPS (*E. coli*; 026B, Difco Laboratories, Detroit, MI, USA) at a dose of 10 mg kg^{-1} ($5 \times LD_{50}$) *via* the external jugular vein and record the blood pressure

Protocol *Continued*

and heart rate. Blood pressure will gradually (3–4 h after injection of LPS) decrease to approx. 40% of the initial value; thereafter the rats might start to die.[2] In parallel with the fall in blood pressure, urine volume decreases significantly and the other parameters of renal function including urinary excretion of sodium and creatinine clearance also decline.

[1] After median incision in the lower abdomen, the catheter is inserted into the abdominal aorta, and the top of the catheter should be placed at the distal position from the bifurcation of the renal arteries. Incision in the lower abdomen should be sutured immediately after cannulation is complete. The carotid arteries are available for monitoring blood pressure instead of abdominal aorta.

[2] Sensitivity of rats to LPS is affected to some extent by the surgical procedures employed. Surgical stress due to abdominal incision can, for instance, increase the lethality of LPS. It is recommended that the LD_{50} of LPS is determined for each experimental protocol with different surgical procedures.

Detection of enhanced production of NO induced by LPS

It is desirable to detect the induction of NO synthase or increased production of NO in each animal model in which the effects of the NO scavenger are evaluated. Some of the methods for measurement of the NO biosynthesis in the rat with endotoxin shock are described below.

Protocol

Quantification of NO_2^- and NO_3^- in blood and urine

Enhanced production of NO induced by the injection of LPS is easily detected by measuring the amount of NO_2^- and NO_3^-, stable oxidized products of NO, produced in the blood and urine[1]. In our laboratory, quantification of NO_2^- and NO_3^- is performed by use of an autoanalyser (e.g. TCI-NOX 1000m, Tokyo Chemical Industry, Tokyo, Japan or ENO-10, Eicom, Kyoto, Japan) in which NO_3^- is reduced to NO_2^- by means of a cadmium-packed column, which is then quantified with the Griess reagent in a flow reactor system.

Identification of NO in biological samples by electron spin resonance (ESR) spectroscopy

Deoxyhaemoglobin as an endogenous NO trap is most conveniently used to detect NO in biological systems. A large amount of NO-haemoglobin is usually detected by ESR spectroscopy in the blood of endotoxin-shock rats. A variety of dithiocarbamate–Fe complexes such as Fe complexes with *N,N*-diethyldithiocarbamate, *N*-methyl-D-glucamine dithiocarbamate, and *N*-(dithiocarboxy)sarcosine have also been used as NO trapping agents. The ESR signals of NO trapped by these dithiocarbamate–Fe complexes are readily measurable because they can be identified by the typical triplet structure under both ambient and frozen conditions. Details of measurement procedures are described elsewhere (3–5).

[1] Because the concentrations of NO_2^- and NO_3^- in the blood and urine vary with each rat depending on the dietary NO_2^- and NO_3^- intake, it is a prerequisite to measure the concentrations of these metabolites in the sample obtained before administration of LPS as a control value for NO_2^-/ NO_3^- in each sample.

Scavenging NO *in vivo*

Protocol

Administration of carboxy-PTIO, an NO scavenger

The NO scavenger should be administered to the rats, after observation of the fall in blood pressure and increased production of NO induced by the injection of LPS, as mentioned above. A decrease in blood pressure can be seen approx. 90 min after injection of LPS; carboxy-PTIO dissolved in PBS (pH 7.4)[1] should then be infused, at a rate of 6 mL h^{-1} by means of an infusion pump, *via* the catheter inserted into the jugular vein.

Effects of the NO scavenger on the rats received LPS

Infusion of carboxy-PTIO at 0.17 mg kg^{-1} min^{-1} inhibits the decline of blood pressure, as shown in Fig. 3. The antihypotensive effect of carboxy-

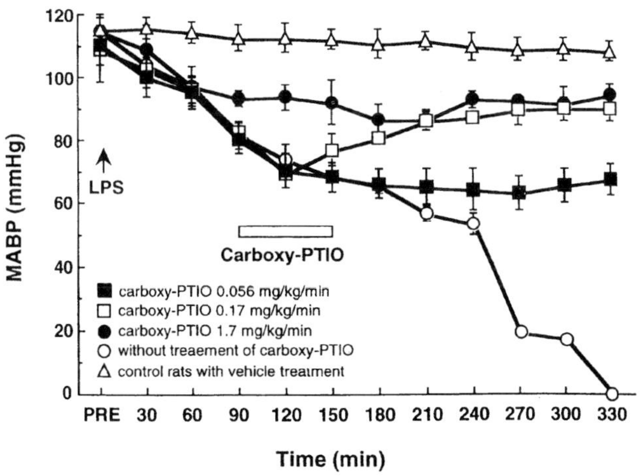

Figure 3. Decline in mean arterial blood pressure (MABP) of rats with endotoxin shock, and the effect of carboxy-PTIO.

Protocol *Continued*

PTIO was observed to be dependent on the dose of carboxy-PTIO administered (0.056–1.7 mg kg^{-1} min^{-1}). Renal failure in the LPS-treated rats, as assessed by decreases in urine volume, creatinine clearance, and urinary excretion of sodium, was also significantly restored by infusion of carboxy-PTIO (2).

Because of the hydrophilic nature of carboxy-PTIO, about 80% of the amount administered was recovered in urine 2–3 h after terminating the infusion. It is, therefore, possible to determine the NO-dependent conversion of carboxy-PTIO to carboxy-PTI by quantifying the amounts of these compounds in the urine by ESR spectroscopy (X-band) at room temperature (Fig. 1).[2]

[1] Because solutions of carboxy-PTIO in pure water are acidic, after preparation of the solution of carboxy-PTIO the pH should be checked.

[2] Because most carboxy-PTIO and carboxy-PTI in biological samples are in the reduced form, an oxidizing agent such as potassium ferricyanide [$K_3Fe(CN)_6$] should be added to the samples before measurement of the ESR spectra.

References

1. Akaike, T., Yoshida, M., Miyamoto, Y., Sato, K., Kohno, M., Sasamoto, K., Miyazaki, K., Ueda, S., and Maeda, H. (1993). Antagonistic action of imidazolineoxyl *N*-oxides against endothelium-derived relaxing factor/.NO through a radical reaction. *Biochemistry*, **32**, 827–32.
2. Yoshida, M., Akaike, T., Wada, Y., Sato, K., Ikeda, K., Ueda, S., and Maeda, H. (1994). Therapeutic effects of imidazolineoxyl *N*-oxide against endotoxin shock through its direct nitric oxide-scavenging activity. *Biochem. Biophys. Res. Commun.*, **202**, 923–930.
3. Akaike, T., Noguchi, Y., Ijiri, S., Setoguchi, K., Suga, M., Zheng, Y. M., Dietzschold, B., and Maeda, H. (1996). Pathogenesis of influenza virus-induced pneumonia: Involvement of both nitric oxide and oxygen radicals. *Proc. Natl. Acad. Sci. USA*, **93**, 2448–53.
4. Kosaka, H., Sawai, Y., Sakaguchi, H., Kumura, E., Harada, N., Watanabe, M., and Shiga, T. (1994). ESR spectral transition by arteriovenous cycle in nitric oxide haemoglobin of cytokine-treated rats. *Am. J. Physiol.*, **266**, C1400–5.
5. Yoshimura, T., Yokoyama, H., Fujii, S., Takayama, F., Oikawa, K., and Kamada, H. (1994). *In vivo* EPR detection and imaging of endogenous nitric oxide in lipopolysaccharide-treated mice. *Nat. Biotechnol.*, **14**, 992–4.

88

Experimental protocols for ROS and RNS: cyclic GMP radioimmunoassay using brain slices

DAISUKE OKADA

Cyclic GMP radioimmunoassay using brain slices

Preparation of 400-μm thick slices of rat cerebellar vermis

Introduction

Methods for preparing rat cerebellar slices using a vibrating slicer (Microslicer DTK 2000 or 3000, DosakaEM, Kyoto) are described (1, 2, 3). The use of brain slices has several important advantages—neuronal circuits, cell-to-cell connections, and intracellular macromolecular complexes remain intact within slices. Although brain-slice preparations enable direct visual access to intact brain structures, and easy control of fluid environment, slice techniques also have disadvantages—slices are by no means 'intact brain'. On the other hand, slices are mixtures of heterogeneous cells in a biochemical sense.

Protocol

1. Anaesthetize the rat with diethyl ether and decapitate at the joint between the occipital bone and cervical vertebrae[1].
2. Remove skin from the head posterior to ears.
3. The occipital bone is removed using scissors. Hold the head upright and insert the sharp tip of the scissors into the foramen magnum. Cut the right and left bones to the level of lambdoid suture, then lift the scissors. Be careful not to damage the brain.
4. Cut off the cerebellum and the brainstem at the occipital cerebrum and rinse in ice-cold Krebs solution[2]. The procedure for obtaining the brain is illustrated in Fig. 1.
5. The rinsed brain is transferred with a spoon to filter paper pre-wetted with cold Krebs solution. Remove the medulla and the cerebrum with a razor blade and replace the cerebellum in the cold Krebs solution by means of a spoon.

Protocol *Continued*

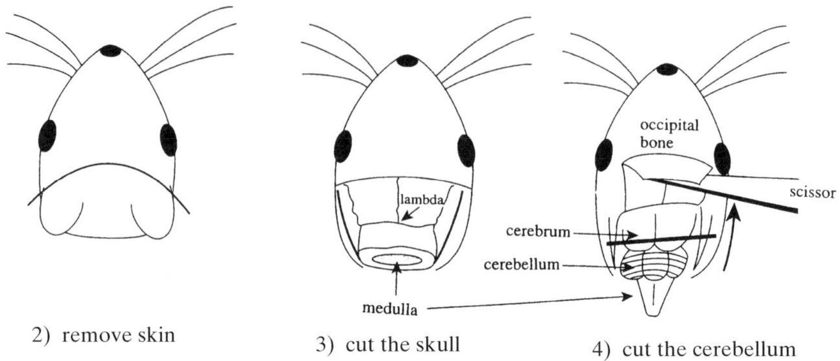

Figure 1. Obtaining the brain: Step 1, removing the cerebellum; Step 2, removing the skin; Step 3, cutting the skull; Step 4, cutting the cerebellum.

6. The top surface of an agar block[3] is spread with glue.
7. The cerebellum is transferred with a spoon to filter paper, to remove excess solution which would prevent attachment to the agar.
8. The cerebellum is attached to the agar block; the brain is supported at the hemispheres by means of bent-tip forceps.
9. Holding the agar block, put the agar block with the cerebellum into cold Krebs buffer while the glue hardens.
10. Place the block on filter paper with the cerebellum facing up.
11. Cut off the hemispheres parasagittally with a blade.
12. One of the cut surfaces of the vermis is glued to the centre of the specimen holder of the microslicer. The procedure for preparing the block is illustrated in Fig. 2.
13. Set the holder in the slicer. Fill the holder and the chamber with cold Krebs solution. Use ice made with Krebs solution to cool the solution in the chamber.
14. Adjust the cutting angle to 10°. Slice the vermis parasagittally at 400-μm intervals. Adjust the cutting frequency and speed so that the sliced part of the vermis does not vibrate while slicing[4].
15. Four slices 400 μm thick can be made from one vermis of an adult rat. They are incubated in a recovery chamber at 35°C for 90 min. Change the Krebs solution several times[5].

[1] Treat animals with care, do not discomfort or surprise them. Excited animals breathe at increased rates, causing deep and sudden anaesthesia, and finally anoxia. Use scissors to decapitate rats younger than 10 weeks old. Do not use a decapitator, because this might cause haemorrhage in the brain because of the high pressure on the neck.

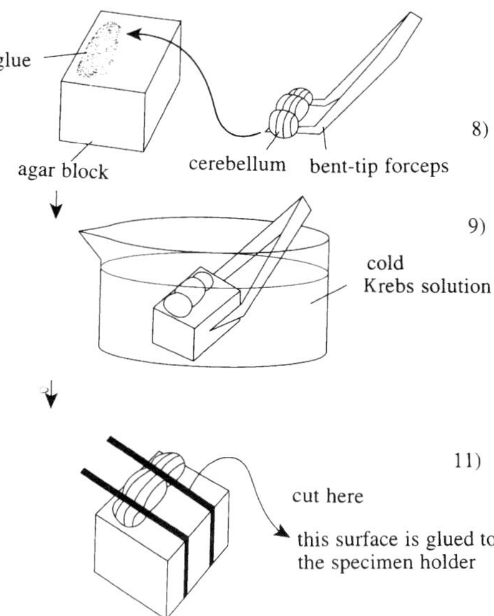

Figure 2. Attaching the cerebellum to the agar block with glue: Step 8, cerebellum supported with bent-tip forceps; Step 9, cold Krebs solution; Step 12, the surface is glued to the specimen holder.

[2] The composition of Krebs ringer solution varies, depending on the experiment. It is convenient to prepare a 10-fold stock solution but because $CaCO_3$ precipitates $NaHCO_3$ should be added after dilution. The standard 10-fold stock is prepared by: (i) dissolving NaCl (17.53 g; final concn 1200 mM), KCl (0.37 g; 20 mM), KH_2PO_4 (0.44 g; 12 mM), $MgSO_4.7H_2O$ (0.74 g; 20 mM) in water (ca 200 mL) (solution A); (ii) dissolving $CaCl_2.2H_2O$ (0.88 g; 24 mM) in water (ca 10 mL) (solution B); and (iii) mixing solutions A and B and diluting to 250 mL with distilled water. This 10-fold stock solution (50 mL) is diluted to 500 mL with water (ultra pure $>$18 Mohm, ex MilliQ), and $NaHCO_3$ (1.09 g; 26 mM) and glucose (0.99g; 11 mM) are added. Bubble with 95% O_2 + 5% CO_2. Check the pH of the solution is approx. 7.4.

[3] The agar block is prepared with 5% agar in 0.9% NaCl. Approximate dimensions of a block are 15 mm $\times$ 10 mm $\times$ 10 mm.

[4] Shaving blades made by Wilkinson, Gillette, or Feather are thin and often used. The cut surfaces of the slices made by Wilkinson blades are smooth, although they sometimes compress the tissue especially adult brain. Wilkinson blades are suitable for cutting brains of young ($<$5 weeks) animals. Although the cut surfaces of the slices made with Gillette blades are coarse, because they cut like a saw, they never compress the tissue. Gillette blades cut adult brains well. Either carbon steel or stainless steel is satisfactory. Remove oil on the blade with acetone before use. Use each blade only once.

[5] It is essential to wait for recovery. To avoid the action of neuroactive compounds such as glutamate leaking from damaged neurons on the cut surfaces, external solution should be changed several times. Handle the slices by means of a fine blush or a 7–8 mm tip diameter pipette. To determine the recovery period, measure the basal levels of the target compounds after recovery. They should be constant for all slices after adequate recovery. The success of experiments on brain slices depends largely on consistent preparation of good quality slices. Prepare slices rapidly and keep the brain (yours and the animal's) cool. Do not mechanically damage the vermis or slices. Slices are very sensitive to pH and anoxia. Molecular layers of a good slice is translucent and gets yellowish and opaque when dying. Purkinje cell bodies are barely visible to the eye when healthy; when dead, Purkinje cells swell and become visible to the eye.

Cyclic GMP radioimmunoassay using cerebellar slices

Introduction

Assay methods for cyclic GMP formation in response to neurotransmitters are described (4, 5). Neuronal activity generates nitric oxide which is a potent activator of soluble guanylate cyclase, a major synthesizing enzyme of cyclic GMP. Cyclic GMP in turn regulates enzymes such as cyclic GMP-dependent protein kinase, cyclic GMP-binding (types 2, 5, and 6) phosphodiesterases, and cyclic GMP-gated cation channels. The NO-cyclic GMP pathway is known to play a crucial role in inducing synaptic plasticity such as long-term depression in the cerebellum.

Protocol

1. Incubate the slice with agonist in the presence or absence of the drugs under investigation.[1]
2. Homogenize the slice in a glass microhomogenizer (Volpi, Italy) containing cold trichloroacetic acid (20% *w/v*, 0.2 mL)[2], then centrifuge the homogenate at 1000 g for 4 min at 4 °C.
3. Dissolve the pellet in NaOH (1 M) and determine the protein concentration (BCA method, Pierce; Bradford method, BioRad; or Lowry method).
4. Collect the supernatant and extract four times with water saturated with diethyl ether. Evaporate traces of ether with N_2 and adjust the pH of the supernatant to 7 with a small volume of NaOH (1 M). Store at –20 °C until radioimmunoassay with a kit[3].

[1] Incubation conditions depend on the experiment. As a standard, we incubate a slice in 1 mL Krebs buffer at 35 °C for 5 min. We include 0.1 mM IBMX, because IBMX inhibits both type 5 and type 1 phosphodiesterases found in Purkinje cells. Incubation with IBMX (100 μM for 5 min) does not affect the basal level of cGMP in the slice, but it makes the agonist-induced changes measurable.
Several concentrations of agonist and the testing drugs should be tested. As a positive control, high concentrations of NO donor should be tested. Sodium nitroprusside at 3 mM results in more than 300 pmol/mg protein of cGMP in a slice. As a negative control, the resting level of cGMP (20–25 pmol/mg protein) should be tested. Because four slices can be obtained from one rat, and two are for these controls, the remaining two slices can be used to test agonist and agonist plus antagonist only. Divergence between trials are therefore largely dependent on slice preparation.

[2] Alternatively, reaction can be terminated by heat, such as boiling slices in hot water or microwave irradiation. However, HPLC analysis performed recently in our hands showed that heating yielded less cGMP than TCA termination.

[3] Amersham provides three different kits for determination of cGMP. Radioimmunoassay using tritium- and iodo-radiolabelled cyclic GMP, and ELISA. Each method has different sensitivity and a different detection principle. The protocols of the assays are described in the manufacturer's instructions.

Comments

In the cerebellar cortex, soluble guanylate cyclase is localized in Purkinje cells, whereas neuronal nitric oxide synthase is in granule, basket, and stellate cells. Although Purkinje cells respond to nitric oxide released from outside of the cell by producing cyclic GMP, cyclic GMP-activating phosphodiesterase (type 5) in the cell hydrolyses cyclic GMP rapidly. This feed-back mechanism makes cyclic GMP increases in the cell very transient. To measure the agonist-induced cyclic GMP increase in the slices, it is essential to include low concentrations of phosphodiesterase inhibitor, otherwise, agonist-induced cGMP increase is not observed. Select the concentration of the inhibitor which does not affect the basal cyclic GMP concentration under the incubation conditions used.

Higher doses and longer incubations are often required to obtain consistent data with drugs applied to slices, because of drug absorption by cellular membranes of broken cells on the cut surface of the slices.

References

1. Dingledine, R. (ed.) (1984). *Brain slices*. Plenum Press, NY.
2. Cassella, J. P., Hay, J., and Lawson, S. J. (1997). *The rat nervous systems. An introduction to preparatory techniques*. John Wiley and Sons, Chichester.
3. Aitken, P. G., Breese, G. R., Dudek, F. F., Edwards, F., Espanol, M. T., Larkman, P. M., Lipton, P., Newman, G. C., Nowak, T. S. Jr., and Panizzoni, K. L. (1995). Preparative methods for brain slices: a discussion. *J. Neurosci. Methods*, **59**, 139–49.
4. Okada, D. (1992). Two pathways of cyclic GMP production through glutamate receptor-mediated nitric oxide synthesis. *J. Neurochem.*, **59**, 1203–9.
5. Okada, D. (1995). Protein kinase C modulated calcium sensitivity of nitric oxide synthase in cerebellar slices. *J. Neurochem.*, **64**, 1298–1304.

89

Oxidative stress models: *E. coli*, cultured cells

QIU-MEI ZHANG, AKIRA TACHIBANA, and SHUJI YONEI

Introduction

Reactive oxygen species occur naturally in aerobic cells, arising from a wide variety of intracellular and environmental sources. Cells maintain strong defence systems against such threats, for example superoxide dismutases, catalases, and peroxidases. In some circumstances, the concentration of reactive oxygen species increases to a level which overwhelms the basal level of such scavenging capacity, giving rise to a condition at oxidative stress (1, 2). Artificial, oxidative stress can be induced by addition of hydrogen peroxide or superoxide-generating drugs, or by increasing the partial pressure of oxygen.

Oxygen stress in *E. coli*

Protocol

Treatment of *E. coli* cells with reactive oxygen species

1. Inoculate LB medium[1] (5 mL) with a single *E. coli* colony.
2. Culture overnight at 37 °C.
3. Dilute the overnight culture 100-fold in a 200-mL Erlenmeyer flask containing LB medium (30 mL).
4. Incubate at 37 °C until the concentration reaches about 2×10^8 cells mL^{-1}.
5. Collect the cells by centrifugation at 6000 rpm for 5 min, wash once and re-suspend in M9 buffer[2].
6. Add hydrogen peroxide or superoxide-generating drugs at different concentrations. Methyl viologen (paraquat), menadione, and plumbagin can be used as superoxide-generating drugs.

7. Incubate at 37 °C for 1 h with vigorous shaking.

Survival

1. Dilute the treated cell suspension appropriately with phosphate-buffered saline[3] (PBS).
2. Spread 0.1 mL on LB plate containing Difco agar (15 g L^{-1}).
3. Incubate at 37 °C for approx. 18 h.
4. Count the number of viable colonies to estimate survival.

Mutation inducing resistance to rifampicin

1. Incubate the cell suspension treated with reactive oxygen species in LB medium at 37 °C for 4 h.
2. Spread 0.1 mL of the cell suspension treated with reactive oxygen species on and LB plate containing rifampicin (100 mg mL^{-1}).
3. Incubate at 37 °C for approx. 40 h.
4. Count the number of mutants on the plate.
5. Calculate the induced mutation frequency by use of the equation: Induced mutation frequency = $(M - M_0)/N$, where M and M_0 are the number of mutants mL^{-1} with or without exposure to reactive oxygen species and N is the number of viable cells mL^{-1} after treatment with the agents.

Induction of *katG::lacZ* and *nfo::lacZ* fusions

1. Collect the treated *E. coli* cells with *katG::lacZ* or *nfo::lacZ* fusion gene.
2. Wash the cells once and re-suspend in LB medium.
3. Incubate at 37 °C for approx. 1 h.
4. Measure the absorbance at 600 nm (A_{600}).
5. Transfer 0.2 mL to Z buffer[4] (1.8 mL).
4. Add SDS (0.1% *w/v*, 0.1 mL) and chloroform (0.1 mL).
5. Vortex-mix for 10 s.
6. Pre-incubate at 28 °C for 10 min.
7. Add *O*-nitrophenyl-β-D-galactopyranoside (ONPG; 4 mg mL^{-1}, 0.4 mL) in KPi buffer (pH 7.0, 0.1 M)
8. Vortex-mix.
9. Incubate at 28 °C until sufficient yellow colour develops.
10. Stop the reaction by adding Na_2CO_3 (1 M, 1 mL; reaction time *t* min).
11. Measure the A_{420} and A_{550}.

Protocol *Continued*

12. Calculate β-galactosidase activity by use of the equation (3): Units/A_{600} = $[1000(A_{420} - 1.75 \times A_{550})]/(0.1 \times t \times A_{600})$.

[1] LB medium is prepared by mixing Bacto tryptone (10 g), Bacto yeast extract (5 g), and NaCl (5 g), diluting with water to 1 L, and adjusting the pH to 7.2.
[2] M9 buffer is prepared by dissolving NaH_2PO_4 (6 g), KH_2PO_4 (3 g), NaCl (0.5 g), and NH_4Cl (1 g) in water (1 L), autoclaving and cooling this solution, and finally adding $MgSO_4$ (0.1 M, 10 mL) and $CaCl_2$ (0.01 M, 10 mL) which have previously been sterilized separately by autoclaving.
[3] PBS contains NaCl (8 g L^{-1}), KCl (0.2 g L^{-1}), NaH_2PO_4 (1.15 g L^{-1}), and KH_2PO_4 (0.2 g L^{-1}).
[4] Z buffer is prepared by mixing KPi buffer (pH 7.0, 0.2 M, 100 mL), KCl (1 M, 2 mL), $MgCl_2$ (1 M, 0.2 mL), β-mercaptoethanol (1 M, 10 mL), and H_2O (87.8 mL). (The β-mercaptoethanol should be added just before the use.) KPi buffer (pH 7.0) is prepared by mixing KH_2PO_4 (0.2 M, 39 mL) and K_2HPO_4 (0.2 M, 61 mL).

Oxygen stress in cultured mammalian cells

Protocol

1. Inoculate 100–400 mouse C3H10T1/2 cells/6-cm plate.
2. Culture at 37 °C for 24 h in BME medium[1].
3. Add drugs (for example methyl viologen and hydrogen peroxide) at 0–20 μg mL^{-1}.
4. Incubate at 37 °C for 24 h.
5. Wash the cells twice with PBS.
6. Incubate the cells at 37 °C for 10–12 days.
7. Stain colonies with Giemsa.
8. Count colonies to estimate survival.

[1] BME medium is prepared by mixing BME–Earle medium (500 mL), glutamine (200 mM, 10 mL), $NaHCO_3$ (10% *w*/*v*, 5 mL), and foetal calf serum (50 mL).

Comments

1. In *E. coli* sodAsodB mutants deficient in both Mn- and Fe-superoxide dismutase activity, oxygen stress can be induced by significantly lower concentrations of drugs.
2. The control experiments should be conducted under anaerobic conditions.
3. Many experiments are reported that induce oxidative stress conditions in bacteria and mammalian cells (2, 4).

References

1. Fridovich, I. (1983). Superoxide radical: An endogenous toxicant. *Annu. Rev. Pharmacol. Toxicol.*, **23**, 239–57.
2. Farr, S. and T. Kogoma. (1991). Oxidative stress responses in *Escherichia coli* and *Salmonella typhimurium*. *Microbiol. Rev.*, **55**, 561–85.
3. Miller, J. H. (1972). *Experiments in molecular genetics*. Cold Spring Harbour Laboratory, New York.
4. Packer, L. and Glazer, A. N. (eds.) (1990). Oxygen radicals in biological systems Part B. In *Methods in Enzymology*, Vol. 186. Academic Press, San Diego.

90

Carcinogenesis model: progression models *in vitro* and *in vivo*

FUTOSHI OKADA and MASUO HOSOKAWA

Introduction

Tumour progression is the final stage of carcinogenesis. In this stage the tumour cell population expands and, more importantly, the tumour cells gradually acquire enhanced malignant properties. Although reactive oxygen species produced at the site of inflammation are known to be one of the major contributors to the carcinogenesis, the role of reactive oxygen species in tumour progression is not fully understood because of a lack of an appropriate animal model. We have established a mouse model in which the process of tumour progression accelerated by inflammation-mediated reactive oxygen species can be observed.

In vitro experimental model of tumour progression

Protocol

Inflammatory cells reactive to a foreign body are collected as follows. All the instruments used during surgical procedures must be autoclaved.

1. Anaesthetize C57BL/6 mice and make a 10-mm cut in the skin of the abdomen after wiping with 70% *v/v* ethanol.
2. Make an 8 mm-cut in the peritoneum.
3. Insert a piece (10 mm × 5 mm × 3 mm) of haemostatic gelatin sponge (Spongel; Yamanouchi Pharm, Japan) into the peritoneal cavity by use of forceps.
4. Close the incision in the peritoneum with a suture, and close the skin with a sterile clip.
5. Harvest peritoneal-exuded cells (which we term inflammatory cells) 5 days later by lavage of the peritoneal cavity after injection of ice-cold PBS (5 mL) supplemented with penicillin G (200 units mL^{-1}) and

heparin sodium (10 units mL^{-1}), then aspiration of the fluid by the use of a syringe with a 23-gauge needle.

6. Repeat the above harvesting procedure twice.
7. Lyse erythrocytes by treating the harvested cell suspension with TRIS-buffered ammonium chloride for a few minutes at 37 °C.
8. Wash the harvested cells three times with MEM medium supplemented with FCS (2% *v/v*) by centrifugation at 1200 rpm for 5 min.
9. Co-culture inflammatory cells (1×10^6) with regressor tumour cells[1] (QR-32, 1×10^4) in a 24-well plate in FCS (8% *v/v*, 2 mL)-supplemented MEM medium for 48 h.
10. Collect supernatant and store at –80 °C.
11. Perform assay for prostaglandin E_2 (PGE_2) in the culture supernatant by ELISA or radioimmunoassay.

[1] The term 'regressor tumour cells' was based on the characteristic spontaneous regression of the cells in normal syngeneic hosts after injection. Regression of tumour cells caused by various modifications such as viral infections, gene transfections, and exposure to chemicals has been demonstrated experimentally, and is known as tumour xenogenization (1).

In vivo experimental model of tumour progression

Protocol

1. Swab a mouse with 70% *v/v* ethanol on the right flank skin around the pelvic region.
2. Make a 10-mm incision in the mouse skin.
3. Make a pocket reaching the thorax between skin and subcutaneous muscle tissue with the tip of a pair of scissors by slowly opening and closing them several times.
4. Insert gelatin sponge (10 mm × 5 mm × 3 mm) into the subcutaneous pocket away from the incision. Foreign bodies are used to induce inflammation. Gelatin sponge is reported to induce acute inflammation and is absorbed spontaneously (Fig. 1). It takes a little more than one month for the gelatin sponge to disappear completely. To induce chronic inflammation in mice, we use a piece of plastic plate (polystyrene) (2). Plastic plate is cut into 10 mm × 5 mm pieces which are sterilized by UV-irradiation. Regressor cells (1×10^4 in 50 μL medium) are seeded on each plastic plate. After incubation for 24 h the plate with attached regressor tumour cells is implanted in the same manner as described for gelatin sponge.

Protocol *Continued*

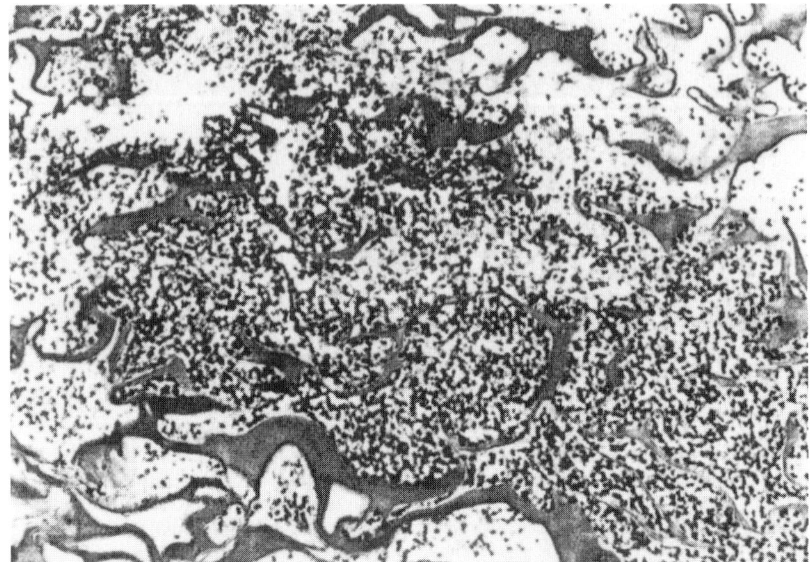

Figure 1. Host acute inflammatory cells penetrate inserted gelatin sponge.

5. Close the incision with clips.
6. Inject QR-32 cells (1×10^5 in 0.1 mL) through a 27-gauge needle at the site of gelatin sponge insertion. Tumours develop at the site of co-implantation.

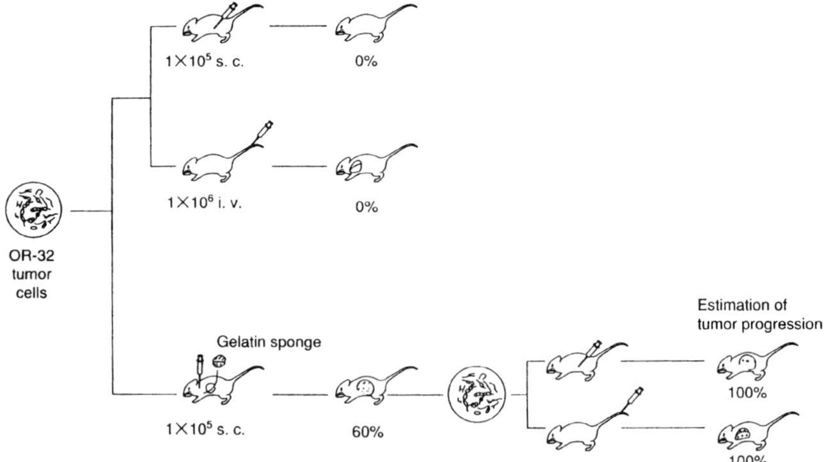

Figure 2. *In vivo* experimental model of tumour progression accelerated by inflammation-mediated reactive oxygen species.

7. Establish culture tumour cell lines from the individually developed tumours.
8. Examine the growth potential of the cultured tumour cells injected s.c. or i.v. into normal syngeneic mice and estimate their progressed phenotypes (Fig. 2).

   Growth of tumours injected into animals can be affected by the presence of host inflammatory cells. To eliminate contamination of established culture cell lines by host cells, culture for at least four passages before re-injection into other normal mice. These procedures are important to ascertain whether tumour progression was actually caused by alteration of the phenotype of the tumour cells rather than host cell contamination. Tumour progression of QR-32 cells is determined depending on whether tumour cell lines exhibit significant increases in either s.c. tumorigenicity or i.v. lung metastases as compared with those of original QR-32 cells (3).

Comments

Tumour progression of QR-32 cells can be determined in either *in vitro* or *in vivo* experiments. When using QR-32 regressor tumour cells *in vitro*, subcutaneous tumorigenicity is reflected in the production of immuno-suppressive prostaglandin E_2 (PGE_2) (1). We can, therefore, use the level of PGE_2 produced by the cultured tumour cells as a surrogate marker of the *in vivo* tumorigenicity of QR-32 cells. We observed that PGE_2 production increased dramatically after co-culture of QR-32 cells with inflammatory cells and somatic mutations of the tumour cells were observed as a result (3–5). Inflammatory-cell-mediated PGE_2 up-regulation and somatic mutations were inhibited by the presence of radical scavengers (5). We also found that *in vivo* progression of QR-32 cells after co-implantation of gelatin sponge was significantly inhibited by induction or administration of anti-oxidative enzymes or antioxidants. We thus confirmed that our *in vitro* and *in vivo* experimental models are useful for investigating the effect of physiologically produced reactive oxygen species on the malignant progression of benign or weakly malignant tumour cells.

References

1. Okada, F., Hosokawa, M., Hasegawa, J., Ishikawa, M., Chiba, I., Nakamura, Y., and Kobayashi, H. (1990). Regression mechanisms of mouse fibrosarcoma cells after *in vitro* exposure to quercetin: diminution of tumorigenicity with a corresponding decrease in the production of prostaglandin E_2. *Cancer Immunol. Immunother.*, **31**, 358–64.
2. Okada, F., Hosokawa, M., Hamada, J.-I., Hasegawa, J., Mizutani, M., Takeichi, N., and Kobayashi, H. (1993). Progression of a weakly tumorigenic mouse

fibrosarcoma at the site of early phase of inflammation caused by plastic plates. *Jpn. J. Cancer Res.*, **84**, 1230–6.

3. Okada, F., Hosokawa, M., Hamada, J.-I., Hasegawa, J., Kato, M., Mizutani, M., Ren, J., Takeichi, N., and Kobayashi, H. (1992). Malignant progression of a mouse fibrosarcoma by host cells reactive to a foreign body (gelatin sponge). *Br. J. Cancer*, **66**, 635–9.
4. Okada, F., Hosokawa, M., Hasegawa, J., Kuramitsu, Y., Nakai, K., Yuan, L., Lao, H., Kobayashi, H., and Takeichi, N. (1994). Enhancement of *in vitro* prostaglandin E_2 production by mouse fibrosarcoma cells after co-culture with various anti-tumour effector cells. *Br. J. Cancer*, **70**, 233–8.
5. Okada, F., Nakai, K., Kobayashi, T., Shibata, T., Tagami, S., Kawakami, Y., Kitazawa, T., Kominami, R., Yoshimura, S., Suzuki, K., Taniguchi, N., Inanami, O., Kuwabara, M., Kishida, H., Nakae, D., Konishi, Y., Moriuchi, T., and Hosokawa, M. (1999). Inflammatory-cell-mediated tumour progression and minisatellite mutation correlate with the decrease of antioxidative enzymes in murin fibrosarcoma cells. *Br. J. Cancer*, **79**, 377–85.

91

Iron-overload model

SHINYA TOYOKUNI

Introduction

Redox cycling is a characteristic of transition metals such as iron. Iron is an essential metal in mammals for oxygen transport by haemoglobin and for the function of many enzymes including catalase and cytochromes. However, the 'free' or 'catalytic' form of iron mediates in the production of reactive oxygen species *via* the Fenton reaction and induces oxidative stress. 'Free' iron is quite cytotoxic, as well as mutagenic and carcinogenic. There is an increasing number of reports of an association between increased body iron stores and increased risk of cancer. Intraperitoneal injection of an iron chelate, ferric nitrilotriacetate (Fe-NTA), induces renal proximal tubular necrosis associated with lipid peroxidation and oxidative DNA damage that finally leads to a high incidence of renal cell carcinoma after repeated administration in rodents (1, 2).

Protocol

1. Mix $Fe(NO_3)_3$ ($Fe(NO_3)_3.9H_2O$; 300 mM; 12.12 g/100 mL H_2O) and NTA (nitrilotriacetic acid disodium salt; 600 mM; 14.11 g/100 mL H_2O) solutions at a volume ratio of 1:2 (molar ratio, 1:4) with magnetic stirring at room temperature[1]. For mice, dilute further with H_2O to furnish an appropriate injection volume.
2. Adjust the pH of the solution to 7.4 with sodium hydrogen carbonate powder (the colour of the solution changes from orange to greenish brown).

Acute study by single injection

3. Inject intraperitoneally into rats or mice. For a 100–125-g male Wistar rat (4–5 weeks), 15 mg iron kg^{-1} (0.27 mL for a 100-g rat) is the sub-lethal maximal dose to be administered. For a 20–24-g male C57BL/6 mouse, 5 mg iron kg^{-1} is the maximal sublethal dose. Dosage should be adjusted for each strain and age of rodent. In general, males are more susceptible to Fe-NTA than females.

Protocol *Continued*

Carcinogenesis protocol for male Wistar rats

4. Use 100–125 g male Wistar rats. Inject intraperitoneally 5 mg iron kg^{-1} Fe-NTA on days 1 to 3, 10 mg iron kg^{-1} on days 4 and 5, and thereafter 10 mg iron kg^{-1} five times a week for 11 weeks. Avoid injection when the animal show marked weight loss (more than 5%).
5. Observe the animals closely, especially after 1 year, and sacrifice the animals when they are found to be dying.

[1] The $Fe(NO_3)_3$ and NTA solutions should be mixed immediately before use.

Results

This is a typical animal model of ROS-induced organ injury and carcinogenesis. The major target organ is kidney—the liver is less affected. Injected Fe-NTA is absorbed *via* the portal vein into the systemic blood flow and goes, *via* the filtration of glomeruli, to the lumina of renal proximal tubules where Fe-NTA acts as a catalyst for the generation of ·OH or its equivalents.

Acute study by single injection

1. Degeneration of the renal proximal tubular cells are observed by light microscopy 3 h after administration. Necrosis is most evident after 24 h (Fig. 1), so the gross appearance of the kidney is whitish.
2. Lipid peroxidation is induced 30 min after administration, and continues for approx. 6 h (peak 1–2 h). Among C_{2-12} aldehydes, 4-hydroxy-2-nonenal shows the greatest increase and malondialdehyde shows the highest abundance (3, 4).
3. Oxidative modification of DNA bases including 8-oxoguanine is increased 3–24 h after administration (5).

Carcinogenesis protocol for male Wistar rats

1. Survival of the animals after 1 year is approximately 70%. Renal cell carcinoma is specific to Fe-NTA administration, and is found after 1 year of incubation. The final incidence is usually more than 60%.
2. Multiple renal cell carcinoma (solely adenocarcinoma, not sarcoma) of low- and high-grade is induced, and high-grade tumour is usually accompanied by invasion and metastasis.
3. The animals die because of peritoneal haemorrhage owing to tumour rupture or because of respiratory failure resulting from massive pulmonary metastasis (Fig. 2).

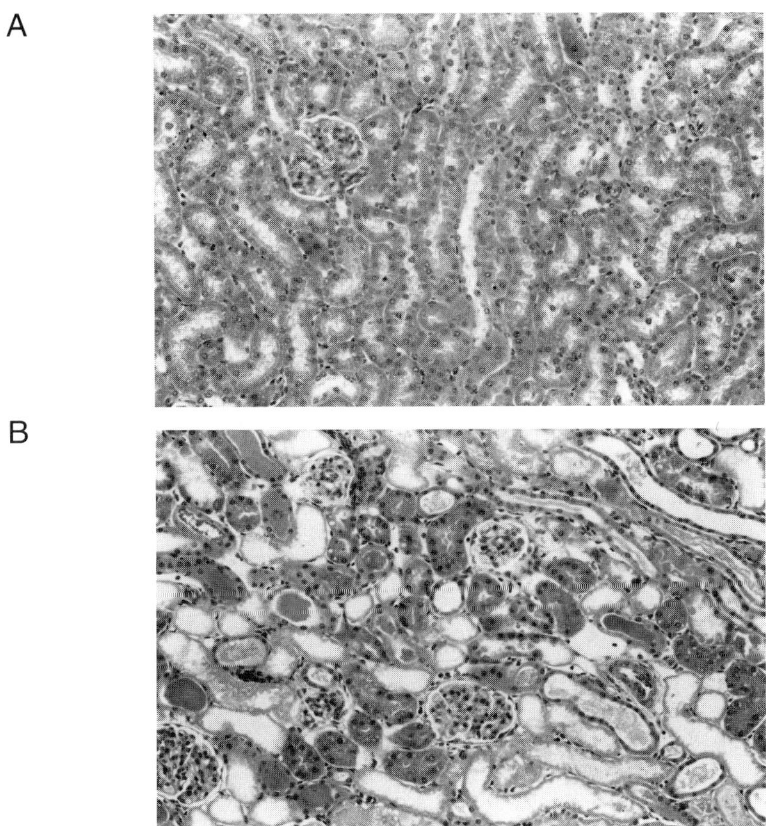

Figure 1. Histology of the kidney of a Wistar rat after administration of Fe-NTA. A. control; B. 24 h after administration. Half of the proximal tubular cells are removed after necrosis. Proteinaceous casts are observed.

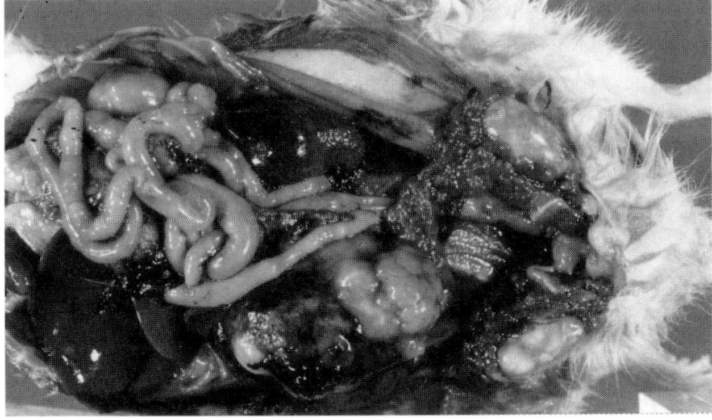

Figure 2. Induced renal cell carcinoma in the right kidney. This animal died of peritoneal haemorrhage. Peritoneal mesothelioma is also observed around the testes.

References

1. Toyokuni, S. (1996). Iron-induced carcinogenesis: the role of redox regulation. *Free Radic. Biol. Med.*, **20**, 553–66.
2. Okada, S. (1996). Iron-induced tissue damage and cancer: The role of reactive oxygen free radicals. *Pathol. Int.*, **46**, 311–32.
3. Toyokuni, S., Uchida, K., Okamoto, K., Hattori-Nakakuki, Y., Hiai, H., and Stadtman, E. R. (1994). Formation of 4-hydroxy-2-nonenal-modified proteins in the renal proximal tubules of rats treated with a renal carcinogen, ferric nitrilotriacetate. *Proc. Natl. Acad. Sci. USA*, **91**, 2616–20.
4. Toyokuni, S., Mori, T., and Dizdaroglu, M. (1994). DNA base modifications in renal chromatin of Wistar rats treated with a renal carcinogen, ferric nitrilotriacetate. *Int. J. Cancer*, **57**, 123–8.
5. Toyokuni, S., Luo, X.-P., Tanaka, T., Uchida, K., Hiai, H., and Lehotay, D. C. (1997). Induction of a wide range of C_{2-12} aldehydes and C_{7-12} acyloins in the kidney of Wistar rats after treatment with a renal carcinogen, ferric nitrilotriacetate. *Free Radic. Biol. Med.*, **22**, 1019–27.

92

Copper-overload model

KEIICHIRO SUZUKI

Introduction

Free copper and iron may facilitate the Fenton reactions to produce reactive oxygen species:

$$O_2\cdot^- + Cu^{2+}\ (Fe^{3+}) \rightarrow O_2 + Cu^+\ (Fe^{2+})$$
$$H_2O_2 + Cu^+\ (Fe^{2+}) \rightarrow OH^- + HO\cdot + Cu^{2+}\ (Fe^{3+})$$
$$O_2\cdot^- + H_2O_2 \rightarrow O_2 + OH^- + HO\cdot$$

Copper or iron overloading can be modelled *in vitro*, but it is difficult to devise a suitable overload model *in vivo*. In this section we introduce LEC rats (Long–Evans with a cinnamon-like coat colour, available from Charles River Laboratories) as an *in vivo* copper overload model (1, 2).

Comments

The LEC rat was established from a closed colony of the Long–Evans strain (2). This rat spontaneously develops hepatitis and eventually hepatoma after 1 year of age. Abnormal copper accumulation in the LEC rat liver (levels ~50 times higher than in control LEA (Long–Evans with an agouti colour) rat or in the normal Wistar rat) was first discovered in 1991 (3) (Fig. 1). Abnormal copper accumulation by the LEC rat has proved useful as a model of copper overload in research on free radicals and antioxidants. For example (Fig. 2), Mn-SOD expression is induced in the LEC rat (4). When D-penicillamine was administered to the LEC rat after birth, the concentration of copper in both the serum and liver decreased, and hereditary hepatitis did not occur. These data suggest that abnormal copper accumulation in the liver was the major cause of hepatitis. Many studies have reported the use of the LEC rat as an animal model of Wilson's disease. Recently, the causative gene for abnormal copper accumulation, which is a rat homologue of the human Wilson's disease gene, was cloned and the defect was demonstrated. The LEC rat is, therefore, a bona fide model of Wilson's disease. Other abnormalities such as immunodeficiency, have also been found in the LEC rat (Fig. 3).

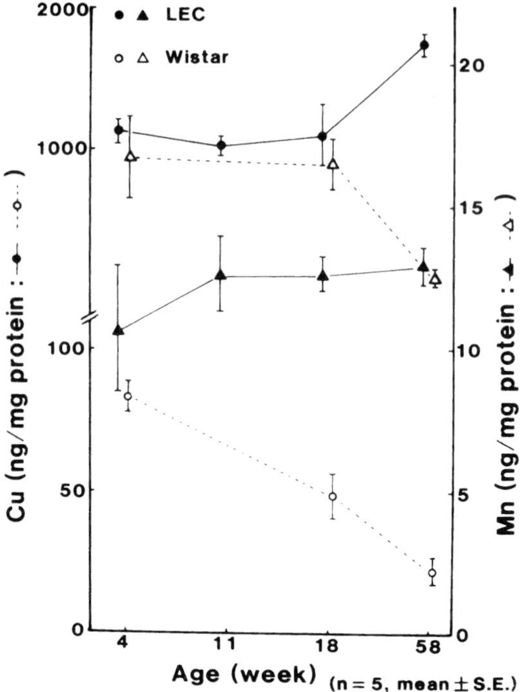

Figure 1. Cu and Mn concentrations in the livers of LEC and Wistar rats (4).

References

1. (1997). The special issue for the LEC Rats, Taniguchi, N. (ed.) *J. Trace Elem. Exp. Med.*, **10**, 47–152.
2. Mori, M., Yoshida, M. C., Takeichi, N., and Taniguchi, N. (ed.) (1991). *The LEC rat*. Springer, Tokyo.
3. Li, Y., Togashi, Y., Sato, S., Emoto, T., Kang, J. H., Takeichi, N., Kobayashi, H., Kojima, Y., Une, Y., and Uchino, J. (1991). Spontaneous hepatic copper accumulation in Long–Evans cinnamon rats with hereditary hepatitis. *J. Clin. Invest.*, **87**, 1858–61.
4. Suzuki, K., Miyazawa, N., Nakata, T., Seo, H. G., Sugiyama, T., and Taniguchi, N. (1993). High copper and iron levels and expression of Mn-superoxide dismutase in mutant rats displaying hereditary hepatitis and hepatoma (LEC rats). *Carcinogenesis*, **14**, 1881–4.
5. Miyoshi, E., Fujii, J., Hayashi, N., and Taniguchi, N. (1997). LEC rats: An overview of recent findings. *J. Trace Elem. Exp. Med.*, **10**, 135–45.

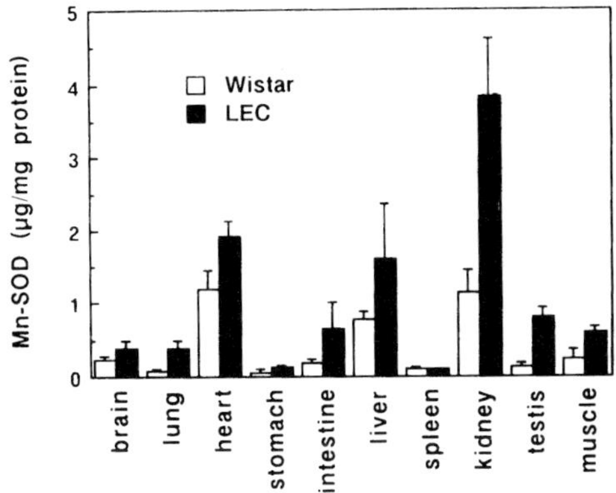

Fig. 1. Tissue distribution of Mn-SOD in LEC and Wistar rats. Values are mean ± SE; $n = 5$.

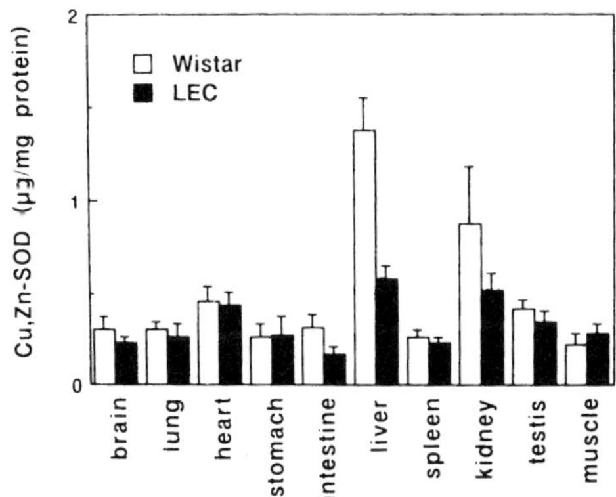

Figure 2. Tissue distribution of Mn-SOD and Cu, Zn-SOD in LEC and Wistar rats (n = 5, mean ± S.E.) (4).

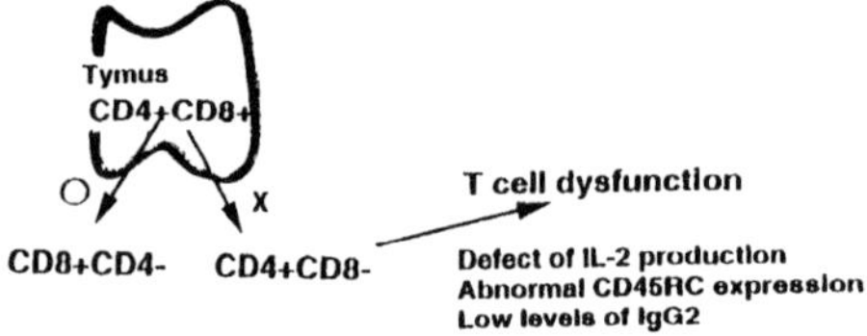

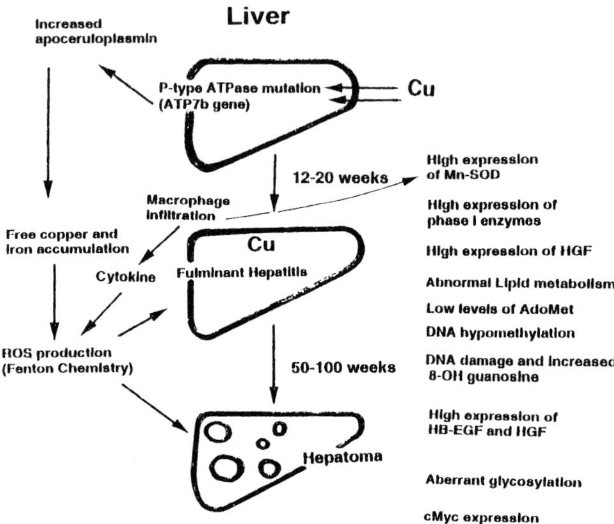

Figure 3. Pathogenesis and biochemical features of LEC rats (5).

Index

Index